CARRIER

A Guided Tour of an Aircraft Carrier

TOM CLANCY'S BESTSELLING NOVELS INCLUDE

The Hunt for Red October
Red Storm Rising
Patriot Games
The Cardinal of the Kremlin
Clear and Present Danger
The Sum of All Fears
Without Remorse
Debt of Honor
Executive Orders
Rainbow Six

CREATED BY TOM CLANCY AND STEVE PIECZENIK

Tom Clancy's Op-Center
Tom Clancy's Op-Center: Mirror Image
Tom Clancy's Op-Center: Games of State
Tom Clancy's Op-Center: Acts of War
Tom Clancy's Op-Center: Balance of Power
Tom Clancy's Net Force

CREATED BY TOM CLANCY AND MARTIN GREENBERG

Tom Clancy's Power Plays: Politika
Tom Clancy's Power Plays: ruthless.com

DON'T MISS

SSN: Strategies of Submarine Warfare

NONFICTION

Submarine:
A Guided Tour Inside a Nuclear Warship
Armored Cav:
A Guided Tour of an Armored Cavalry Regiment
Fighter Wing:
A Guided Tour of an Air Force Combat Wing
Marine:
A Guided Tour of a Marine Expeditionary Unit
Airborne:
A Guided Tour of an Airborne Task Force
Carrier:
A Guided Tour of an Aircraft Carrier

Into the Storm: A Study in Command

TOM CLANCY

CARRIER

A Guided Tour of an Aircraft Carrier

BERKLEY BOOKS, NEW YORK

DISCLAIMER

The views and opinions expressed in this book are entirely those of the author, and do not necessarily correspond with those of any corporation, military service, or government organization of any country.

This book is an original publication of The Berkley Publishing Group.

CARRIER

A Berkley Book / published by arrangement with
Rubicon, Inc.

PRINTING HISTORY
Berkley trade paperback edition / February 1999

For information address: The Berkley Publishing Group,
a member of Penguin Putnam Inc.,
375 Hudson Street, New York, New York 10014.

The Penguin Putnam Inc. World Wide Web site address is
http://www.penguinputnam.com

ISBN: 0-425-16682-1

BERKLEY®
Berkley Books are published by The Berkley Publishing Group,
a member of Penguin Putnam Inc.,
375 Hudson Street, New York, New York 10014.
BERKLEY and the "B" design
are trademarks belonging to Berkley Publishing Corporation.

PRINTED IN THE UNITED STATES OF AMERICA

10 9 8 7 6 5 4 3 2 1

It's hard when you lose friends. Especially those who were close or important to what you have been doing. This last year was especially tough, because we lost four people special to our efforts. To these men we dedicate this book:

Dr. Jeffery Ethell, Ph.D. An aviation historian, pilot, commentator, and friend with unparalleled credentials, who died in June 1997 while flying a vintage P-38 Lightning in Oregon.

Mr. Russell Eggnor. Director of the Navy Still Photo Branch at the Pentagon, he lost a fight to cancer in June 1997. Though Russ did not write the words in our books, the office and organization that he built supplied images and stories for every volume in this series.

Lieutenant Colonel Henry Van Winkle, USMC. The Executive Officer of VMFA-251, he was a constant source of wisdom and truth in the "Dirty Shirt" mess aboard USS *George Washington* (CVN-73). "Rip" Van Winkle died as a result of a midair collision in the Persian Gulf while flying an F/A-18 Hornet on February 6th, 1998.

Lieutenant General David J. McCloud. Head of the Alaskan Air Command and U.S. Forces in Alaska, Dave McCloud was an old and trusted friend of ours. When he and another flier died on July 26th, 1998, in the crash of a small aerobatic aircraft, his friends and the nation lost a treasure, which will not easily be replaced. We will miss you, "Marshall."

Contents

Acknowledgments

As we finish up the sixth book in this series, it is once again time to give credit where it is due. I'll start with my longtime friend, partner, and researcher, John D. Gresham. Once again, John met the people, took the pictures, spent nights aboard ship, and did all the things that make sure readers feel like they are there. We also have again benefited from the wisdom, experience, and efforts of series editor Professor Martin H. Greenberg, Larry Segriff, and all the staff at Tekno Books. Laura DeNinno is here again with her wonderful drawings, which have added so much to this book. As well, Tony Koltz and many others all need to be recognized for their outstanding editorial support that was so critical and timely.

Carrier required the support of many senior sea service personnel in a number of sensitive positions. In this regard, we have again been blessed with all the support that we needed and more. At the top were Admiral Jay Johnson and our old friend General Chuck Krulak. Both of these officers gave us their valuable time and support, and we cannot repay their trust and friendship. Their boss, Secretary of the Navy John Dalton, gave us critical support as well. Elsewhere around the Washington Beltway, we had the help of other influential leaders. Folks like Rear Admirals Dennis McGuinn and Carlos Johnson, and Captain Chuck Nash made it possible to get the information that we needed. This year, our home-away-from-home was the ships of the *George Washington* battle group, and they took us to some really exciting places. Led by Rear Admiral Mike Mullen, this unit is key to helping keep us safe in a dangerous world. Running the *GW* was an extraordinary crew led by Captains ''Yank'' Rutheford and Mark Groothausen, as well as Commander Chuck Smith. These men took us under their wings, and kept us warm and fed. Thanks also to Captains Jim Deppe of USS *Normandy* and Jim Phillips of USS *Vella Gulf* for sharing insights and time and letting us break bread with them. For the thousands of other unnamed men and women of the *GW* group who took the time to show us the vital things that they do, we say a hearty ''Thanks!''

Another group that is always vital to our efforts consists of the members of the various military public and media offices (PAOs) that handled our numerous requests for visits, interviews, and information. Tops on our list were Rear Admirals Kendall Pease and Tom Jurkowsky in CHINFO at the Pentagon. Also at CHINFO were our project officers, Lieutenants Merritt Allen and Wendy Snyder, who did so much to keep things going. Over in the office of the Chief of Naval Operations was Captain Jim Kudla, who coordinated our interview requests. Down with the Atlantic Fleet in Norfolk, Virginia, Commander Joe Gradisher, Lieutenant Commander Roxy Merritt, and Mike Maus ably assisted us. Then there were the folks of the *GW*'s PAO shop, led by the outstanding Lieutenant Joe Navritril. Along with Joe, an excellent young crew of media-relations specialists took us on some memorable adventures. Finally, we want to thank the special folks at the Navy Still Photo Branch, who have serviced our needs for so many years. They include Lieutenant Chris Madden and an incomparable staff of photographic experts. We thank them for their efforts as friends and professionals.

Again, thanks are due to our various industrial partners, without whom all the information on the various ships, aircraft, weapons, and systems would never have come to light. Down at Newport News Shipbuilding, we were allowed a look that few outsiders have ever had. Thanks are owed to Jerri Fuller Dickseski, Bill Hatfield, Mike Peters, Mike Shawcross, the folks from the U.S. Navy SUSHIPS office, and literally thousands of others. At the aircraft manufacturers, there were Barbara Anderson and Lon Nordeen of Boeing, Joe Stout, Karen Hagar, and Jeff Rhodes of Lockheed Martin, and finally, our old friend Bill Tuttle of Boeing Sikorsky. We also made and renewed many friendships at the various missile, armament, and system manufacturers, including: Tony Geishanuser and Vicki Fendalson at Raytheon Strike Systems, Larry Ernst at General Atomics, Craig Van Bieber at Lockheed, and the eternal Ed Rodemsky of Trimble Navigation. We also received an incredible amount of help from Dave ''Hey Joe'' Parsons and the fine folks at Whitney, Bradley, & Brown, Inc.

We owe thanks for all of our friends in New York, especially Robert Gottlieb, Debra Goldstein, and Matt Bialer at William Morris, as well as Robert Youdelman and Tom Mallon, who took care of the legal details. Over at Berkley Books, our highest thanks go to our series editor, Tom Colgan, as well as David Shanks, Kim Waltemyer, and the staff of Berkley Books. To old friends like Matt Caffrey, Jim Stevenson, A. D. Baker, Norman Polmar, and Bob Dorr, thanks again for your contributions and wisdom. Thanks also to the late Jeff Ethell and Russ Eggnor, who gave so much of themselves to us and the world. And to all the folks who took us for rides, tours, shoots, and exercises, thanks again for teaching the ignorant how things *really* work. As for our friends, families, and loved ones, we again thank you.

Foreword

"Where are the carriers?" This has been the likely first question asked by every President of the United States since World War II when faced with a developing international crisis that involves U.S. interests. It was probably also asked by Admiral Isoroku Yamamoto (the Commander in Chief of the Japanese Combined Fleet) after the Japanese attack on Pearl Harbor initiating World War II. This same question was always a top concern of the Soviet leadership throughout the Cold War. It drove an inordinate amount of their military expenditures, as well as many of their operational planning decisions.

More recently, in March of 1996, two U.S. aircraft carrier battle groups (CVBGs) were dispatched to the Taiwan Straits after the People's Republic of China launched a program of ballistic missile exercises close to Taiwan. The presence of the two aircraft carrier groups so close to the mainland of China defused the crisis, and prevented a Chinese escalation or miscalculation of our resolve.

The following year saw the latest in a series of crises with Iraq over Saddam Hussein's refusal to meet United Nations inspection criteria over his weapons of mass destruction. This was responded to by sending two more CVBGs to the Persian Gulf, this time to prepare for possible strikes on Iraqi targets had that been necessary.

Clearly, the flexibility, mobility, and independence of these versatile and forward-deployed assets will keep them center stage as our nation leads the world in the transition to a free-market system of democracies.

The rapid development and growth of airpower as the primary enabling capability for military operations represents one of the true military revolutions of the 20th century. At the close of this century, with manned space exploration and earth-orbiting satellites commonplace, it is hard to conceive that just ninety-five years ago, the Wright brothers made their first flight at Kitty Hawk, North Carolina. That historic first effort traveled less distance than the wingspan of a modern jumbo jet. However, things began to rapidly progress with the coming of the First World War. With the start of the Great War visionaries around the world realized the potential significance of aviation capabilities on military operations. By 1914, then-Secretary of the Navy Josephus Daniels had announced "that the point has been reached where aircraft must form a large part of our naval forces for offensive and defensive operations." It was an insightful thought.

The ensuing twenty-five years before our entry into World War II saw the United States developing the assets and vision to take airpower to sea in a way unmatched by any other nation. As a maritime nation dependent on the sea lines of communications for its economic and national security interests, the United States would need the edge provided by Naval aviation to win the greatest over-water military campaigns ever conducted. The history of the Second World War in the Pacific documents the great debt of gratitude our nation owes to the early pioneers of naval aviation. These were legendary men like Glenn Curtis, Eugene Ely, Theodore Ellyson, John Towers, John Rogers, Washington Chambers, Henry Mustin, and many more too numerous to mention.

However, it was at Pearl Harbor on December 7th, 1941, with the war cry of "*Tora . . . Tora . . . Tora!*" and our own lax state of readiness, that Japan brought home to the world the impact of carrier aviation.[1] The fact that none of our three Pacific-based aircraft carriers were in port that fateful morning may have been the single most significant factor in our eventual victory during the Great Pacific War. At the time of our entry into World War II, the U.S. Navy had just seven big-deck aircraft carriers in commission: *Saratoga*, *Lexington*, *Ranger*, *Yorktown*, *Enterprise*, *Wasp*, and *Hornet*. These "seven sisters" would take the war to our enemies from Casablanca and Malta to Midway and Guadalcanal.

Clearly, Admiral Yamamoto knew that Japan had awakened a "sleeping giant," and he believed a prolonged war would go in favor of the United States. He knew the potential productivity of American industry and its people, something that he had witnessed personally while on naval attaché duty in Washington. Thus it was that Japan, needing a quick decisive victory over the U.S. Navy in the Pacific, set in motion the great sea battle off Midway Island in mid-1942.[2] Yamamoto mustered an overwhelming naval armada, designed to take Midway and hand the U.S. Navy and their carrier groups a crushing defeat. However, when the Battle of Midway was over, the tide had turned in the Pacific, though not in the favor of Japan. Thanks to the raw courage and aggressive tactics of the U.S. carrier pilots as well as superb intelligence, four Japanese carriers and a cruiser were sunk. In the process, Japan's ability to project naval air power throughout the vast Pacific was crippled forever.

---

1 "*Tora . . . Tora . . . Tora*" is Japanese for "*Tiger . . . Tiger . . . Tiger.*" This was the radio call indicating a fully successful strike on the Hawaiian air and Naval bases.

2 Midway and the Battles of the Coral Sea, Eastern Solomons, and Santa Cruz Islands were all fought between Japanese and American carrier groups in 1942. They were unique in being the first battles where the major opposing forces never actually sighted each other, and the majority of the damage was inflicted by air strikes instead of gun or torpedo fire. In these actions, six Japanese and three American flattops were sunk.

The U.S. carrier groups and their courageous aviators had, on paper, no right to win. But win they did. The cost was not insignificant; fifteen of fifteen aircraft and twenty-nine of thirty aircrew in Torpedo Squadron 8 alone were lost. Along with scores of American aircraft and their crews, the USN lost the *Yorktown* and a destroyer.[3] However, finding a way to win in the face of adversity is a naval aviation tradition.

Today, U.S. carrier aviation is inextricably tied to the concept of United States forward presence and power projection; the "From the Sea" doctrine. Since the end of the East/West conflict, the United States military has withdrawn from the majority of its overseas bases. Consequently, America's ability to exercise a forward military presence and provide military forces depends on a combination of naval power and power projection from the continental United States. This means that in the complex post-Cold War world, where the majority of the world's major population centers are within two hundred miles of the open ocean, naval forces are increasingly relevant, and able to influence all manner of events that shape regional stability. The fact that this can be done with little or no land-based support and with no host nation support is a tremendous advantage for our national interests.

The independence, sustainability, and staying power of naval units often makes them the forces of choice for our National Command Authorities. This includes protecting the sea-lanes for a global free-market economy, reinforcing and supporting American embassies, and executing non-combatant evacuations of American citizens overseas. These and many other missions are ideally suited to our forward-deployed naval forces. This has been continuously demonstrated in places like the Taiwan Straits, the Persian Gulf, Somalia, Albania, the Central African Republic, Liberia, Zaire, and Sierra Leone. America is an island nation, dependent upon the seas for our economic prosperity and security. There was good reason why our founding fathers determined the need for the nation to maintain naval forces and raise an army. We should occasionally remind ourselves of this reality, since it is the geopolitics, not the geography of the world, that has changed over time.

Unfortunately, aircraft carriers and naval forces in general have often been seen as both provocative and vulnerable. Many critics who do not understand the science of modern naval operations have claimed that advances in space systems and missile technology make the carrier/naval forces excessively vulnerable to air and missile attacks. Certainly technology has increased the threat from these systems, but far less so than that faced by fixed land bases and ground forces from terrorism and ballistic missile attacks.

For starters, there is the challenge to any would-be enemy who would try to find a CVBG in the open ocean. Naval units are highly mobile and the world's oceans are a big, dynamic place. Trying to coordinate sophisti-

3 In the spring of 1998, oceanographer and adventurer Dr. Robert Ballard led an expedition that located the sunken *Yorktown* on the ocean floor north of Midway Atoll. Upright on the bottom, she is in excellent shape, with her guns still trained out, as if ready for action.

cated long-range targeting solutions onto a target that can move thirty nautical miles in any direction in just one hour, or up to seven hundred nautical miles in a day, is a tough business. Clearly, a CVBG is not an easy target. The inherent mobility, together with sophisticated CVBG electronic-warfare-deception packages (radar "blip" enhancers, target decoys, etc.), combined with the air defenses provided by our Aegis-equipped escorts (*Ticonderoga*-class [CG-47] cruisers and *Arleigh Burke*-class destroyers [DDG-51]) as well as the CVN's own organic aircraft, make the vulnerability quite manageable.

The threat of theater ballistic and cruise missiles is also a matter of concern for the CVBG, and work is rapidly progressing to increase our defenses against these classes of weapons. The Aegis combat system is being improved and extended to be able to provide theater-wide defense from the sea, for both land and sea forces. Survivability from these threats will always be greater from a mobile bastion at sea than a fixed base on land. Arriving along with this new capability are new aircraft, ships, and even new carrier designs, which will help keep the CVBG credible long after the last manned-aircraft designs are retired. However, one does not have such naval forces for purely defensive purposes.

The real strength of CVBGs is offensive, making them a threat to the very despots and enemies that might themselves wish ill to the carrier group. Able to generate hundreds of air and missile attack sorties day and night, the modern CVBG is a powerful tool that requires no permission of ally or foe to do its job. Today, when the challenge is to get the most return for our limited defense dollars, it is significant to note that since the end of World War II, we have not lost *any* carriers to enemy action or geopolitical changes.

This is hardly true in the case of our overseas land bases. In such countries as Iran, Libya, Vietnam, and the Philippines to name just a few, we not only lost the airfields that the U.S. paid for, but also the costly infrastructure devoted to support, maintenance, and quality-of-life issues. There also is the fact that we pay a high monetary and often unacceptable political price for even restricted access to foreign military land and air bases. As recently as 1997, the U.S. was not allowed to place the desired number of USAF aircraft in Saudi Arabia, where the U.S. presence was already established. From this viewpoint, the aircraft carrier, which has a forty-five-year life cycle and remains free from such entanglements, is a relative bargain for our scarce defense dollars.

As a new crop of world economic and potential military superpowers emerge in the coming years, the value of aircraft carriers to U.S. foreign policy goals will dramatically increase. One of the unchallenged realities of modern warfare is that you cannot be victorious in any conflict on the ground or at sea without air/space superiority. In an era of sophisticated precision weapons, including cruise and ballistic missiles, this is the medium that enables our land and sea forces to operate with acceptable risk. Air superiority is even more essential for forward-deployed forces that are shaping the battlespace, trying to create stability and prevent conflict from occurring through their own forward presence. In more and more cases, this flexible combat power will have to be provided by forward-deployed carrier and amphibious groups. This is a reality since the world's surface is 70% covered by water, and our free-market economy depends on open access to the sea lines of communication.

Naval forces are more than just ships, planes, and weapons. What I hope this book conveys is the quality and dedication of the people it takes to provide the nation the kind of flexibility and fighting punch packaged in our modem CVBGs. The carriers, Aegis cruisers, and destroyers, together with their aircraft and fast-attack submarines, would be nothing without the people who make them work. Operating a high-usage airport in day and night operations, while moving at thirty knots on the open seas, is one thing. However, to provide all the organic support to do this for extended periods of time at a great distance from a home base is another thing all together.

A modern *Nimitz*-class (CVN-68) carrier is the equivalent of a small American city packaged into just four-and-a-half acres. This city not only operates an airport on its roof, but also can move over seven hundred nautical miles in any given day. It also provides full medical support, machine shops, jet engine test cells, food service operations, computer support, electrical generation, and almost everything else that you can imagine.

Now picture the carrier as a business, a company that has a net worth of six to seven billion dollars and employs over six thousand people. The average age of the six thousand employees is less than twenty-one years. On top of this, the Chairman of the Board (Admiral and Staff), the President and Chief Operating Officer (Captain and Air Wing Commander), all the Vice Presidents (Department Heads), and *every* other employee rotates out of the company every two to three years. Common sense would dictate that you could *never* make a profit with *any* business under those conditions. Yet the U.S. Navy operates successfully under these very conditions, and the profit is freedom, and protection of our national interests.

This dedication of young Americans, the symphony of their teamwork, and the indomitable spirit of the American sailor make this all possible. We owe them our respect and gratitude, and must never take the service or sacrifices they and their families make for granted. It was my privilege to be a shipmate with these great Americans for over thirty-seven years. For this I salute the American Sailors, Marines, Soldiers, Airmen, and Coast Guardsmen of every generation who have protected our freedom at home and around the world.

—Leon A. ''Bud'' Edney
Admiral, USN (Retired)
Former Commander, U.S. Atlantic Command &
NATO Supreme Allied Commander, Atlantic

Introduction

Presence, influence, and options. In these three words are the basic rationale for why politicians want carrier battle groups, and have been willing to spend over a trillion U.S. taxpayer dollars building a dozen for American use. That was hardly the original reason, though. Back in the years after the Great War, naval powers were trying to find loopholes in the first series of arms-control treaties (which had to do with naval forces). With the numbers and size of battleships and other vessels limited by the agreements, various nations began to consider what ships carrying aircraft might be able to contribute to navies. At first, the duties of these first carrier-borne aircraft were limited to spotting the fall of naval shells and providing a primitive fighter cover for the fleet. Within a few years, though, aircraft technologies began to undergo a revolutionary series of improvements. Metal aircraft structures, improved power plants and fuels, as well as the first of what we would call avionics began to find their way onto airplanes. By the outbreak of World War II, some naval analysts and leaders even suspected that carriers and their embarked aircraft might be capable of sinking the same battleships and other surface ships that they had originally been designed to cover.

The Second World War will be remembered by naval historians as a conflict dominated by two new classes of ships: fast carriers and submarines. The diesel-electric submarines were a highly efficient force able to deny navies and nations the use of the sea-lanes for commerce and warfare. Unfortunately, as the German Kriegsmarine and Grand Admiral Karl Donitz found, you do not win wars through simple denial of a battlespace like the Atlantic Ocean. Victory through seapower requires the ability to take the offensive on terms and at times of your choosing. This means being able to dominate vast volumes of air, ocean, and even near-earth space. Without a balanced force to project its power over the entire range of possibilities and situations, one-dimensional forces like the U-boat-dominated *Kriegsmarine* wound up being crushed in the crucible of war.

By contrast, the carriers and their escorts of World War II were able to project offensive power over the entire globe. From the North Cape to the islands of the Central Pacific, carrier-based aircraft dominated the greatest naval war in history. Along the way, they helped nullify the threat from Germany's U-boats and other enemy submarines, as well as sweeping the seas of enemy ships and aircraft. While the eventual Japanese surrender may have been signed aboard the battleship *Missouri* in Tokyo Bay, it occurred in the shadow of a sky blackened by hundreds of carrier aircraft flying overhead in review. Called ''Halsey's Folly,'' the flyover was the final proof of the real force that had ended the second global war of this century. Despite the claims of Air Force leaders who pronounced navies worthless in an era of nuclear-armed bombers, when the next shooting conflict erupted in Korea, it was carrier aircraft that covered the withdrawal to the Pusan Perimeter and the amphibious landings at Inchon. They then dropped into a role that would become common in the next half-century, acting as mobile air bases to project combat power ashore.

Despite the best efforts of the former Soviet Union to develop a credible ''blue-water'' fleet during the Cold War, the U.S. Navy never lost control of any ocean that it cared about. One of the big reasons for this was the regular presence of carrier battle groups, which took any sort of ''home-court advantage'' away from a potential enemy. Armed with aircraft that were the match of anything flying from a land base, and flown by the best-trained aviators in the world, the American carriers and their escorts were the ''eight-hundred-pound guerrillas'' of the Cold War naval world. This is a position that they still hold to this day. However, their contributions have taken on a deadly new relevance in the post–Cold War world.

One of the tragic truths about America's winning of the Cold War was that we did it with anyone who would help us. This meant that the U.S. frequently backed any local dictator with a well-placed air or naval base and a willingness to say that Communism was bad. The need to contain the ambitions of the Soviet Union and their allies took a front seat to common sense and human rights. The result was a series of alliances with despots ranging from Ferdinand Marcos to Manuel Noriega. However, there was a war to win and we did win it. The price, however, is what we are paying today. Around the world, Americans are being asked to please pack up their aircraft, ships, and bases and please take them home. We should not be offended; we did it to ourselves. The continuing legacy of squalor in places like Olongapo City in the Philippines and other ''outside the gate'' towns was more than the emerging democracies of the post–Cold War era could stand. When you add in our continued interference in the internal politics of the countries that hosted our bases, it is a wonder that we have any friends left in the world as the 20th century ends.

Our poor foreign policy record aside, the United States and our allies still have a number of responsibilities in the post–Cold War world. This means simply that to wield military force in a crisis, we now have just a few options. One is to ask nicely if a friendly host nation might allow us to base personnel, aircraft, and equipment on their soil so that we can threaten their neighbors with military force. As might be imagined, this can be a tough thing to do in these muddled times. George Bush managed to do it in the Persian Gulf in 1991, but Bill Clinton failed in the same task in 1997 and 1998. Even with a dictator like Saddam Hussein, most regional neighbors would rather tolerate the bully than risk the death and destruction that occurred in Kuwait in 1990 and 1991. This leaves just two other credible options; to base military power at homeland bases or aboard sovereign flagged ships at sea. The first of these options means that fleets of transport ships and aircraft must be maintained just to move them to the place where a crisis is breaking out. It also takes time to move combat aircraft and ground units to the places where trouble may be brewing. This is why having units forward-based aboard ships is so incredibly important to us these days.

Time in a crisis is more precious than gold. As much as any other factor, the time delay in responding to a developing conflict determines whether it results in war, peace, or a distasteful standoff. While we may never know for sure, there is a good chance that Saddam Hussein stopped at the Saudi border in 1990 because of the rapid flood of U.S. and coalition forces into the Kingdom. However, it would be a tough act to duplicate today. One of the benefits of our military buildup in the late years of the Cold War was the ability to do both of these things well. Along with lots of continental-based forces with excellent transport capabilities, we usually had a number of carrier and amphibious groups forward-based to respond to crises. However, these rich circumstances are now just happy memories.

Today the U.S. Navy considers itself lucky to have retained an even dozen carrier battle groups, along with their matched amphibious ready groups. By being able to keep just two or three of these forward-deployed at any time, the United States has managed to maintain a toehold in places where it has few allies and no bases. The recent confrontation with Iraq over United Nations weapons inspectors, had it led to war, would have been prosecuted almost entirely from a pair of carrier groups based in the Persian Gulf. With the 1990/91 allied coalition splintered over each country's regional interests, almost nobody would allow U.S. warplanes and ground forces onto their soil. This is a 180° change from 1990/91, when the majority of Allied airpower was land-based.

This brings us back to the three words at the beginning of this introduction: presence, influence, and options. Naval forces generally provide presence. Carrier groups, though, dominate an area for hundreds of miles/kilometers in every direction, including near-earth space. While a frigate or destroyer impresses everyone who sees it, a carrier group can change the balance of military and political power of an entire region. A weak country backed by an American carrier group is going to be much tougher to overthrow or invade for a local or regional rogue state or warlord. That is the definition of international presence these days. Finally, there is the matter of options.

In the deepest heart of every politician, there is a love of options. Having choices in a tough situation is every politician's greatest desire, and carrier groups give them that. It is one of the oddities of national politics that until they become President or Prime Minister, politicians frequently and publicly view large military units like carrier groups as a waste of taxpayer money. However, let the politicians hit the top of a nation's political food chain, and they sing another tune entirely. It is almost a matter of national folklore that every Chief Executive will, at some time in their Presidency, ask those four famous words: "Where are the carriers?" It certainly has been the case since Franklin Roosevelt haunted the halls of the White House. Today, in fact, the use of forward-deployed forces afloat may be the only option open to a national leader.

Understanding aircraft carriers and their associated aircraft and battle group escorts is not an easy task. Focusing only on the flattop is like tunnel vision, since the carrier's own weapons are purely defensive and quite short-ranged. To fully understand what effects a carrier group moving into your neighborhood is going to have, it is necessary to look beyond the carrier's bulk and dig deeper. You must look into the embarked air wing with its wide variety of aircraft and weapons, as well as the escorts. These range from Aegis-equipped missile cruisers and destroyers, to deadly nuclear-powered attack submarines. Armed with surface-to-air missiles and Tomahawk cruise missiles, they not only protect the carrier from attack, but have their own mighty offensive punch as well. To see it all takes a wider, deeper look than you are likely to find on the nightly news or in your daily newspaper. To do that requires that you spend time with people. Lots of people. These include the Navy's leaders, who make the policy decisions and have the responsibility of keeping our Navy the best in the world. You also need to spend some time with the folks who build the ships, aircraft, and weapons that make the force credible and dangerous. Finally, you have to know the thousands of people who run the battle groups and sail them to the places where they are needed across the globe.

I hope as you read this book that you get some sense of the people, because it is they that are the real strength of the carrier groups, and our nation. While you and I stay home safe and warm in the company of our families and loved ones, they go out for months at a time to put teeth into our national policies and backbone into our words. It is they who make the sacrifices and perhaps pay the ultimate price. I hope you see that in these pages, and you think of them as you get to know the "heavy metal" of the U.S. Navy up close. If you do, I think that you will gain a real perspective on their difficult, but vital, profession.

—Tom Clancy
July 1998

Naval Aviation 101

"Where are the carriers?"

Every American President since
Franklin Delano Roosevelt

Aircraft carriers stretch perceptions. First of all, they're *big*—bigger than most skyscrapers—skyscrapers that can move across the sea at a better than fair clip. And yet, despite their great size, when you watch flight operations on the flight deck (usually as busy as a medium-sized municipal airport), you can't help but wonder how so much gets done in such a tiny space. They not only stretch perceptions, they stretch the limits of the nation's finances and industrial capacity; and they stretch credibility. It's hard to find a weapon that raises more controversy.

Controversy has troubled naval aviation from the early days of the century, when primitive airplanes originally went to sea. At first, airpower was seen as a useless diversion of scarce funds from more pressing naval requirements like the construction of big-gun battleships. Later, after naval aviation became a serious competitor for sea power's throne, bitter infighting arose between gunnery and airpower advocates. Today, as the acknowledged "big stick" of America's Navy, the aircraft carrier is under attack from those who claim to have better ways to project military power into forward areas. Air Force generals plug B-2A stealth bombers with precision weapons (so-called "virtual presence"). Submariners and surface naval officers hawk their platforms carrying precision strike missiles. A good case can be made for all of these. Still, in a post–Cold War world that becomes more dangerous and uncertain by the week, aircraft carriers have a proven track record of effectiveness in crisis situations. Neither bombers nor "arsenal ships" can make that claim.

Question: What makes aircraft carriers so effective?

The USS *George Washington* (CVN-73) operating her embarked carrier air wing One (CVW-1). Battle groups based around aircraft carriers are the backbone of American seapower.
OFFICIAL U.S. NAVY PHOTO

Answer: Carriers and their accompanying battle groups can move freely over the oceans of the world (their free movement is legally protected by the principles of ''Freedom of Navigation''), and can do as they please as long as they stay outside of other nations' territorial waters.

A nation's warships are legally sovereign territories wherever they might be floating; and other nations have no legal influence over their actions or personnel. Thus, an aircraft carrier can park the equivalent of an Air Force fighter wing offshore to conduct sustained flight and/or combat operations. In other words, if a crisis breaks out in some littoral (coastal) region, and a carrier battle group (CVBG) is in the area, then the nation controlling it can influence the outcome of the crisis.[4] Add to this CVBG an Amphibious Ready Group (ARG) loaded with a Marine Expeditionary Unit-Special Operations Capable (MEU (SOC)), and you have even more influence.[5] This, in a nutshell, is the real value of aircraft carriers.

Such influence does not come without cost. Each CVBG represents a national investment approaching US $20 billion. And with over ten thousand embarked personnel that need to be fed, paid, and cared for, each group costs in the neighborhood of a billion dollars to operate and maintain annually. That's a lot of school lunches. That's a lot of *schools*! Add to this current United States plans to maintain twelve CVBGs. And then add the massive costs of the government infrastructure that backs these up (supply ships, ports, naval air stations, training organizations, etc.), as well as the vast

4 ''Littoral'' regions are defined geographically as those areas lying within several hundred miles/kilometers of a coastline. Since the majority of the world's population, finance, industry, and infrastructure reside in littoral regions, the sea services focus on operations there.

5 For more on the ARG and MEU (SOC), see: *Marine: A Guided Tour of a Marine Expeditionary Unit* (Berkley Books, 1996).

commercial interests (shipbuilders, aircraft and weapons manufacturers, etc.) necessary to keep the battle groups modern and credible. And then consider that not all twelve battle groups are available at one time. Because the ships need periodic yard service and the crews and air crews need to be trained and qualified, only two or three CVBGs are normally forward-deployed. (There is usually a group in the Mediterranean Sea, another in the Western Pacific Ocean, and another supporting operations in the Persian Gulf region.)

Is this handful of mobile airfields worth the cost? The answer depends on the responses to several other questions. Such as: How much influence does our country want to have in the world? What kind? How much do we want to affect the actions and behavior of other countries? And so on.

Sure, it's not hard to equate the role of CVBGs with "gunboat diplomacy" policies of the 19th century. But doing that trivializes the true value of the carriers to America and her allies. Among the lessons the last few years have taught us is one that's inescapable: The United States has global responsibilities. These go far beyond simply maintaining freedom of maritime lines of communications and supporting our allies in times of crisis. Whether we like it or not, most of the world's nations look to America as a leader. And these same nations (whether they want to say so officially or not) see us as the world's policeman. When trouble breaks out somewhere, who're you going to call? China? Russia? Japan? Not in this decade.

Sure, it's not always in the best interest of the United States to give a positive answer to every request for support and aid. But when the answer *is* positive, there is the problem of how to deliver the needed response. Once upon a time, our network of overseas bases allowed us to project a forward presence. No longer. Over the last half century, a poorly conceived and ill-executed American foreign policy has allowed us to be evicted from something over 75% of these bases. Add to this the limited resources available following the recent military drawdowns, and the National Command Authorities are left with very few options. Most of these are resident in the CVBGs and ARGs that make up the forward-deployed forces of the United States Navy.

At any given time, there are usually two or three CVBGs out there on six-month cruises, doing their day-in, day-out job of looking out for the interests of America and our allies, with adventure and danger only a satellite transmission away. Thanks to the support of service forces (fuel tankers, supply ships, etc.), a well-handled CVBG's only limitations are the durability of machinery and the morale of the people aboard. Given the will of a strong nation to back it, CVBGs can be parked off any coast in the world, and sit out there like a bird of prey.

That is the true meaning of "presence."

Rationale: Why Aircraft Carriers?

So why does America *really* need aircraft carriers? We've seen the theoretical, "policy" answer to that question. But what's the practical, real-world answer? What value does a ninety-year-old military concept have in an age of satellite surveillance and ballistic missiles? How does a relative handful of aircraft based aboard Naval vessels actually effect events on a regional scale? Finally, what does this capability give to a regional CinC or other on-scene commander? All of these questions must be explored if the real value of carriers and CVBGs is to be fully understood.

Aircraft Carriers: An Open Architecture

In less than a hundred years, we've passed from the first heavier-than-air test flights to deep-space probes. During that same time, after over five centuries of preeminence, we have seen the demise of gunnery as the measure of Naval power. The decline of naval guns and the rise of airpower were not instantly obvious. In fact, in the early 1900's, to suggest it would have invited a straitjacket. The first flying machines were toys for rich adventurers and stuntmen, their payload and range were extremely limited, and their worth in military operations was insignificant. The technology of early manned flight was derived from kites, bicycles, and automobiles. Structures were flimsy and heavy, and the engines bulky and inefficient.

Though the First World War did much to improve aircraft technologies, and made many military leaders believers in the value of airpower, the world powers had just made a staggering investment in big-gun dreadnought-type battleships that Naval leaders had no appetite to replace. Thus, Naval airpower wound up being limited by arms treaties or shuffled to the bottom of the funding priorities. Even so, though few saw this then, the future of Naval airpower was already a given. There are two reasons for this:

First—Aircraft soon proved they could carry weapons loads farther than guns could shoot, and with greater flexibility.

Second—An aircraft carrier can more easily accommodate upgrades and improvements than an armored ship with fixed-bore guns.

In order to retrofit a larger gun to deliver a larger shell, you have to replace the turrets and barbettes. And to do that, you have to completely rebuild a battleship or cruiser. By comparison, for an aircraft carrier to operate a new kind of aircraft, bomb, or missile, you only need to make sure that the new system fits inside the hangars and elevators. You also need to make sure that it's not too heavy for the flight deck, and (if it's an aircraft) that it can take off and land on the deck. Simply put, as long as an aircraft or weapons *fits* aboard a carrier, it can probably be employed successfully. In modern systems terminology, the carrier is an "open architecture" weapons system, with well-understood interfaces and parameters. Much like a computer with built-in capabilities for expansion cards and networking, aircraft carriers have a vast capability to accept new weapons and systems. Thus, some battleships built at the beginning of the First World War were scrapped after less than five years service, while modern supercarriers have planned lives measured in decades.

Sure, gun-armed warships can still hurt aircraft carriers. And in fact, during World War II, several flattops found themselves on the losing end of duels with surface ships. Today, missile-armed ships and submarines pose an even greater hazard to flattops, as they do to all vessels. However, all things being equal, the range of their aircraft is going to give carriers a critical edge in any combat. Carrier aircraft can hold an enemy ship or target at a safe distance, and then either neutralize or destroy it. The word for this advantage is "standoff." By "standing off" from an enemy and attacking him from over the horizon, you greatly reduce his ability to counterattack the carrier force, making defense much easier. In fact, just finding a CVBG is harder than you might think, as the Soviet Union discovered to its great chagrin on more than one occasion during the Cold War. If—as now seems likely—the next generation of American flattops incorporates stealth technology, then you can plan on aircraft carriers serving well into the next century.

Some Propositions about Sea-Based Airpower

The "real-world" effects of "sea-based" naval aviation (that is, aircraft based aboard ships at sea) and the principles by which battle group commanders ply their intricate and difficult trade are many, varied, and complex; and learning these takes years. What follows is no substitute for those years. Still, knowledge of some of the basic propositions about sea-based airpower that guide the plans and actions of our Naval leaders can't help but be useful:

- **Control of the Total Littoral Battlespace Is Impossible without Airpower—**While it cannot realistically win a battle, campaign, or war by itself, no victory is possible without airpower. Broadly defined as the effective military use of the skies—airpower is vital to controlling the "battlespace" of the littoral regions. One only need look back at British operations in the Falklands in 1982 to see how much can go wrong when a fleet operates within range of enemy land-based aircraft without proper air cover. As a result, their victory in that war was "a very near thing."
- **Sea-Based Airpower Involves a Variety of Systems—**Naval forces bring a variety of systems and sensors to the littoral battlespace. To name a few: fighter jets and transport helicopters; submarine-hunting helicopters and aircraft; surface-to-air (SAM) missiles defending against aircraft and ballistic missiles; and cruise missiles. This functional diversity means that a CVBG commander can bring any number of systems and employment options to bear, greatly compounding the defensive problem of an adversary. Properly utilized and supported, sea-based airpower can provide enabling force and muscle for any number and type of military operations.

An F-14D Tomcat taxies through catapult steam on the deck of the USS *Carl Vinson* (CVN-70). The four-and-a-half-acre flight deck is one of the busiest and most dangerous workplaces in the world. It also is the place where carriers prove their worth in the real world.
OFFICIAL U.S. NAVY PHOTO

Examples of this functional diversity include: deterring the use of ballistic and cruise missiles in a regional conflict, supporting amphibious and airborne operations, providing cover for a non-combatant personnel evacuation, or firing land-attack missiles and controlling unmanned aerial vehicles from submarines.

- **Sea-Based Airpower Is Inherently Flexible and Mobile—**Because they are based aboard ships, sea-based aviation assets are highly mobile. Modern CVBGs can easily move five hundred nautical miles in a day, which means that they can redeploy almost anywhere in the world in just a few weeks. And with a little warning, a forward-deployed force can be in a crisis zone in days, sometimes even in hours. Because they are not directly tied to a land-based command structure, the personnel and units embarked aboard the ships are equipped and trained to work on their own. Finally, because sea-based air units pack a lot of power into very small packages, they have great agility in an uncertain, fast-moving crisis or combat situation.
- **Sea-Based Airpower Is Inherently Offensive—**While airpower has powerful defensive capabilities, it is best used in offensive operations, thus allowing its full power to be focused and timed into blows of maximum power and efficiency. The ability to rapidly shift position, for example, allows sea-based units to change their axis of attack, and makes the defensive problem of the enemy much more difficult. By simply moving into an area, sea-based aviation units fill the skies with their presence, affecting both the military situation and the mind-set of a potential enemy. Should combat operations be initiated, sea-based air units are prepared to launch sustained strikes against enemy targets for as long as required. Even if the enemy forces choose to strike back at the naval force, the

mere act of the attacking fleet units degrades the hostile air and naval units involved.

- **Sea-Based Airpower Provides Instant Regional Situational Awareness—**A battle group entering an area provides a wide variety of intelligence-collection capabilities for a regional CinC. Along with the air and shipborne sensors organic to a naval force, the unit commanders have a number of regional and national-level intelligence-collection capabilities that can rapidly fuse the data into a coherent situational analysis. This makes the job of deciding upon future action and committing follow-on forces much less uncertain. As a further benefit, the staying power of the naval force means that minute-to-minute changes in the military and political situation in a crisis/combat zone can be watched, and trends and developments can be tracked over time, allowing a deeper and wider understanding of the regional situation.
- **Sea-Based Airpower Is Protected from the Effects of International Politics—**Unlike land-based air and ground units, which can't operate without the approval of a regional ally or host country, naval forces (and air units in particular) are not affected by such issues. They are also less vulnerable to attack by enemy forces or acts of terrorism. Shielded by the international laws covering freedom of navigation, sea-based units are free to act independently. Since each ship and aircraft is the sovereign territory of the owning country, any attack or intrusion becomes a potential act of war and a violation of international law. Since few nations have the will to violate these accords, this makes naval aviation a force that does not have to ask permission to act.
- **Sea-Based Airpower Provides Long-Term Presence and Power—**Maritime nations have long made allowance for resupply and support of their forces at sea. As long as proper sea lines of communications can be maintained, and replacement ships and aircraft can be rotated, ships and sea-based air units can be sustained almost indefinitely on station, and mission durations of months or even years can be supported. This is a key attribute of great maritime nations, and the addition of sea-based air units to their force mix greatly enhances the power and presence they can generate. Recent examples of this kind of forward naval presence are the naval embargoes of Iraq and the Balkans, and the lead-up to the 1991 Gulf War.
- **Sea-Based Airpower Can Conduct Multiple Missions at the Same Time—**Since naval forces are designed with robust command-and-control capabilities, and sea-based aircraft are multi-mission-capable by necessity, sea-based air units are capable of many types of missions, and can conduct them simultaneously. Thus, attack aircraft can conduct suppressive missions on enemy air defenses, while other units are engaging in precision cruise-missile strikes, armed helicopters are securing the battlespace around the naval force, and SAM-equipped ships are conducting defensive operations against enemy ballistic- and cruise-missile strikes. Such flexi-

The launch of a BGM-109 Tomahawk cruise missile from the guided-missile destroyer *Laboon* (DDG-58) during Operation Desert Strike in 1996.
OFFICIAL U.S. NAVY PHOTO

bility gives naval leaders a critical edge when fast-breaking, rapidly changing crisis and combat situations are in play.

- **Sea-Based Airpower Can Generate a Wide Variety of Effects**—A naval force generates reactions that range from coercion to terror. Sea-based air units add to this power, by adding a wide variety of weapon and mission effects, ranging from the use of surveillance aircraft and the delivery of special operations forces to more traditional results like the aerial delivery of munitions onto targets. Yet even here, variety is the watchword. Because naval air units are based at sea, there are no restrictions upon the munitions they can carry and employ. This means that an enemy can expect to face everything from precision-guided penetration bombs to cluster munitions—or even a nuclear strike. Such threats can often deliver the most useful of all weapons effects, deterrence from acting with hostile force against a neighboring nation.
- **Sea-Based Airpower Keeps Threats Far Away**—America's Navy has historically displayed its greatest value by keeping the threat of enemy military action on the other side of the world's oceans. In fact, no hostile military force of any size has intruded upon our territory since the War of 1812. Today, our sea services continue this mission, and sea-based airpower provides our naval forces with much of the muscle that makes it possible. By keeping the enemy threats against our homeland at arm's length, sea-based airpower keeps our nation strong, and our people safe in an otherwise uncertain world.

Milestones: The Development of a Modern Weapon

It goes without saying that institutions as large, diverse, and powerful as naval aviation do not just happen overnight. They evolve over time, and are the product of the forces and personalities that impact upon them. In fact, naval aviation grew to maturity surprisingly quickly, and most of the critical events and trends that shaped it happened in the roughly five decades stretching from 1908 through the mid-1950's. During that time, the basic forms and functions that define carriers and their aircraft today were conceived and developed. Let's take a look at a few of the most critical of these events and trends. We'll start with the first act in the birth of the world's most powerful conventional weapons system.

Eugene Ely's Stunt

Our journey begins in 1908, just five years after the Wright brothers' first flight, when Glenn Curtiss, an early aerial pioneer, laid out a bombing range in the shape of a battleship, and simulated attacking it. Though the U.S. Navy took notice of Curtiss's test run, it took no action. Several years later, after word reached America of a German attempt to fly an airplane from the deck of a ship, the U.S. Navy decided to try a similar experiment. They built a wooden platform over the main deck of the light cruiser *Birmingham* (CL-2) and engaged Eugene Ely, a stunt pilot working for Curtiss, to fly off it. At 3 P.M. on the afternoon of November 14th, 1910, while *Birmingham* was anchored in Hampton Roads, Virginia, Ely gunned his engine, rolled down the wooden platform, and flew off. He landed near Norfolk several miles away. A few months later, Ely reversed the process and landed on another platform built on the stern of the armored cruiser *Pennsylvania* (ACR-4), which was then anchored in San Francisco Bay. Soon afterward, Congress began to appropriate money, the first naval aviators began to be trained, and planes began to go to sea with the fleet. It was a humble beginning, but Eugene Ely's barnstorming stunt had started something very much bigger than that.

The First Flattop: The Conversion of the USS *Langley* (CV-1)

Stunts were one thing, but making naval aviation a credible military force was something else entirely. During World War I, U.S. naval aviation was primarily seaplanes used for gunnery spotting and antisubmarine patrols. However, the British achieved some fascinating results using normal (wheeled) pursuit aircraft (fighters) launched from towed barges, and later from specially built aircraft carriers converted from the hulls of other ships. These aircraft attacked German Zeppelin hangars and other targets.[6] The benefits of taking high-performance aircraft to sea were so obvious to the British that the Royal Navy rapidly set to converting further ships into aircraft carriers. This move did not go unnoticed by other Naval powers after World War I. By 1919, the Japanese were also constructing a purpose-built carrier, the *Hosho*. Meanwhile the British continued their program of converting hulls

6 The primary Zeppelin base for operations against England and the North Sea fleets was at Tondern near Whelimshaven (on the German/Danish border). In July of 1917, seven Sopwith Camels flying from the flying-off deck of HMS *Furious* attacked the Zeppelin sheds there; three Zeppelins were destroyed in their hangar.

Eugene Ely flies off of the USS *Pennsylvania* at 3 P.M. on November 14th, 1910. This was the moment of birth for naval aviation.
Official U.S. Navy Photo

into aircraft carriers, and began work on their own from-the-keel-up carrier, the *Hermes*.

These programs spurred the General Board of the U.S. Navy to start its own aircraft carrier program. In 1919, the board allocated funds to convert a surplus collier, the USS *Jupiter*, into the Navy's first aircraft carrier, the USS *Langley* (CV-1)—nicknamed the "Covered Wagon" by her crew. For the next two decades, the little *Langley* provided the first generation of U.S. carrier aviators with their initial carrier training, and offered the fleet a platform to experiment with the combat use of aircraft carriers. When World War II arrived, the slow little ship was converted into a transport for moving aircraft to forward bases, and was sunk during the fighting around the Java barrier in 1942. However, the *Langley* remains a beloved memory for the men who learned the naval aviation trade aboard her.

The Washington Naval Treaty: The Birth of the Modern Aircraft Carrier

While the *Langley* was primarily a test and training vessel, her initial trials led the Navy leadership to build larger aircraft carriers that could actually serve with the battle fleet. The problem was finding the money to build these new ships. The early 1920's were hardly the time to request funds for a new and unproved naval technology, when the fleet was desperately trying to hold onto the modern battleships constructed during the First World War. The solution came after the five great naval powers (the United States, Great Britain, Japan, France, and Italy) signed the world's first arms-control treaty at the Washington Naval Conference of 1922. Though the treaty set quotas and limits on all sorts of warship classes, including aircraft carriers, a bit of

The USS *Langley* (CV-1), the U.S. Navy's first aircraft carrier. She was converted from the collier *Jupiter*. She served as a floating laboratory for U.S. naval aviation into the 1930s, and was subsequently sunk in 1942 during the Battle of the Java Sea.

OFFICIAL U.S. NAVY PHOTO FROM THE COLLECTION OF A. D. BAKER

fine print provided all the signatories with the opportunity to get "something for nothing."

At the end of the war, several countries were constructing heavy battleships and battle cruisers,[7] which were still unfinished in the early 1920's. Meanwhile, the 1922 Washington Naval Treaty set limits on the maximum allowable displacement and gun size of individual ships, as well as a total quota of tonnage available to each signatory nation (the famous 5:5:3 ratio).[8] Even after scrapping older dreadnought-era battleships, the nations within the agreement were left with no room for building new battleships and battle cruisers (which were classed together because of gun size). However, the treaty allowed the signatories to convert a percentage of their allowable carrier tonnage from the hulls of the uncompleted capital ships. What made this especially attractive was that the new carriers could be armed with the same 8-in/203mm gun armament as a heavy cruiser. Thus, even if the aircraft carriers themselves proved to be unsuccessful, those heavy cruiser guns would still make the ships useful.

The British had already converted their tonnage quota with the *Furious*, *Courageous*, *Glorious*, and *Eagle*, while the Japanese converted their new carriers from the uncompleted battle cruiser *Akagi* and the battleship *Kaga*. The American vessels, however, were something special. The U.S. Navy wanted its two new carriers to be the biggest, fastest, and most capable in the world. The starting points were a pair of partially completed battle cruiser hulls. Already christened the *Lexington* and *Saratoga*, they were converted into the ships that the fledgling naval air arm had always dreamed of. When commissioned in 1927, the *Lexington* (CV-2) and *Saratoga* (CV-3) were not only the largest (36,000-tons displacement), fastest (thirty-five knots), most powerful warships in the world, (most important) they could operate up to

7 Battle cruisers, a British invention, combined a large hull and power plant with a battleship's armament. While as fast as a cruiser (twenty-five-plus knots) and as heavily armed as a battleship, they lacked the armor protection of a traditional dreadnought. This made them vulnerable to enemy fire in a gunnery duel, though they could normally run away from a stock battleship.

8 The "5:5:3 ratio" represented the allowable naval tonnage under the treaty for the U.S., Great Britain and Japan respectively. The treaty held until the 1930s, when the run-up to World War II began.

The aircraft carriers *Saratoga* (CV-3, in the foreground) and *Lexington* (CV-2, in the background) together near Diamond Head, Hawaii. At the time this was taken, the two converted battle cruisers were the largest, fastest, and most powerful warships in the world.

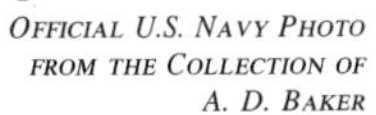

OFFICIAL U.S. NAVY PHOTO FROM THE COLLECTION OF A. D. BAKER

ninety aircraft, twice the capacity of the Japanese or British carriers.[9] The *Lexington* and *Saratoga* also featured a number of new design features (such as the now-familiar ''island'' structures, which contained the bridge, flight control stations, and uptakes for the engineering exhausts), which greatly improved their efficiency and usefulness. The treaty-mandated gun turrets were placed in four mounts fore and aft of the island structure.

With the commissioning of the *Lexington* and *Saratoga* (and parallel rapid strides in naval aircraft design), the U.S. Navy took the world lead in naval aviation development. Virtually all of the American leaders who commanded carriers and air units during the Second World War served their early tours of duty aboard the two giant carriers. In addition, the series of fleet problems (war games) involving the *Lexington* and *Saratoga* led to the tactics America would take into the coming Pacific war with Japan.

The Taranto Raid and the Sinking of the Battleship *Bismarck*

Always leaders in the development of naval aviation technology and tactics, the British had planned for and assimilated the aircraft carrier into their fleet long before the opening of the Second World War. This was not merely institutional integration, for there were also plans for potential wartime carrier operations. One of these plans, devised in the 1930s, involved a surprise strike against the Italian battle fleet based at Taranto harbor in southern Italy: A carrier force would approach at night, launch torpedo bombers, and sink the Italian battleships at their moorings.

---

9 Though it would be two decades before practical experience would prove it, the single most important characteristic of carrier design is aircraft capacity. No other factor, including speed, antiaircraft armament, or armor protection is so desirable as the ability to carry and operate *lots* of aircraft. The British found this out the hard way, when they sacrificed aircraft capacity for armor protection in their *Illustrious*-class carriers, which could only carry about thirty-six planes (while the American *Yorktown* (CV-5) and Japanese *Shokaku*-class carriers could carry ninety).

The opportunity to implement the plan came soon after the Italian declaration of war on Great Britain (in June of 1940) and the fall of France (later that summer). Despite the highly aggressive efforts of the British Mediterranean Fleet under their legendary commander, Fleet Admiral Sir Andrew Cunningham, the fleet was in trouble from the start. It was outnumbered and split by Fascist Italy, since the Italian peninsula more or less bisects the Mediterranean. By the fall of 1940, Italy had six modern battleships, while Cunningham only commanded a pair. His only real advantages were a few ships equipped with radar, the British intelligence ability to read Axis cryptographic (code and cipher) traffic, and a pair of aircraft carriers—the old *Eagle* and the brand-new armored deck flattop HMS *Illustrious*. Doing what he could to make the odds more even, Cunningham ordered his staff to plan a carrier aircraft strike on the Italian fleet base at Taranto. Though they had no real-world experience to work from, and only sketchy data from old fleet exercises about how to proceed, with typical British aplomb they began training aircrews and modifying their aerial torpedoes so they would run successfully in the shallow water of Taranto Harbor. Meanwhile, a special flight of Martin Maryland bombers began regular reconnaissance of Italian fleet anchorages. By November of 1940, they were ready to go with Operation Judgment.

Though the original Operation Judgment plan called for almost thirty Swordfish torpedo bombers from both *Eagle* and *Illustrious*, engine problems with *Eagle* and a hangar fire on *Illustrious* cut that number considerably. In the end, only *Illustrious*, along with an escort force of battleships, cruisers, and destroyers, set out to conduct the attack. On the night of November 11th, *Illustrious* and several escorting cruisers broke off from the main force, and made a run north into the Gulf of Taranto. Later that night, *Illustrious* launched a pair of airstrikes using twenty-one Swordfish torpedo bombers (only a dozen of which carried the modified shallow-water torpedoes). The two strikes sank three of the six Italian battleships then in port and damaged several smaller ships and some shore facilities.[10] In just a few hours, the brilliantly executed strike had cut the Italian battleship fleet in half, and changed the balance of naval power in the Mediterranean.

While most of the world's attention was focused at the time on the Battle of Britain, the eyes of naval leaders were turned on Operation Judgment. Even before the Italians began salvage operations, naval observers from around the world began to pour into Taranto to view the wreckage, and write reports back to their home countries. Most of these reports were quietly read and filed away, or else were read and discounted (such was still the potency of the battleship myth). In Tokyo, however, the report of the Japanese naval attaché was read with interest. This report eventually became the blueprint

10 Of the three battleships that sank to the bottom of Taranto Harbor, the *Littorio* and *Caio Duillo* were eventually raised and returned to service. The third vessel, the *Conte de Cavor,* was not repaired prior to the Italian Armistice in 1943.

for an even more devastating carrier raid the following year, when over 360 aircraft launched from six big carriers would make the strike. The target would be entire U.S. Pacific Fleet at Pearl Harbor. Out of the tiny strike on Taranto emerged the decisive naval weapon of the Second World War.

Less than six months after the Taranto raid, battleship enthusiasts got a shocking dose of reality with the sea chase and sinking of the German battleship *Bismarck*, one of the most powerful warships in the world. After the *Bismarck* broke out of the Baltic Sea into the North Atlantic, she sank the British battle cruiser HMS *Hood*. Outraged at this defeat (and humiliation), Prime Minister Winston Churchill ordered the *Bismarck* to be sunk at all costs. Though she was damaged enough during her fight with the *Hood* to need repairs in port, and her British enemies were in hot pursuit, *Bismarck* was still a dangerous foe, and was able to slip away from her pursuers and make for a French port.

She might well have escaped, but for the efforts of two British aircraft carriers. A strike by Swordfish torpedo bombers from the carrier *Victorious* slowed down the German monster, while another strike from the carrier *Ark Royal* crippled her. The following day, *Bismarck* was finally sunk by shellfire from the British battleships *King George V* and *Rodney*. In the celebration that followed, the contributions of the Swordfish crews from *Victorious* and *Ark Royal* generally went unnoticed—again. However, naval observers took note and wrote their reports home; and naval professionals around the world began to wonder if aircraft from carriers *might* do more than just hit ships in harbor. One of the most modern and powerful ships in the world had been crippled by a single torpedo dropped by a nearly obsolete, fabric-covered biplane in the open ocean.

Before the end of 1941, further proof that the age of battleships had passed came with the Japanese attack on Pearl Harbor and the sinking a few days later of the British battleship *Prince of Wales* and the battle cruiser *Repulse* by land-based aircraft. While battleships would continue to play an important part in World War II, it was naval aircraft flying from carriers that would win the coming naval war. The strike on Taranto and the crippling of the *Bismarck* had seen to that.

Task Force 34/58: The Ultimate Naval Force

Now that the new weapon was proven, the next stage in its evolution was to work out its most effective use. This came during 1943. That year saw a period of rebuilding for both the United States and Japan. After the vast carrier-verses-carrier battles (Coral Sea, Midway, Eastern Solomons, and Santa Cruz) that had dominated the previous year's fighting, the two navies had reached something like stalemate and exhausted their fleets of prewar carriers. Meanwhile, in the Solomons, on New Guinea, and in the Marshall Islands in the Central Pacific, Allied ground forces were conducting their first amphibious invasions on the road to Tokyo.

On January 1st, 1943, the first of a new generation of American fleet carriers, the *Essex*-class (CV-9), was commissioned. Over the next two years, almost two dozen of these incomparable vessels came off the builder's ways. Utilizing all the lessons learned from earlier U.S. carriers, the *Essex*-class vessels were big, fast, and built to take the kinds of punishment that modern naval combat sometimes dishes out. Their designs also gave them huge margins for modifications and systems growth. So adaptable were *Essex*-class ships that a few were still in service in the 1970's, flying supersonic jets armed with nuclear weapons!

The ships of the *Essex*-class were just the tip of the America carrier production iceberg in 1943, for the U.S. Navy also approved the conversion of nine cruiser hulls into light carriers (with a complement of thirty-five aircraft). Though small and cramped, they were fast enough (thirty-three knots) to keep up with their *Essex*-class siblings. Known as the *Independence* class (CVL-22), they served well throughout the remainder of the war.

Along with the fast fleet carriers, the United States also produced almost a hundred smaller escort, or "jeep," carriers. Built on hulls designed for merchant vessels, they could make about twenty knots and carry around two dozen aircraft. While their crews joked wryly that their ships were "combustible, vulnerable, and expendable" (from their designator: CVE), the escort carriers fulfilled a variety of necessary tasks. These included antisubmarine warfare (ASW), aircraft transportation, amphibious support, close air support (CAS), etc. This had the effect of freeing the big fleet carriers for their coming duels with the Imperial Japanese Navy.

As the new fleet carriers headed west into the Pacific, they would stop at Pearl Harbor for training and integration into carrier forces. Together with a steady flow of fast, new battleships, cruisers, destroyers, and other support ships, they would be formed into what were called "task groups." Experience gained during raids on various Japanese island outposts in 1943 showed that the optimum size for such groups was three or four carriers (additional carriers tended to make the groups unwieldy), a pair of fast battleships, four cruisers, and between twelve and sixteen destroyers.

On those occasions when larger forces were called for, two or more task groups were joined into a "task force." These were commanded by senior Naval aviators, and were assigned joint strike missions, refueling assignments, and even independent raids. Though it took time to pull this huge organization together and find the men capable of leading it, by the winter of 1943/1944, what became known as Task Force 34/58 was ready for action.[11] Task Force 34/58, the most powerful naval force in history, put the lid on the Japanese Navy's coffin, and nailed it shut.

---

11 Unlike the Japanese, who tended to keep their warriers in combat until they died, the United States developed a rotation system to rest and replenish its combat personnel at all levels—even admirals. Thus, the fast carrier fleet had two sets of commanders and staffs: the 3rd Fleet under Admiral Halsey, and the 5th commanded by Admiral Raymond Spruance. After each operation, the two fleet staffs would switch, allowing the outgoing group to rest and plan the next mission. Thus, when Halsey was in command, the fleet was known as Task Force 34; and when Admiral Spruance took over, the carrier force was known as Task Force 58.

The ships of the fast carrier force at Ulithi Atoll in 1944.
Official U.S. Navy Photo from the Collection of A. D. Baker

In February of 1944, now composed of four task groups with twelve fast carriers, Task Force 58, under Vice Admiral Marc Mitscher, raided the Japanese fleet anchorage at Truk, wrecking the base and driving the Imperial Fleet out of the Central Pacific. Mitscher, a crusty pioneer naval aviator, aided by his legendary chief of staff Captain Arleigh Burke, ran Task Force 58 like a well-oiled machine. By the end of May, preparations had been completed for an invasion of the Marianas Island group, just 1,500 nm/2,800 km from Tokyo (thus within range of the new B-29 heavy bombers). Since these islands were essential to the defense of the home islands, the Japanese had to fight for them. The largest carrier-versus-carrier fight of the war resulted.

As soon as the invasion forces of Admiral Spruance's 5th Fleet hit the beaches of Saipan in early June, the entire Japanese battle fleet sortied from their base in northern Borneo to counterattack. When they arrived on June 19th, the nine carriers of the revitalized Japanese carrier force (three large, three medium, and three light fleet carriers) got in the first strike, launching their planes against Task Force 58 (now with seven large and eight light fleet carriers). That was their final hurrah; for the Japanese strike simply fell apart against the radar-directed fighters and antiaircraft fire of the American task groups. Of the 326 Japanese planes launched against the American fleet, 220 were shot down. Not a single U.S. ship was sunk or seriously damaged.

The next day, the U.S. fleet found the Japanese carrier force and launched a counterstrike. Blasting through the surviving Japanese planes, they sank the carrier *Hiyo* and several vital fleet oilers, and damaged numerous other ships before returning to Task Force 58.[12] The next day, the decisively beaten Japanese force withdrew to Japan. So great were the losses to Japanese

12 In addition to the loss of the *Hiyo*, the Japanese also lost the two large fleet carriers *Shokaku* and *Taiho* to submarine attacks.

air crews that their carriers would never again sortie as a credible force. When the U.S. 3rd Fleet invaded the Philippines in October of 1944, the four Japanese carriers that took part in the Battle of Leyte Gulf were used purely as decoys, and sunk by air attacks from Task Force 34.

The Revolt of the Admirals, the USS *United States* (CV-58), and the Korean War

When Japan surrendered in September of 1945, the United States had over a hundred carriers in commission or being built. Within months, the Navy had been slashed to a fraction of its wartime peak. Only the newest and most capable carriers and other warships were retained in the tiny Navy that remained. Part of this massive force reduction was a consequence of the simple fact that the war had ended and the naval threat from the Axis nations had been eliminated. But that was not the only rationale for cutting the fleet and other conventional forces.

The major reason for the cut was the development of the atomic bomb. Specifically, the leadership of the new United States Air Force (USAF) had convinced the Truman Administration that their force of heavy bombers armed with the new nuclear weapons could enforce the peace, protect the interests of the United States, and do it without large conventional ground and naval forces. This was a debatable point, which events were soon to prove hugely wrong. But the immediate result was a mass of hostility that broke out between the Navy and USAF in the last years of the 1940's.

The hostility did not start then, however. It had its roots in the 1920's in the battles over airpower between the Navy and Brigadier General Billy Mitchell. Mitchell, an airpower zealot and visionary, was not an easy man to like. He had already fought a losing battle to convince Army leaders of the virtues of airpower. Meanwhile, the small corps of Army aviators saw the developing strength of Naval aviation, which some of them saw as taking funds and support that should have been theirs. To set right this (perceived) imbalance, Mitchell and his fliers (against orders) sank the captured German battleship *Ostfriesland*, an act that did not sit well with the Navy. In 1925, fed up with Mitchell's stings and barbs, his superiors brought him up before a court-martial, where Mitchell, ever unrepentant, stated that airpower made the navies of the world both obsolete and unnecessary. Not surprisingly, the Navy (and others) publicly defended themselves against these charges, and they did it so effectively that Mitchell's professional career was finished. Mitchell's supporters never forgot or forgave that. The result was a multi-decade blood feud.

The Navy/Air Force war reached its peak during the 1949 fight for new weapons appropriations. Then as now, new weapons systems were expensive. Then, as now, the Navy and the Air Force saw it as a zero-sum game: You win/I lose (or vice versa). Practically speaking, the fight was over whether the nation's defense would be built around the new B-36 long-range bomber (armed with the H-bomb), or a new fleet of large aircraft carriers (called supercarriers) armed with a new series of naval aircraft that could carry atomic weapons. There was only enough money in the defense budget for one of these systems, and the Navy lost. The first supercarrier, the USS *United States* (CV-58), was canceled by Secretary of Defense Louis Johnson just days after her keel had been laid at Newport News, Virginia.

Outraged, the Navy's leadership made their case for Naval aviation in a series of heated (some would say fiery) congressional hearings that called into question the capabilities of the B-36 and the handling of the matter by Secretary Johnson and the Air Force. Johnson did not accept this "Revolt of the Admirals" patiently; the Navy's leadership suffered for their rebellion against him. Many top admirals were forcibly retired, and the Navy paid a high price in personnel and appropriations.[13] However, it did manage to win some fiscal support for modernization of older fleet carriers and development of new jet aircraft.

This turned out to be a godsend, for the fiscal frugality of the Truman Administration came to a crashing halt with the outbreak of the Korean War in 1950, which caught the U.S. and the world with their military pants down. Except for some Air Force units in Japan and a few of the surviving aircraft carriers and their escorts, there was little to stop the North Korean forces from overwhelming the South. Built around the USS *Valley Forge* (CV-45) and the British light carrier *Triumph*, Task Force 77 was sent by the United Nations to interdict the flow of North Korean supplies and men. Eventually, Task Force 77 grew to four *Essex*-class carriers, and would become a permanent fixture not only during the Korean Conflict, but also throughout the Cold War.

For the next three years, carrier-based fighter-bombers rained destruction on the forces of North Korea and (after they entered the conflict) the People's Republic of China. Korea was not a glamorous war. For the pilots and crews of the carriers and their escorts, it was a long, cold, drudgery-laden, never-ending fight in which victory always seemed distant. What glory there was went to the "jet-jocks" flying their USAF F-86 Saber jets up into "MiG Alley" to duel with the Korean, Chinese, and Russian pilots in their MiG-15's. But for the Navy and Marine pilots on the carriers, Korea meant blasting the same bridges and railroads they had hit last week, and would hit again next week. Still, Korea answered any question of America's need for Naval aviation to protect its far-flung interests during the Cold War.

With the end of the Korean Conflict, and the inauguration of a new President, the answer took physical shape in the completion of the aircraft carrier development cycle. Within just a few years, the first of four new *Forrestal*-class (CV-59) supercarriers would be built, setting a model for every new American carrier built ever since. Despite improvements in every

13 One of the few survivors of the purge was Admiral Arleigh Burke, later to become—arguably—the Navy's greatest modern leader.

system imaginable (from nuclear power plants to radar-guided SAM systems), the *Forrestals* have defined the shape of U.S. carriers for almost forty years. Meanwhile, the development of aircraft like the F-4 Phantom II, E-1 Tracer, S-2 Tracker, and others, led to the present-day structure of American carrier air wings. And at the same time, the roles and missions of carriers and their battle groups—their moves as pieces on the Cold War chessboard—were fixed in the minds of the politicians that would use them. The model set by the *Forrestal* and her jet-powered air wing was an almost perfect mix for the Cold War. With some improvements in Naval architecture and aircraft design, it has stayed on and done a great job.

Critical Technologies: Getting On and Off the Boat

What things make carrier-based Naval aviation possible? Actually, a surprisingly few critical technologies set carrier and carrier-capable aircraft design apart from conventional ship and land-based aircraft designs. Most have to do with getting on and off of the ship, and being tough enough to do it over a period of decades.

The Need for Speed: Chasing the Wind

Other than being a lot of fun, speed is essential for aircraft carriers . . . for two reasons:

- High speed generates artificial wind over the flight deck to assist in the launching and landing of aircraft.
- High sustained speed allows carriers to rapidly transit from one part of the world to another.

Wind over the deck allows some influence over an aircraft's "stall speed"—that is, the minimum speed at which an aircraft can still be controlled without falling out of the sky. The lower an aircraft's stall speed, the easier it will be to launch and land (a consideration that's especially important on the pitching deck of an aircraft carrier). You get wind over the deck, first of all, simply by steering the carrier *into* the wind. Every knot of wind over the bow acts as a knot of airspeed for an aircraft trying to take off or land, which is why carriers always come into the wind to conduct flight operations. You get even more wind over the deck by cranking up the speed of the carrier. Thus, if you have a fifteen-knot wind and steam into it at twenty-five knots, you can effectively launch and land aircraft at forty knots *under* their normal stall speed. Putting wind over the deck also maximizes aircraft payload and return weight and reduces stress on the flight deck. All of this means that carriers will be using their maximum speed more often than other ships.

Carriers need more than just a high maximum speed (for launching and

A prototype F/A-18E Super Hornet prepares for a test launch from a catapult aboard the USS *John Stennis* (CVN-74). The plane handler is guiding the pilot to the catapult shuttle, which will launch the aircraft.

OFFICIAL U.S. NAVY PHOTO

recovering aircraft); they need to maintain a high transit speed so CVBGs can move quickly across the oceans. The whole point of forward presence is to have it available *now*. Building a high, sustained speed into a ship is not easy. While many ships may be capable of "dashing" for short times at high speeds, they are normally designed to cruise at more sane and economical rates. The twelve-knot cruising speed of your average merchant ship is fine for transporting cars or athletic shoes, but it just won't do if you want to move a CVBG in a few days from the South China Sea (say) to the Persian Gulf. That means carrier power plants have to be durable enough to cruise at high speeds for days or weeks at a time, without having to put in for repairs or overhaul. This is one of the reasons why nuclear power plants and their highly reliable machinery have been the gold standard for carriers for going on three decades. Just how fast is fast enough? Most naval analysts believe that carriers require minimum battle/flank speeds of thirty-three knots/ sixty-one kph to operate aircraft in the widest possible wind and weather conditions, and sustained speeds of at least twenty knots/thirty-seven kph to allow for rapid transits to crisis areas.

Catapults and Wires: Getting On and Off the Boat

Though aircraft carriers are *very big*, there is still *very little* room on the flight deck to support takeoffs and landings. Since a carrier operates as many aircraft as a small regional airport on just a few acres of flat space (about 4.5 acres on a *Nimitz*-class (CVN-68) ship), it makes sense to take advantage of some mechanical muscle to assist the aircraft on and off the flight deck. To this end, carrier designers have for many years depended upon the tried-and-true technologies of catapults (to give aircraft the speed to take off) and arresting wires (to give the drag to land).

The current generation of carrier catapults are basically nothing but steam-powered pistons . . . steam-powered pistons that can throw a Cadillac half a mile (one kilometer). That's a *lot* of power! But when you're trying to fling a fully loaded aircraft like an F-14 Tomcat or E-2C Hawkeye off a carrier deck, you need *that* much power. This is how it works. Simply described, the catapult is a pair of several-hundred-foot-long tubes built into the deck, with an open slot along the top (at deck level) that's sealed by a pair of overlapping synthetic rubber flanges. A ''shuttle'' running above the deck is attached (through the flanges) to pistons at the rear of the tubes; and the nosewheel towbar of the aircraft is attached to the shuttle when it is launched. To accomplish the launch, high-pressure steam, drawn from the carrier's propulsion plant pressurizes the tubes behind the pistons. When the proper pressure is reached, a lock is released, a small, disposable fastener called a ''holdback'' (it holds the nosewheel to the shuttle) breaks loose, and the pistons (and attached shuttle) fling the aircraft down the deck. At the end of the deck the towbar releases from the shuttle, and the aircraft is airborne. The piston and shuttle assemblies are then run aft (back to the rear of the tubes) in order to prepare for the next launch.

Catapults are high-maintenance, complex, high-risk pieces of equipment that have the ugly habit of failing or breaking if they are not treated with loving care. This is one of the reasons why some nations have chosen to forgo them in their carriers and employ instead vertical/short takeoff and landing (V/STOL) aircraft (like the Harrier/Sea Harrier jump jet), which do not require catapults to operate from ships. Though the technology behind a carrier catapult is relatively simple, the size of the tubes and the magnitude of the forces involved make designing and building them hugely difficult. Very few nations have either the technical or industrial skills to build them. Thus, the very proud and competitive French (who don't like to admit to being second in *anything* military) are buying American catapult units for their new supercarrier, *Charles de Gaulle*. The Soviets, after a generation of trying, failed to devise a reliable catapult unit for their carrier, the *Kuznetzov*.

While taking off from a carrier is difficult, landing on one is almost appalling! Setting down a CTOL (Conventional Take Off and Landing) aircraft like an F/A-18 Hornet strike fighter, for example, has been compared to taking a swan dive out of a second-floor window and hitting a postage stamp on the ground with your tongue. During the Vietnam War, scientists made a study to find out when naval aviators were under their greatest stress during a mission. Their cardiac monitors told the scientists that getting shot at in a bomb run was not even close to the stress of a night carrier landing in heavy weather. In order to make carrier landings easier and less fearsome, the Navy has developed a series of automatic and assisted landing aides to help pilots get their aircraft onto the heaving, pitching deck. But once you're there, how do you stop thirty or forty tons of aircraft that have just slammed down at something over a hundred knots?

Well, you attach a hook to the tail of your aircraft (the famous ''tailhook'') and ''trap'' it on one of a series of cables set across the deck. These cables are woven from high-tensile steel wire, which are stretched across the after portion of the ship. Usually four of these cables are laid out along the deck. The first is placed at the very rear of the carrier (called the ''ramp''

A prototype F/A-18E Super Hornet about to "trap" a landing wire during trials aboard the carrier *John Stennis* (CVN-74).
BOEING MILITARY AIRCRAFT

by naval aviators); the second a few hundred feet forward of that; and so on. The last goes just behind the angle that leads off the port (left) side of the ship. This creates a box into which the pilot must fly the aircraft and plant his tailhook onto the deck.

What happens if a pilot misses the wires? Well, that is another issue entirely. CTOL carrier landing decks are angled to port (left), about 14° off the centerline. This is so that if an aircraft fails to "trap" a wire, then it is not headed forward into a mass of parked aircraft. Instead, the aircraft is now headed forward to port. This is the reason why on every landing, as soon as they feel their wheels hit the deck, pilots slam the engine throttles to full power. Thus, if they do not feel the reassuring tug of the wire catching the hook (more of a forward slam actually), they can just fly off the forward deck (a "touch and go") and get back into the pattern for another try. This is known as a "bolter," and most naval aviators make a lot of these in their careers.

Generally, hitting the rearmost (or "number one") wire is considered dangerous, since by doing that you're risking coming in too low and possibly hitting the stern (fantail) of the carrier (which is known as a "ramp strike"). So too is catching the last one ("number four"). Because you don't have much room to regain airspeed in the event of a "bolter," you risk a stall and possible crash while trying to climb back into the pattern. Catching the number-two wire is acceptable. But catching the number-three wire (called an "OK Three" by the air crews) is optimum, for it allows maximum room from the fantail and maximum rolling distance to regain speed and energy in the event of a bolter. Catching the "number three" is evidence of great professionalism and skill. In fact, if there is not a shooting war around to test your abilities and courage, then a consistent string of "OK Three" traps is considered the best path to promotion and success for a carrier pilot.

So what comes next? You have hit an "OK Three" trap, your aircraft's tailhook has successfully caught a wire, yet you are still hurtling forward at a breathtaking speed and may fly off the forward deck edge of the "angle" at any moment if all doesn't go well. In other words, the excitement isn't over. Each end of the arresting wire runs though a mechanism in the deck down to a series of hydraulic ram buffers, which act to hold tension on the wire. When the aircraft's tailhook hits the wire, the buffers dampen the energy from the aircraft, yanking it to a rapid halt. Once the aircraft stops, the pilot retracts the hook, and is rapidly taxied out of the landing zone guided by a plane handler. While this is happening, the wires are retracted to their "ready landing" position, so that another aircraft can be landed as quickly as possible. When it is done properly, modern carriers can land an aircraft every twenty to thirty seconds.

Aircraft Structures: Controlled Crashes

Any combat aircraft is subjected to extraordinary stresses and strains. However, compared with your average Boeing 737 running between, say, Baltimore and Pittsburgh, carrier-capable aircraft have the added stresses of catapult launches and wire-caught landings that are actually "controlled crashes." That means your average carrier-capable fighter or support aircraft is going to lug around a bit more muscle in its airframes than, say, a USAF F-16 operating from a land base with a nice, long, wide, concrete runway. This added robustness of carrier aircraft (compared with their land-based counterparts) is a good thing when surface-to-air missiles and antiaircraft guns are pumping ordnance in their direction. But it also means that carrier aircraft, because of their greater structural weight, have always paid a penalty in performance, payload, and range compared with similar land-based aircraft.

This structural penalty, however, may well be becoming a thing of the past. Today, aircraft designers are armed with a growing family of nonmetallic structural materials (composites, carbon-carbon, etc.), as well as new design tools, such as computer-aided design/computer-aided manufacturing (CAD/CAM) equipment. They have been finding ways to make the most recent generation of carrier aircraft light and strong, while giving them the performance to keep up with the best land-based aircraft. This is why carrier-capable aircraft like the F/A-18 Hornet have done well in export sales (Australia, Finland, and Switzerland have bought them). The Hornet gives up nothing in performance to its competitors from Lockheed Martin, Dassault-Breguet, Saab, MiG, and Sukhoi. In fact, the new generation of U.S. tactical aircraft, the JSF, may not pay any "structural" penalty at all. Current plans have all three versions (land-based, carrier-capable, and V/STOL) using the same basic structural components, which means that all three should have similar performance characteristics. Not bad for a flying machine that has to lug around the hundreds of pounds of extra structure and equipment that allow it to operate off aircraft carriers.

All of these technologies have brought carrier aircraft to their current state of the art. However, plan on seeing important changes in the next few years. For example, developments in engine technology may mean aircraft with steerable nozzles that will allow for takeoffs and landings independent of catapults and arresting wires. Whatever happens in the technology arena, count on naval aircraft designers to take advantage of every trick that will buy them a pound of payload, or a knot of speed or range. That's because it's a mean, cruel world out there these days!

Hand on the Helm: An Interview with Admiral Jay Johnson

Guiding Principals: Operational Primacy, Leadership, Teamwork, and Pride.

Admiral Jay Johnson, *Steer by the Stars*

During the long history of the U.S. Navy, there have been many inspirational examples of individuals coming out of nowhere at the time of need to lead ships, planes, and fleets on to victory. During the American Civil War, for example, a bearded, bespectacled gnome of an officer named Lieutenant John Worden took a new and untried little ship named the *Monitor* into battle. When Worden faced the mighty Confederate ironclad ram *Virginia* at Hampton Roads in 1862, his actions with the *Monitor* saved the Union frigate *Minnesota*, the Union blockade fleet, and General George McClellan's army from destruction.[14] More importantly, his inspired use of the little turreted ironclad forever changed the course of naval design technology, and made the wooden ship obsolete forever. There are other examples.

A mere half century ago, the United States Pacific Fleet was nearly destroyed by the Japanese at Pearl Harbor. Within days of the raid that brought the United States into World War II, a gravelly-voiced, leather-faced Texan named Chester Nimitz was picked to lead what was left of the Pacific Fleet against the powerful forces of Imperial Japan. Nimitz's early Naval service (mostly spent quietly in the ''pig boats'' that the U.S. Navy passed off as submarines in those days) gave no indication that he was the man for the job. Nor did his later career in virtually invisible jobs at obscure (to ordinary folks) places like the Bureau of Navigation add much to that aura. When he was made CINCPAC (Commander in Chief of the Pacific), few Americans outside of his friends in the Navy even knew the man's name.

14 The CSS *Virginia* is more widely, though incorrectly, known as the *Merrimac*, after the Union ship that she was built from.

With fleet morale shattered by the events at Pearl Harbor, he hardly seemed an inspiring choice.

That opinion began to change almost immediately, when Nimitz retained many of the staff officers present at Pearl Harbor, rather than cashiering them and bringing in his own people. The men responded with total loyalty, and many were instrumental in the subsequent Allied victory in the Pacific. His action in retaining these officers, even though some commanders would have gotten rid of them for their perceived "responsibility" for the disaster, proved to be the first of an unbroken string of brilliant personnel, planning, and operational decisions. These eventually brought Nimitz to the deck of the USS *Missouri* (BB-63) in 1945 as the Navy's representative to accept the Japanese surrender.

Though the Navy has been blessed with many fine leaders in its illustrious history, all the successes of the past are meaningless unless it can serve effectively today and in the future. The late 1980s and early 1990s have tested the faith of even the most fervent U.S. Navy supporters. Following what some felt was a mediocre performance during Desert Storm in 1991, the Navy suffered a string of public relations "black eyes" that included the infamous 1991 Tailhook scandal. There was worse to come. In the spring of 1996, after a media frenzy and an intense round of public criticism over both his handling of personnel matters and his own character, the popular Chief of Naval Operations (CNO), Admiral Mike Boorda, died of a self-inflicted gunshot wound. The suicide of this much-admired sailor cast a pall over the entire fleet; and many in and out of the Navy began to question the quality of Navy leadership. Clearly, it was time for a top-notch leader to step up and take the helm. The man selected to take over as Chief of Naval Operations was actually much closer at hand than some would have thought—in fact, just a few doors away in the office of the Vice Chief of Naval Operations. Admiral Jay Johnson would soon start the Navy back on the road to excellence.

Johnson, a career naval aviator and fighter pilot, has quietly served his country and his Navy for more than three decades. A slim and trim officer who looks years younger than his age, Johnson is a quiet and sometimes shy man. But the quiet demeanor is something of a smoke screen. This man is a "doer," who has chosen to make the hard decisions that will give the U.S. Navy a real future in the 21st century. Johnson is a passionate man, one who cares deeply about his country, his Navy, and the sailors who serve under him. He channels all that emotion into one goal: to build the U.S. Navy into a superb fighting machine, an organization that is once again the envy of military officers everywhere in the world.

Jay L. Johnson came into the world in Great Falls, Montana, on June 5th, 1946. The son of a soldier in the Army Air Corps, he spent the bulk of his youth in West Salem, Wisconsin. Let's let him tell the story of his journey into naval service:

Admiral Jay L. Johnson, USN
Official U.S. Navy Photo

Tom Clancy: Could you please tell us a little about your background and Navy career?

Admiral Johnson: I was born in Montana. My dad was serving there at the time. I didn't stay there long—only about a year. I spent the rest of my youth in Wisconsin, in a little town with a lake near it, not far from the headwaters of the Mississippi River. That's the total exposure to water that I had in my early years.

Tom Clancy: What made you choose the Navy as a career?

Admiral Johnson: I'd been intrigued by the military service academies as I was growing up. I had a distant relative who had gone to West Point, and was thinking about applying there myself. Then I went to a Boy Scout National Jamboree out in Colorado Springs, Colorado, in what is now the Black Forest, just down the road from the Air Force Academy. It was in 1960, I believe, about a year after the Air Force Academy had come into being. As part of our stay, we were invited to a tour there. We also got to see a show by the Thunderbirds [the Air Force precision-flight demonstration team]. As I watched that performance, and looked at that academy, I said to myself, "I can do this!" When I returned home, I decided that I'd apply to the Air Force Academy. Before I did so, I found that I had an option to go to the Naval Academy at Annapolis. I looked into it, found out a bit about carrier aviation, and decided that was what I wanted. I took that opportunity, and here I am.

Admiral Jay Johnson, in his Pentagon office with the author.
John D. Gresham

Tom Clancy: Did you have any particular ''defining'' experiences while at the Academy?

Admiral Johnson: Well . . . I got to watch Roger Staubach [the great Naval Academy and Dallas Cowboys star quarterback] play football. On a more serious note, the most striking thing I remember about my time there is how close my company mates and I became. To this day, we're inseparable. A lot of them are still in the Navy today. Admiral Willie Moore, who is the USS *Independence* [CV-62] battle group commander, was a company mate of mine. My former roommate is the Naval attaché to India. Rear Admiral Paul Gaffney, who is the Chief of Naval Research, was also in my company. These are just a few of the people I met at the Academy who are special to me personally.

Tom Clancy: Were there other notable members of the Academy classes while you were there?

Admiral Johnson: Guys like Ollie North and Jim Webb [the former Secretary of the Navy]—and of course Roger Staubach from the class of '65. I have always admired him. Even then, he was a man of great integrity, courage, and superb physical prowess. What I see of Roger today matches exactly what I saw then. It's nice to see a guy who is that solid early in his life, remain so through a highly visible career, retirement, and new career.

Tom Clancy: You graduated during the depths of the Vietnam conflict [1968]. Were you immediately sent out to flight school and into the Replacement Air Group [RAG]?

Admiral Johnson: Well, they did move us through at a nice pace, though I don't remember it being any kind of "rush" job. I went through flight training in pretty much a normal time frame. I got my wings in October of 1969. From there I headed out to San Diego and NAS Miramar to learn to fly the F-8 Crusader.

Tom Clancy: You must have been there with some living legends, men like "Hot Dog" Brown and Jim "Ruff" Ruffelson, right?

Admiral Johnson: Yes, they were there. Being one of the F-8 "MiG Killers" was kind of the unusual for a new guy back then. It was the time when a lot of the guys fresh out of the Academy were getting orders to F-4's [Phantom IIs], and most of us were lined up to get into the Phantom community because they were new and they were hot! More than a few of us wound up flying F-8's though, and in retrospect it was the best thing that ever happened to me. The F-8 was an awesome airplane. And, as good as the airplane was, the community of people who flew and supported it was even better. We're all still pretty tight. We have F-8 Crusader reunions every year.

Tom Clancy: Could you tell us a little about your experiences in the Crusader?

Admiral Johnson: I had about a thousand hours in the Crusader. I did two combat cruises to Vietnam in VF-191 aboard the USS *Oriskany* [CVA-34], in 1970 and 1972. As I recall, we went out for a long cruise, came back for a short time, and then did an even longer cruise. In all that time, I only had one backseat ride in a Phantom. I think I may be one of the few naval aviators of my generation who has never flown an F-4. From the Crusader I went straight into the F-14 Tomcat.

Tom Clancy: From your record, it looks like you spent the majority of your career in the Tomcat community.

Admiral Johnson: That's right. I did my department head tour and my squadron command tour in Tomcats. However, when I went to become an air wing commander, I tried to fly most of the air wing airplanes. The planes I flew back then included the A-7 Corsair, which is like a stubby-nosed cousin to the F-8 without an afterburner. I also flew the A-6 Intruder. Later, on my second CAG [Commander, Air Group—the traditional nickname for an Air Wing Commander dating back to the beginnings of carrier aviation], on my battle group command tour, I wound up flying the F/A-18 Hornet. I still remember flying the F-8, though. Your first jet assignment is like your first love. It's where everything is defined for you.

Tom Clancy: Following your time in F-8's, you seem to have spent most of your time in the East Coast units. Is that correct?

Admiral Johnson: It's correct, but it really wasn't a conscious decision on my part. I guess it just worked out that way. Initially, when I learned to fly the Tomcat, I headed back out to the West Coast and went through the F-14 RAG [Replacement Air Group], VF-124. Then I was moved back to the East Coast, where I have pretty much stayed ever since.

Tom Clancy: Obviously, you spent an eventful couple of decades with the fleet in the 1970's and 80's. Can you tell us a few of the things that stand out in your mind?

Admiral Johnson: The Vietnam experience stands out, of course. The operations against Libya in the 80's were interesting—Operations Prairie Fire and Eldorado Canyon [the bombing of Libya in April 1986]. I was in and out of there several times during that period. I also remember the day that Commander Hank Kleeman and the guys from VF-41 [the Black Aces] ''splashed'' two Libyan Sukhois back in [August] 1981. I was sitting in flight deck control [on the USS *Nimitz* [CVN-68], getting ready to man up and recycle one of the combat air patrol [CAP] stations. The plan was to land the first pair of F-14 Tomcats. Then I was going to be part of the second ''go'' of the day. It was announced over the 1MC [the master public address system on board the ship] that something ''big'' had just happened. When the two F-14's that had shot down the two Libyan fighter-bombers got back aboard, everyone wanted to look at the planes and see what had happened.

Tom Clancy: You came into this job [as Chief of Naval Operations, or CNO] at a time of great crisis and turmoil for the Navy. Among other issues, Admiral Boorda's death was a great blow to the Navy. What were the important things that you had to do when you arrived?

Admiral Johnson: It was important to me to make sure, because of Admiral Boorda's reputation as a sailor in the fleet, that the officers and sailors in the fleet knew that things were going to be ''O.K.'' I sent out an ''all hands'' message to that effect, and spent the next eight or nine months traveling around the world to get the message out to the people in the fleet.

Tom Clancy: As CNO, you seem to have a unique working partnership with Secretary of the Navy John Dalton, and the Commandant of the Marine Corps, General Charles ''Chuck'' Krulak. Can you tell us about that relationship?

Admiral Jay Johnson eating a 1997 holiday meal with sailors aboard ship in the Persian Gulf.
Official U.S. Navy Photo

Admiral Johnson: As you know, before I got here, Secretary Dalton made the decision to relocate the Commandant of the Marine Corps and most of his staff from the old Navy Annex up the hill to the E-Ring of the Pentagon. So now Secretary Dalton's office is bracketed by the Commandant's office on one side, and the CNO's on the other. He's got us in stereo! The decision to move the Marine Corps Commandant was a powerful one, in my opinion. The relationship between Secretary Dalton and Chuck Krulak was already in place even before I arrived. When I got here as Vice CNO, and particularly as I made the transition to CNO, both men were very understanding, supportive, and helpful. I could not have asked for a better welcome.

Tom Clancy: It sounds like the three of you have forged a special working relationship on this end of the E-Ring corridor. Is that true?

Admiral Johnson: The short answer is *yes*! These relationships work very well due to a number of factors. First of all, Chuck Krulak and I are friends. He and I are close personally, as are our wives. That's a good start to a professional relationship, but there's more to it than that. We share some important common goals. For example, we are both making a concerted effort to lead our sailors and marines to work well together in this age of cooperation and coordination between the various branches of the military. I mean, how the hell are you going to do that, if the top sailor and marine can't get along?

The relevance of the sea services, both the Navy *and* the Marine Corps, is that we're the forward presence for our country in virtually any military operation. We're there first, and we're out last. It's essential that we coordinate our forces to do the job as well as it can be done. We're proud of our mission, proud of our people, and proud of our ability to do the job together. That's the strength that we give to the country.

Now, just because we're trying to work together on our various

Admiral Jay Johnson relaxing in his Pentagon office during his interview with the author.

John D. Gresham

missions does not mean that the job of coordinating the Navy and the Marine Corps is an easy one, either for Chuck and me, or for the other officers and enlisted soldiers on our staffs. We work with some *very* challenging issues, and we aren't always able to agree completely on every point we discuss. As in any working relationship, there are occasional conflicts.

But we're committed to working through them and formulating solutions. The principles that underlie our working partnership and the friendship between Chuck and me girds it all and makes it possible for us to work through those hard decisions. This benefits both services. Both Chuck and I have the support and guidance of Secretary Dalton as well. I think we have a pretty good team.

Tom Clancy: As we all know, it's been a challenging decade for the Navy. In addition to issues like Tailhook and Admiral Boorda's death, there were real problems that had been building for over two decades. You were placed in charge of a Navy whose ships had been run hard during the Cold War years. Can you tell us a bit about the state of the fleet today?

Admiral Johnson: Despite the many challenges we're had to endure, the Navy has carried on wonderfully, in my view, in terms of reacting to the requirements that have been levied upon it. Our mission as the nation's forward-deployed force means we have to be prepared to respond at all times to any situation in which we are needed. The relevance of that mission will not change as we go into the 21st century. I believe we are ready. That's what we do, seven days a week, 365 days a year. I think that one of the greatest challenges that we face in the Navy is reassuring the American people of the level of our commitment to the mission to serve and protect them. This is important, because for a lot of people, what we do is sort of ''off of the radarscope.''

Tom Clancy: Given what you have just said, how is the fleet bearing up under this extremely high Operations Tempo [Optempo]?

Admiral Johnson: That's a question that requires an answer on more than one level. There is no denying that our sailors, by the very nature of their work, spend time away from their homes and families. Some of the things that we are looking at are ways to make sure that we don't overstretch ourselves.

Right now we have a policy that says that ships will have no more than six months forward-deployed at sea, from portal to portal. We're also maintaining a ratio of two-to-one for time at home port to deployed time, and no more than fifty percent of time out of home port when you are off deployment.

We are adhering to that policy, and I am the only one who can waive it for any reason. In fact, whether we are standing by that policy is one of my own measures of whether we are ''stretching the rubber band too tight,'' where people are concerned. So right now, we're OK with that situation. Now, we have had a couple of exceptions to this rule last year because of problems with ship maintenance in a yard that closed down. The result is that in terms of readiness and execution, the fleet is ''answering on all bells.'' I want to make sure as you walk back from looking at deployment issues, that everyone is getting enough training to get ready, but not so much that their home lives suffer. We also want to make sure that the right equipment is available during training, so that the fleet fights with the same gear it trains with.

Tom Clancy: How is retention of personnel holding up?

Admiral Johnson: Retention right now is good, though there are pockets of concern in that situation. If you look, for instance, at pilot retention numbers, the aggregate numbers, they're great. They're not even worth talking about today. There is *no* problem there. Within that community, though, if you ''peel that onion'' back a layer, we're beginning to see that we need to pay attention to the attrition rates of some kinds of air crews.

In my view, these situations are not developing just because the airlines are hiring. The airlines are *always* hiring, and will continue to hire. That's a reality that we can't change. But I do think that part of this softness in community retention is based upon the ''turnaround'' and non-deployed side of a Naval career.

In particular, we need to make sure that we're not keeping people too far from home for too much time doing temporary kinds of assignments. We need to make sure that we don't have backlogs in aircraft and equipment depot maintenance, so that our crews have enough airplanes to fly during turnarounds and workups. We also have to pay attention to the matter of funding enough flying hours to keep our people sharp. Let's face it, junior officers [JOs] *never* get enough flight hours. I know that I didn't as a young aviator, and I don't know anyone who did. We've still got some work to do in that area.

These "soft" community areas are not just limited to naval aviation. We've got some year groups in the submarine community that we're watching carefully, as well as some in the surface warfare professionals. Overall, though, we're OK. On the enlisted side the numbers are excellent, and most significantly, the high quality of personnel is there.

These days, we're having to work very hard to get that quality, and it's a real challenge. The goal of our recruiting is to have ninety-five-percent high school graduates, with sixty-five percent of those recruits in the *top* mental group in their classes. When we achieve that, it's good for the fleet, and we're committed to achieving that.

However, the competition for that part of the labor market is really intense out there. Given the pressures of a healthy economy, I think that it's going to be more and more of a challenge. The *really* good young men and women out there—the ones who are really smart and talented—everybody wants them. Frankly, while I can offer them a lot, there are other folks who can offer them more of things like money. Still, there are wonderful and patriotic young folks who take up the challenge, and we work hard to find them and keep them in the fleet.

Trust me when I say that the recruiting challenges will not go away. Remember, back in the Cold War we had to bring around 100,000 new recruits a year into the fleet to fill our needs. Today, even in a time of relative peace, we still need between 45,000 and fifty thousand new sailors every year to keep our force healthy and running.

Tom Clancy: You just mentioned the end of the Cold War. Can you tell us something about the challenges that you and the fleet have faced in light of the end of the East/West conflict?

Admiral Johnson: I think that one of the biggest challenges that the Navy has is to make sure that our nation still has an appreciation for the value of the sea services, especially within the citizenry and the Congress. I think we need to educate the public to understand that while we have a great Navy now, it takes a lot of effort and money to keep it that way. Another challenge the Navy has been faced with in the last few years has been overcoming the public perceptions left by Tailhook. I believe that we've made great strides since then.

Tom Clancy: Do you feel that the challenges that the Navy had to face as a result of Tailhook and other incidents have helped the sea services deal better with the issues of women in the force?

Admiral Johnson: Yes, I do. Since we were the first of the services forced to confront the gender-related issues that the other military departments face right now, I hope and truly believe that we have learned from those hard experiences and are better for them. We needed to change the ways that we were doing business in many respects, and I think that we have, and I'm proud of that. I believe that we have a much better and stronger force today than we did before Tailhook.

Tom Clancy: Let's talk a little more about the roles and missions that the Navy is undertaking in the post-Cold War world. For example, with the decline of the Russian fleet, what have you got the submarine force doing?

Admiral Johnson: We actually have some exciting things coming up for the submarine force. The New Attack Submarine [NSSN] program is underway, and *Seawolf* [SSN-21] has been commissioned. As far as the submarine mission is concerned, it is much more diverse than during the Cold War. Their main mission is still undersea warfare and antisubmarine warfare [ASW]. Today that mission is less predictable than it was during the Cold War, but challenging nonetheless.

The "big water" ASW mission is still a part of our lives, but these days littoral [inshore] ASW is an even bigger and emerging piece of that mission. I might add that inshore operations bring with them a whole new set of challenges. Our submarine force is today involved in strike warfare, reconnaissance, special operations, and lots of other things. The community is being reshaped to reflect all of these new missions, and remains a valuable and viable part of the fleet. And even though today's Russian submarine force is not the Soviet fleet of the Cold War era, I feel strongly that we must maintain sufficient capability to deal with it. With all of that factored in, our force of submarines is still going to shrink. We have around seventy SSNs today, and we'll probably drop to around fifty-five in the next few years.

Tom Clancy: The doctrinal move of the sea services to specialize on the littoral regions has been going on for some time now and the force seems to have adapted well. Can you please tell us your views on how the transition from a "blue water" Navy to an inshore focus has gone?

Admiral Johnson: The transition has gone extremely well, though the Navy has always concentrated on that mission to some degree. You have to remember that the majority of the world's capitals and much of its population reside close to the shores of the world's oceans. Because of that, the Navy has always been tasked for littoral warfare. We kept the littoral missions at the forefront of our planning and preparation throughout this century.

During the Cold War years that mission was somewhat overshadowed by open ocean missions. With the demise of the Soviet Union, what we call "blue water" missions have declined in importance somewhat. Overall, I think we're ideally tasked to meet the challenges of the new century with the force that we have today and the force that we're building for tomorrow. This includes systems like the land attack destroyer variant of the SC-21 escort design that we're currently designing.

Tom Clancy: So does this mean that the mission of the U.S. Navy in the 21st century is going to be like that of the Royal Navy in the 19th century? In other words, showing the flag, keeping the peace, and letting the locals know that we're there?

Admiral Johnson: There's certainly a lot of that in our future plans. I think that the way that we would describe our mission in the Navy is that we plan to shape the environment or battle space. We will do that through forward presence. I might also add that we will do that while carrying the full spectrum of weapons, sensors, and other tools that we need, so that the national leadership does not have to wait for the action to be joined. That is critical because, as I like to say, there is no substitute for being there.

The Navy has to be there and ready to trigger whatever kind of response might be required by a rapidly developing situation. That response might be little, it might be big. In one context we're the enabling force for follow-on units, and in another we're the striking power all by ourselves for whatever might be going on in a particular area. Our mission is always situational. There are times when we'll do port visits and paint schoolhouses. That's a part of our job. But we are also ready to kick somebody's ass if that's required. So to that degree, your Royal Navy analogy is quite valid.

Tom Clancy: One major change in how the military is doing business today compared to the past is that, unlike the CNOs of just a decade ago, you don't actually command ships and planes anymore. Under Goldwater-Nichols, the various branches of the military and their assets are combined into organization "package" forces for regional commanders in chief [CinCs] to use as required by the National Command Authorities. Under the new system that came into effect in the late 1980's, the world has been divided into regions and warfare specialties, with a joint unified command and a commander in chief [CinC] assigned to each. The CinCs package forces into joint task forces [JTFs], which are the basic working

Admiral Jay Johnson making a point during his interview with the author.

John D. Gresham

unit of joint warfare. Could you tell us a little about how that process works?

Admiral Johnson: As far as the Navy goes, I'm the ''organize, train, and equip guy.'' I get the forces ready by making sure that the Navy is well staffed, that our personnel know what they're doing, and that the machinery to support them is of the right kind and in good working order. At the appropriate time, I turn over these forces to the warfighting CinCs. While those ships and aircraft are assigned to numbered fleets and squadrons, we often have to package them in some rather unique ways, depending upon the situation and the requirements.

Tom Clancy: By that do you mean contingencies like Haiti back in 1994? I recall that you stripped two big-deck aircraft carriers of their air wings, and replaced them with a couple of aviation brigades with helicopters and troops from XVIII Airborne Corps.

Admiral Johnson: Well, since I was the Deputy Joint Task Force [JTF] Commander and naval component commander of that operation, let me give you my two cents' worth on how that all worked out. It was really interesting and, I might add, the right thing to do for that situation. I was the commander of the Second Fleet at the time, and was looking at new ways to use carriers. I took some abuse from some of my Naval aviation pals at the time, who said, ''Well, there you go, Johnson. You've sold naval aviation down the river. Next thing you know you'll be putting army helicopters on aircraft carriers.''

The truth of the situation is that I was not threatened by that at all. If you look at what the Navy and our joint service partners were asked to do in Haiti, and you put it into the context of that particular place, at that time, against that threat, and that total scenario, what we did was damned near perfect. It really was. Almost as close to perfect as you can get. Now, the next time, in a different place and situation, doing something like that may be the dumbest idea in the world. For Haiti, though, converting the carriers was as "right on" as anyone could have asked.

Tom Clancy: Once the need for the helicopters during Operation Restore Democracy was finished, how long did it take to get the regular air groups back on board and operating normally?

Admiral Johnson: Well, let me square you on this whole process. The *Eisenhower* [CVN-69] had the aviation brigade from the 10th Mountain Division, and the *America* [CV-66] had the folks from the Army's 160th Special Operations Aviation Regiment at Fort Campbell. When we were ready to change back to normal, here's what happened. First the helicopters and their personnel cleared off to the Haitian mainland, and *Eisenhower* turned away. I was there, and watched this with my own eyes. The carrier battle groups left Port-au-Prince harbor, motored out past the island of Guni, and before they left the operations area, both had "trapped" their Tomcats, their A-6 Intruders, and their E-2C Hawkeye support aircraft. By the time that they headed north, both carriers had fully reconstituted air groups, and made the transit home mission-ready. It was a great use of carriers, in my view.

Tom Clancy: So would you say that one of the real challenges of this post-Cold War transition has been trying to adapt the minds and thinking of people in the Navy to new ideas and concepts? Making people say "Why not?" as opposed to "Are you out of your mind?"

Admiral Johnson: Absolutely. It's hard to adjust to change. And we're all guilty of resisting it sometimes, as it turns out. I'm as bad as the rest of them, even though I like to think that I'm open-minded. I *am* open-minded, until you start messing with one of *my* ships. . . .

Tom Clancy: Could it not be said that your willingness to be "adaptive" with those two carriers may very well help to justify continued aircraft carrier development and procurement?

Admiral Johnson: It could indeed. I can tell you for a fact that the Navy gained a whole lot of new friends in the U.S. Army as a result of that exercise. Especially when their troops found out that they did not have to eat MREs [Meals, Ready to Eat] during the mission. In fact, here's an interesting piece of trivia from the commanding officers [COs] of those two carriers. It turned out that the soldiers from the two Army aviation units liked Navy chow so much, and ate so much of it, that we had to retool the resupply schedule. The soldiers were just shoveling down all this food on board the ships. Navy chow is good!

Tom Clancy: Taking the Haiti example a bit further, it is fairly clear that since the end of the Cold War, the Navy has been used for a wide variety of roles and missions—everything from blockades and strike warfare [Persian Gulf] to rescues and humanitarian relief [Balkans and Somalia]. Given that you already do such a wide variety of things so well, what else do you want the Navy to be capable of doing in the 21st century?

Admiral Johnson: You're right, the Navy's pretty flexible! In the future, I think that you're going to see us doing some new things with the Marine Corps. We're finding new ways to organize and structure our forces to accommodate new roles and missions. One specific area that I know we'll be developing is Theater Ballistic-Missile Defense [TBMD], using our Aegis cruisers and destroyers. That's new and exciting stuff that ten years from now will be everyday business, though today it's all leading-edge technology.

Tom Clancy: Especially in the absence of a "blue water" threat, has the Navy gotten down to developing a real doctrine to go with the move to littoral warfare?

Admiral Johnson: The answer is yes, but I qualify that answer by saying that we're just at the leading edge of getting it done. At the Naval Warfare Doctrine Command, they're looking at how we can take the earlier "blue-water" doctrine of the Cold War, and embed it in a very solid way into this new reality of littoral warfare. We're trying hard to build new linkages with our various Naval academic institutions like the War College [in Newport, Rhode Island], the Postgraduate School [in Monterey, California], and even the Naval Academy [in Annapolis, Maryland], as well as in the tactical and operational sides of the fleet.

Tom Clancy: Once upon a time, not so long ago, the Navy was seen as not being a good partner in the joint warfare arena. Can you tell us, from the Navy point of view, how you view your corporation and participation in joint warfare these days?

Admiral Johnson: Frankly, I don't see *any* friction today. I think that's old news. As far as I'm concerned, the Navy is on the leading edge in the joint warfare business these days. In fact, we're committed to it at all levels. Here's a case in point. When we do our carrier battle group [CVBG] and amphibious ready group [ARG] workups, that's all joint. What we used to call a FLEETEX [Fleet Exercise] in the old days is now the JTFEX [Joint Task Force Exercise]. Of course, we still work within our fundamental core sea service [Navy and Marine Corps] competencies during training. But once we get into the JTFEX, it's units like the 2nd Fleet CVBG, the II MEF [Marine Expeditionary Force] MEU [SOC], the XVIII Airborne Corps, the 8th Air Force, and our allies all together. So we are absolutely committed to the joint warfare arena, right down to training within the Joint Training Matrix. This is not the way it was during Desert Storm where the Navy was still "fighting the feeling."

That does not mean that we have solved all of our challenges. Full utilization of CTAPS [the joint theater air planning tool] and distribution of the ATO [Air Tasking Order] is still giving us problems, but by and large, we're on board in the joint arena. I might add that we're proud to be part of it, because that's the way that we're going to be fighting in the future as a nation.

Tom Clancy: One of the most interesting joint training exercises that has been run recently is Operation Tandem Thrust, down in Australia. Can you tell us about it?

Admiral Johnson: You have to remember that we have a "special" relationship with Australia, one that has been critical to both countries in this century. Tandem Thrust is just another classic example of that relationship. We just came back from Operation Tandem Thrust. It was *huge*, involving over 22,000 U.S. Army, Navy, Marine Corps, and Air Force personnel. We accomplished our objectives and I think everybody learned a great deal.

When you are running a large military exercise, one of the biggest considerations is the matter of finding new range spaces for the joint forces to exercise and train in. If you talk with Chuck Krulak, he'll tell you about his interest in using some of the range facilities in Australia. They are *beautiful*! And the Royal Australian Navy and the rest of their forces are just superb to work with. They are *wonderful* allies. Australia is an amazing country—just eighteen million people on a land mass the size of the continental United States. You see that when you fly over the place. You just fly for hours and hours and see nothing but open space.

Tom Clancy: Talk a little more about modernization if you will. Every couple of generations, there seems to be a CNO who, because of timing and circumstances, defines the U.S. Navy for a period of decades. Elmo Zumwalt filled that role in the 1970's, since so much of what the Navy uses today was defined, designed, or built during his tenure. You seem to be in a similar situation today in the 1990's. Given this notion, what kinds of things do *you* want this Navy to do?

Admiral Johnson: I think that what we're trying to cast for tomorrow and the future is to be able to say five, ten, twenty, even twenty-five years from now, that *this* Navy is really relevant. We need to know that the Navy is giving the country a presence force that can still respond across the full spectrum of crises or requirements that the country asks them to respond to. We don't even know for sure what kinds of crises we'll be facing in that distant future. But the decisions we make today will have a direct impact on our readiness tomorrow.

In general terms, we know exactly where we're going. The new equipment we're building and the new shaping of the force that we are currently going through are very important to us, as is the way we push ourselves into the next century. It's very exciting, though somewhat daunting, to be in this job at a time when the infrastructure is under development to this degree, but I think we're building a marvelous future for the Navy. In my opinion, the future Navy will still be anchored in the carrier battle group with its air wing, in the amphibious ready group and the embarked Marine Expeditionary Unit. These are the two core assets that the sea services give to the country. I want that to be clearly conveyed as we move forward into the next century.

Tom Clancy: Let's talk a little more about that issue of "forward presence." Several years ago when we interviewed General Krulak [the current Commandant of the Marine Corps], he described it as: "A native in a canoe is able to reach out and touch the gray-painted hull of an American warship in *his* territory." How does that match up with your vision for American presence in the 21st century?

Admiral Johnson: The strength of our forward presence is exactly that. The recent Quadrennial Defense Review provided for a strategic vision that carried with it the three elements or phases of military power that our nation requires. These are shaping, responding, and preparing. We've talked a lot in this interview about responding and preparing. So let's take a little time to talk about shaping the world's military situation.

That's what we do every day. That's why we have 350 ships afloat in the world's oceans right now. That's the guy in the canoe who touches the side of *our* gray-hulled ships. We believe that's a tremendously powerful mission, both for our Navy and the country, because of what it means to the rest of the world. You know, even if that man in the canoe can't touch our ship, but can only see it and watch it come and go as it pleases, then that sends a message of great strength to him and to all the other people who see what we can do. Because we're out there, the world is changed every day.

Tom Clancy: Let's talk a little about the material side of the Navy these days. All the ships, aircraft, and other things that were bought during the Reagan Administration are now almost fifteen years old. Military spending has been significantly reduced in recent years. Are you having problems modernizing and reconstituting the Navy for the 21st century?

Admiral Johnson: I would not categorize the Navy's needs at this stage as problems. I think of them as opportunities, and I would say that the future looks promising. I'm just sorry that I'm not going to be a JO [junior officer] to take advantage of all the things we're going to be getting in the future.

If you look at the programs that we've got on the boards for the next decade, it's a long list. There are the DDG-51-class Aegis destroyers, which we are continuing to build at a rate of between three and four a year. We're getting those ships at between $800 and $900 million a copy, depending upon whose numbers you use, which is quite a bargain. I know that sounds like a *lot* of money for a tin can [the traditional nickname for destroyers], but it's a pretty impressive tin can!

After the Aegis, the next class of surface combatant will be the Surface Combatant-21 [SC-21], which we're just coming to clarity on right now. The first phase of that program will give us what we call a "land attack" destroyer or "DD-21." Downstream from that will probably be a group of those ships that will begin to replace the early units of the Aegis fleet.

We can be sure of one thing—SC-21 is going to have to be much more affordable than the DDG-51's. That's the bottom-line challenge in all this. That's why we're invested in something called "Smart Ship" [the USS *Yorktown* [CG-48], which is being outfitted]. We want to see what we can learn about making these ships not only less expensive to buy, but to operate and maintain as well.

But they'll never be cheap. You have to remember that Navy combatants are not cruise ships. They need to have combat capability *all* the time. How you make the different trade-offs for crew size, displacement, engineering plants, weapons, sensors, and other things is very, very important. One day, lives may depend on how well we make our decisions now.

Tom Clancy: What other new classes of warships do you have on the horizon? I know that the first of the new-generation ships will be the *San Antonio*-class [LPD-17] amphibious ships, which are under construction right now.

Admiral Johnson: The *San Antonio*-class [LPD-17] amphibious ship replaces four different classes of older ships in just one hull. It's an important ship to me, as well as to Chuck Krulak [the Commandant of the Marine Corps]. As you know, the ARG [Amphibious Ready Group] of the 21st century is going to be a three-ship force. There will be a big deck aviation/amphibious ship like a *Tarawa* [LHA-1] or *Wasp* [LHD-1], one of the *Whidbey Island* [LSD-41] or *Harpers Ferry*–class [LSD-49] dock ships, and a *San Antonio*. That *San Antonio*-class ship is going to be the inshore fighter, which will launch the new AAAV amphibious tractors, as well as air-cushioned landing craft and helicopters.

The design and mix of the ARG and these new ships will give us the ability to fight both in the littorals and in the ''blue water'' of the open oceans. It's going to be an awesome platform. That ship is coming along well, as well as CVN-77, which we see as a transition carrier to take us to some technological developments on our way to the next generation of carrier, the CVX.

Tom Clancy: Tell us some more about the CVN-77, if you would.

Admiral Johnson: Some of the improvements we contemplate for it are not unlike what we're doing with the Aegis cruiser *Yorktown*, which we're adding a number of different automation systems to for things like assistance on the bridge, damage control monitoring, and a fiber-optic local area network [LAN] backbone. These improvements are designed to reduce the manning of the Aegis platforms, if it proves practical. We want to see what technology can do for us as a practical matter on future combatants. Once we've been to school on that, then we will do the same kinds of things with CVN-77.

We think technological improvements will help us a lot on the road to our future carrier designs, especially with regards to things like size, shape, and manning, which are some of the critical design factors that determine the costs of new ships. So the plan right now is that CVN-77 will indeed be a transition ship to take us to CVX. We feel that it is the right thing to do. We're going to make it just different enough through a ''Smart Ship/Smart Buy'' concept. What we're trying to do is to leverage technology to do things differently and with fewer people, and let technology make the Naval platforms of the next century even smarter and better than the ones we have right now.

Tom Clancy: If you were going to sit here today and describe what CVX will become, what would be your vision of that carrier when it arrives sometime in the 21st century?

Admiral Johnson: Let me do this based upon my own experience. I started on the *Orisknay* [CV-34, a modernized World War II-era *Essex*-class (CV-9/SCB-27C) carrier], and I've flown on and off of everything from the *Midway* [CVA-41] to a number of the *Nimitz*-class [CVN-68] nuclear carriers. I would tell you that what I want CVX to provide is the same kind of flexibility as you can get out of a *Nimitz*-class carrier. I also want it to be able to deliver many of the same kinds of services and benefits that we already get from carriers right now. I especially want it to be able to move around the same way.

This ship has implications from the strategic level all the way down to tactical implications—like whether I can crank up enough wind over the deck to be able to land an aircraft with the flaps stuck in the "up" position. So we need tremendous flexibility out of this platform, including areas like berthing, data networks, sensors, and tactical systems.

The CVX will also need to be an "open architecture" ship, so that we can "net" it into the new kinds of "network centric" battle forces that we want to build in the 21st century. We will want to have distributed sensor and firepower capabilities spread throughout the battle group in ways that allow us to have situational awareness on every platform, both ships and aircraft, and not just the carrier. The carrier is still going to be the core ship of the CVBG. Therefore, it will still need to have flexibility on the flight deck, in the systems that it carries, and in habitability, to ensure a decent quality of life for the crew that will man it. I believe that the Navy in the 21st century will continue to be a forward-deployed force, and given that reality, this ship is a blank sheet of paper in every way.

Tom Clancy: Does that mean that you see every feature of the CVX as being open for new ideas?

Admiral Johnson: As far as I'm concerned, yes. Propulsion, sensors, catapult systems—they are all open to new and innovative ideas, should they be offered or presented. Now, when we talk about a CVX-type carrier, we're talking about a ship that will arrive at a time where the dominant aircraft it will carry will be the new Joint Strike Fighter [JSF], the F/A-18E/F Super Hornet, and something we call the Common Support Aircraft [CSA]. So this ship will have to be optimized to our vision for operating those future aircraft, none of which are operational today.

Tom Clancy: I've heard some of the people involved in the design and development of CVX call this the first non-Navy or "CinC's" [regional commander in chief's] carrier. Given your own use of carriers during Haiti in 1994, would you concur with this view?

Admiral Johnson: We're saying the same thing. As I mentioned earlier, we're looking for open architecture and connectivity to be able to deal with operations ashore, as well as the Joint Task Force [JTF] commanders in the field, and to handle whatever other circumstances may arise. When you're trying to shape the battlespace and respond to emerging situations, then a battle group commander is going to have to be responsible for a full spectrum of crises. Whether it's a little bitty event or the biggest situation, a commander needs a carrier that can respond on the spot. That's what we need to embed in the CVX design.

It will be very exciting to see the kinds of things that we'll be coming out with in areas like catapult and arresting gear technology, combat systems upgrades, and other new systems. We'll be looking at the proper air wing aircraft mix, including V/STOL [vertical/short takeoff and landing] or STOL [short takeoff and landing] kinds of airplanes, for this new platform. Everything is wide open right now.

Tom Clancy: While I know that your first passion is naval aviation and carriers, I also know that you are passionate about modernizing the submarine force as well. Tell us, if you would, a little about *Seawolf* [SSN-21] and the New Attack Submarine [NSSN] programs?

Admiral Johnson: I recently took a ride on *Seawolf,* and it is awesome. The best submarine that has ever been built in the world, period. The *Seawolf* is truly, truly a magnificent submarine—and remember, I'm a fighter pilot saying this! I took some submariners with me on the *Seawolf,* and watched their reactions, listened to their comments, and made my own observations. All of that convinced me that this is an awesome platform. I can't wait to get it into the fleet, as well as the two others that come behind it.

After the *Seawolf,* we move into NSSN, where we're going to use a special teaming arrangement between General Dynamics Electric Boat Division and Newport News Shipbuilding. The idea is to try and get the cost down so we can afford to buy them in the numbers that we'll be needing to replace the *Los Angeles*-class [SSN-688] boats when they retire.

Tom Clancy: Let's talk about aircraft procurement. It's been a really tough decade for the Navy with regards to new aircraft procurement. There hasn't been a single new tactical aircraft for the sea services in more than two decades. Are you comfortable with the current Navy aircraft development and procurement strategy?

Admiral Johnson: Yes. We've made some workable plans to upgrade our aircraft. Though I must point out that if you were to look at a graphic depiction of the last twenty years, it would tell you that we're coming out of something that looks like a bathtub with regards to new aircraft deliveries. I know that we need to buy new airplanes, the plans are in place to begin to acquire them, and I think that we have the platforms and programs that can deliver in a way that makes sense for Naval aviation.

The current plan covers the V-22 Osprey for the Marines, the strike fighters we've already talked about, T-45 trainers for our undergraduate training programs, and H-60 airframes for ASW and fleet replenishment. I know that sounds like a lot of aircraft, but we're working our way out of a period when we were lucky to buy more than just a couple of airplanes a year.

Tom Clancy: Since money is going to be the determining factor in making these procurement plans into reality, one wonders how well the Congress is receiving your message about the value of naval aviation. Just how well are you getting that message across?

Admiral Johnson: You'd have to ask *them* how well we're doing. But from my perspective, when I go talk or testify to Congress, I see a *lot* of support.

Tom Clancy: If you don't mind, let's run down those aircraft programs one at a time and get a comment on each from you.

Admiral Johnson: **F/A-18E/F Super Hornet**—From my standpoint, this is a model program. The aircraft is meeting or exceeding every milestone and specification that we've put out there. It's a wonderful airplane. I've flown it, and though it's bigger than the F/A-18C/D Hornet, it flies ''smaller.'' I say this publicly and I mean it. This plane is the cornerstone of our future Navy air wing. Over the next two decades, they will first replace our fleet of F-14 Tomcats, and eventually our older F/A-18's. By the end of the next decade, we will have three squadrons [with twelve aircraft per squadron] of these aboard every carrier.

Joint Strike Fighter (JSF)—This bird will eventually replace the newest of our F/A-18C Hornets and Marine AV-8B Harrier IIs, which we are buying right now. Initially, each carrier air wing [CVW] will have a single squadron of JSFs, with fourteen aircraft per squadron. When CVX-78 arrives, this will give it a total of 36 F/A-18E/Fs and 14 JSFs. We expect the concept demonstration and fly-off between Lockheed Martin and Boeing to happen in 2001.

V-22 Osprey—Even though this is technically a Marine Corps airplane with *Marine Corps* painted on the side if it, it's part of *our* budget, and a part of the Navy/Marine Corps forward-presence force. So it's as important to us as it is to Chuck Krulak. Whether it has a role in the U.S. Navy, I'm frankly not smart enough to answer that at this time. If I had to give you an answer, I'd probably have to say yes. Right now, though, those V-22 derivatives are not what I'm focusing on. That's only because the *total* focus of our effort for V-22 *must* be to get them into service to replace those H-46's that are older than the men and women who are flying them.

Helicopter Programs—We're necking down into just the H-60 series. The H-60R airframe is going to be what we use for everything within the battle groups, from ASW [with the SH-60R] to logistics and vertical replenishment [VERTREP with the CH-60R].

Tom Clancy: Could you summarize the major focus of the Naval aircraft procurement for the next few years?

Admiral Johnson: Right now, our focus and effort within Naval aviation is clearly with the Super Hornet and what that takes us to with JSF. Those are the two main tactical aircraft programs. The EA-6B Prowler and E-2C Hawkeye are also important to us. The F-14's are vital to us surely, but we are anxious to get the Super Hornets into the fleet to replace the Tomcats in an orderly flow and fashion. Over the next fifteen years or so, if everything goes as planned, what you will see is Super Hornet replacing Tomcats as well as some of the oldest regular F/A-18 Hornets; then JSF will come in and replace the rest of the F/A-18Cs. So, by around 2015, the combat "punch" on carrier flight decks is going to be filling up with *Super Hornets* and JSFs. That's the vision that we have.

Tom Clancy: Does this mean that you are going to be leveraging the remaining life in existing airframes like the F-14 Tomcat, EA-6B Prowler, and S-3 Viking, to buy time to get those new airframes into service?

Admiral Johnson: Yes. The S-3's are integral to the CVWs right now, and their replacement is part of the CSA program that we discussed earlier. The S-3's, the ES-3's, and EA-6B's are all part of that effort. The Prowlers are of particular value to us, since they are now national assets, due to an understanding with the Marine Corps and Air Force.[15] We're completing the buy of Prowlers right now at 125 aircraft. When we're finished filling out that force, they will be well employed until we decide exactly what the Prowler follow-on will be. If you had to ask me today what that will be, I'd have some expectation of a two-seat variant of the Super Hornet with an automated jamming system. The Wild Weasels may rise again.

Tom Clancy: Over the last fifty years, one of the most important parts of Naval aviation has been the medium-attack squadrons, which used to fly the A-6. With the retirement of the last of the Intruders, has that community more or less died?

Admiral Johnson: Well, I guess because the A-6 is gone that you can say that, but their people and missions have been integrated into other com-

---

15 In 1995, the Air Force signed an agreement with the Navy and Marine Corps to retire their fleet of EF-111A Raven electronic warfare/jamming aircraft for a series of joint squadrons composed of EA-6B Prowlers. These joint squadrons, which have personnel from all three services, have been formed to provide suppression of enemy air defense (SEAD) services for joint component commanders, and deployed CVWs.

munities. Places like the Hornet and Tomcat communities as well as other places. Even the EA-6B Prowler and S-3B Viking squadrons are gaining the experience of former Intruder crews and personnel. The name per se may be gone, but the people and mission live on.

I might add that the new Super Hornet is going to be taking on a lot of the jobs that the Intruder used to do for us. In fact, not too long ago the test crews at NAS Patuxent River [the Navy's test facility in Maryland] launched a Super Hornet loaded up at over 65,000 pounds, which is a thousand pounds more than the Intruder used to fly at. The Super Hornet flies with a full kit of precision guided munitions [PGMs], including the new GBU-29/30/31/32 JDAMS, AGM-154 JSOW, and AGM-88E SLAMER.

Tom Clancy: You just talked about the kinds of weapons that you're going to be carrying and dropping from the Super Hornet and JSF. Is it a safe statement to make that if a target is valuable enough for a carrier-based aircraft to hit it, then that aircraft will use some kind of precision or other tailored munitions to do the job?

Admiral Johnson: I guess my answer to that would be that it would depend on the target set. Generally, I would say yes, that's a fair thing to say. The new things that we're developing in JDAMS and JSOW are really going to help us with our combat punch.

Tom Clancy: You also have strike weapons that aren't launched from aircraft, like Tomahawk and a future series of standoff battlefield support munitions on the horizon. Could you tell us more about them?

Admiral Johnson: We're going to embed some quite remarkable combat power in the CVBG of tomorrow. For example, look at our new SC-21 escort design, which we mentioned earlier. The first variant of that is a land-attack destroyer that will have vertically loading guns and vertical missile launchers loaded with all of the new and improved land-attack missiles that you mentioned.

Tom Clancy: Isn't the Navy about to deploy the first TBMD [Theater Ballistic Missile Defense] system aboard the Aegis ships, even ahead of the Army and Air Force?

Admiral Johnson: Yes, but keep in mind that I am really in competition with *time*. I'm not in competition with the Army and Air Force. I firmly believe that the fleet of Aegis cruisers and destroyers that we have out there is absolutely the optimum place to embed that capability, because of the mobility and flexibility that it gives to the National Command Authorities. So we're full speed ahead on our area-wide, lower-tier system, as well as the theater-wide, upper-tier system. It's going to be an awesome capability.

As you know, the top priority of the Department of Defense [DoD] is to get the various area systems on line as quickly as possible. Those are the Army Patriot PAC-3 and the Navy Aegis Area systems. It's looking good right now, and we're planning to have it shipborne in just a few years. That's really a lot of what we're trying to do Navy-wide these days. Doing things ''leaner,'' but more effectively. That's what we need to do to ''punch through'' into the 21st century.

Tom Clancy: Would it be a fair statement, based upon what you just said, that you're trying to get more out of existing systems and people, rather than start from scratch on new systems?

Admiral Johnson: Yes. We want to harness and focus the technologies that are out there, and embed them in these new systems in ways that give us maximum combat power and flexibility in new and exciting ways. We also want to have the ships and systems manned by fewer people. I believe that, with the right equipment, we can do that and still maintain our effectiveness.

We have to be careful how we flow into all that. But you know about our ''Smart Ship'' program, which is teaching us a lot about how to do these things. We're learning a lot, really focusing on what makes sense for us on a combat platform in terms of downsizing the number of people we need aboard. For instance, the ''mark on the wall'' that we have for the SC-21 land-attack destroyer is that we want that ship to be manned by ninety-five people or less. That's a ship the size of an *Arleigh Burke*-class [DDG-51] guided-missile destroyer, but with a crew about one-third the size. That's where we are going.

Tom Clancy: We talked a lot about the ships, aircraft, and things that you have to buy to give the Navy power. But people make those things work. Obviously, just like the rest of the services, you've had to draw down the size of your personnel pool. You're saying that in the future you want to be able to man your ships with fewer people, each of whom will have to do more. Tell us about the young people you want in the Navy of the future, and what you expect from them?

Admiral Johnson: People *are* our Navy. But the Navy is going to have to become leaner and more capable. The Navy has very high recruiting standards. As we mentioned earlier, we have a ''crossbar'' of ninety-five percent high school graduates and sixty-five percent in the upper mental group as recruiting standards. We believe that gives us the quality of sailor that we need to operate our new systems and take us into the next century. I don't see that changing.

Admiral Jay Johnson speaking to officers in the Middle East.
Official U.S. Navy Photo

But the competition for those young men and women is very intense. It's the same corner of the personnel market that private industry, my Joint Chiefs brethren, and everyone else is going for these days. So far, we've been holding our own in the recruiting process. We will build from that pool of great young men and women a Navy that is reshaped into the proper size and structure for the future. We will give them the best tools for their jobs and the quality of life that they deserve.

We accept the reality that says the Navy must get smaller. The caution in all of that is that if the Navy gets smaller and our requirements don't change, we run the risk of having to ask our people to do more with less. I've told my Navy that *right now*, we're out of the "do more with less" business. We don't do that anymore. What we're going to do is reshape ourselves in such a way that we'll be sized for tomorrow, and then do the missions that we are called to do while maintaining a proper optempo, so we don't operate on the backs of our sailors.

Let me tell you, that's a very tough thing to do. That's what I tell my sailors. It's a much easier thing to say than to do. Our policy of six months deployment portal to portal, two-to-one turnaround ratio, and fifty-percent minimum in-port time over a five-year period, gives us a set of standards and policies that I think the Navy can live with. The CNO is the only one who can waive that policy, and we've only done it a total of five times in the last year. I might add that four of those five waivers were written for ships in out-of-home-port maintenance. So we're holding well to that policy.

Tom Clancy: You've been saying all along that you're going to be trying to man your new generation of ships with fewer sailors doing more jobs than on older vessels. This means that you're probably going to have to raise the crossbar when it comes to getting new sailors trained. Chuck Krulak has much the same plans for his Marines, and has instituted the

Crucible program to help form and toughen his recruits. Are you going to do something similar for Navy recruits?

Admiral Johnson: It's a work in progress. We have upped our own crossbar. Let me give you a couple of quick examples. I talked earlier about the young men and women who come into the Navy from the upper parts of the demographic profile. These are really smart, well-schooled young folks. What we do with them then is send them into a recruit training experience that is a very different, very positive, and very challenging experience.

Now, I'm not too proud to admit that we have liked what we have seen of the programs that you have mentioned from General Krulak, including the *Crucible*. We now have a "final battle problem" exercise evaluation instituted at Great Lakes Training Center. This is a Navy version of a *Crucible*-like evolution. We call it "*Battle Stations*," and it's a very arduous, physically demanding fourteen-hour damage-control problem/scenario requiring stamina, ingenuity, and teamwork from the recruits to pass.

We just came back from Great Lakes, where we observed pieces of the pilot version. We think that this is an extremely good and powerful program. The way that we treat our recruits and the things that we indoctrinate them with—heritage, core values, tradition, and pride—lets us groom them into very strong sailors when they leave Great Lakes.

Then we have what we call the Basic Military Training Continuum, which takes them into the fleet and builds on what they have learned in boot camp. We also have embedded throughout the Navy something we call the Leadership Training Continuum. Now, I'm only the implementer of this program, not the inventor. The program was Admiral Frank Kelso's idea. Kelso was CNO before Admiral Boorda, who also worked on it.

It's powerful! It consists of four two-week training blocks for officers and enlisted personnel, and provides formalized leadership training throughout their careers. That's the basic framework, and we'll build on that later.

Right now I'm interested in getting these four basic blocks instituted throughout the Navy. And mark my words: If you plan on being in the Navy as a career and want to advance, you *will* take these training blocks! The Navy has made an institutional investment in formalized leadership training. I'm convinced, based on just the early feedback training and what I've seen thus far, that when you and I are gone from this world, this Navy will be a stronger at all levels because of it.

Tom Clancy: Obviously, the Navy has had a rough and rocky time integrating women into the force. Yet, one gets the feeling that the Navy is farther through the process than perhaps the other services and that you've paid a high price to reach that goal. Is it your opinion that the first-stage initiatives for fully integrating women into the combat force have been successfully completed?

Admiral Johnson: Absolutely. We're through that. As a good example, the CVWs and carriers are already fully integrated. CVW-11 just came back off deployment on the *Kitty Hawk* [CV-63] fully integrated, and it was a marvelous deployment for them. Our surface combatant integration program is going well, though the pacing item is that we want the ships to be properly built or modified so that the habitability standards we have established for the Women at Sea Program are followed. In addition, the crew must be shaped the right way, so that the proper critical mass and makeup of female personnel is maintained. There's a right way and a wrong way to do that, and we've learned how to do that. We're a little over halfway through that initiative right now, and it's going well.

Keep in mind though that Women at Sea issues are not the only things that drive our overhauls. Environmental ''Green'' upgrades, as well as improvements to combat, habitability, and other systems are just as important. Our ship overhauls are the ultimate fifty thousand-mile checkup, and happen every five years that a ship is in the fleet.

Tom Clancy: As you go out into the fleet today, are the sailors having fun doing their jobs?

Admiral Johnson: I think that, overall, the forward-deployed forces are having fun. They're working hard, making a contribution; they're at the tip of the spear executing their missions, and they're doing the things that they came into the Navy to do. On the non-deployed side, we're doing pretty well, but we've got some work to do, some taking care of business. We owe those personnel a reasonable pace when they're not deployed and we owe them ships and airplanes that are properly maintained. Those are the challenges that I'm working on right now. The ''tip of the spear'' is doing great. The non-deployed part of the force is doing well too, but I think that I owe them a bit more than they're getting right now.

Tom Clancy: Obviously, the last ten years have been a roller-coaster ride for senior leaders in the services. Could you look into your crystal ball, and tell us what new roles and missions that you see the Navy taking on as it moves into the 21st century?

Admiral Johnson: Well, to start with, I don't want to lose any of the core skills that we have right now. I think that we would be very shortsighted to lose any of those capabilities. ASW is a classic example. A lot of people think that you can "take your pack off" now and not worry about it. *I do not concur!* We're putting great focus and effort into undersea warfare and specifically ASW. We're the only ones in the world who can do that. That's *Navy* stuff! That gets back to my operational primacy guidestar: "We can never take our eyes off of that ball." The truth of it is, those core combat skills are things that we need to maintain. You've asked what is new. I give you one word: TBMD. That's something fundamentally new and different from what we are doing now. It's a brand-new capability that will reside in our fleet.

Tom Clancy: To wrap things up, I'd like to give you the opportunity to speak your mind about your vision for the Navy. What would you like to say to the readers, sir?

Admiral Johnson: I think that we've touched on the big things already in this interview. One point that I would hope to make is that the capability that CVBGs and the Navy in general give to the country and the world is vital. We've talked a lot about the equipment, and that *is* vital. But I think more than anything, we've got to really represent all the people in the Navy. That's the story. When you go out and "tie on" with one of those groups, you'll see that people are the magic that makes it all happen.

I'd also like to say that we need to make the American people see the need for maintaining the greatest Navy in the world. There *still* is a need. The lessons of history tell us that. So our commitment to them is that we will *never* "take our packs off." Operational primacy will stay as one of our guiding stars as we head into the new century, and we'll do it with leadership, teamwork, and pride.

For the first time in almost a decade, the Navy seems to be on a steady course, with a plan, and with stable leadership to guide it through the uncertain waters between the 20th and 21st centuries. Like the early mariners who navigated from star to star, Admiral Johnson has found a constellation for the Navy to follow to the future. Along the way, he has proven himself a quiet but effective warrior. In a time when the Navy needed a champion and hero for the wars on the banks of the Potomac River, they seem to have found a winner—a steady hand on the helm, to guide the Navy into a new millennium.

Wings of Gold: A Naval Aviator's Life

"Why is America lucky enough to have such men? They leave this tiny ship and fly against the enemy. Then they must seek the ship, lost somewhere on the sea. And when they find it, they have to land on its pitching deck. Where do we get such men?"

The Bridges at Toko-Ri (James A. Michener, 1953)

When James A. Michener wrote these words almost forty-five years ago, carrier decks were straight and made of wood, and the first generation of jet naval aviators were still learning to fly off them. Carriers, jets, and piloting have changed greatly since then, yet the words ring as true today as they did then.

Naval aviators are a national treasure. They are, first of all, America's front-line combat aviators. Much like their Marine Corps brethren, when there is trouble out there, they expect to be the first called. Though this is an attractive challenge for some people, there is more to the naval aviation profession than just being first in line to be shot at. Flying for the sea services requires unique dedication and skills (such as exceptional eyesight and hand-eye coordination under stress), and demands sacrifices that other military pilots don't even have to imagine—all of which has endowed naval aviation with a (mostly) well-justified mystique.

Flying on and off aircraft carriers is a big part of that mystique. There is an old saying among pilots that flying is not inherently dangerous, just very unforgiving. Though there are no truer words, there are also notable exceptions—"trapping" aboard a rolling and pitching aircraft carrier deck on a stormy night, for instance. It is this skill—landing aboard a moving flight deck in all sorts of conditions—that most clearly differentiates naval aviators from all other pilots. There is simply no way to compare flying from a runway on a land base with the stress and responsibility that sea service pilots have to contend with every time they launch. Every time you take off from a carrier, you leave knowing that you might not find your way back onto the "boat" and will have to eject into a hostile ocean. Clearly, there is more at stake than just a $50 million airplane (and a career). Mastering the stress and responsibility of such flying requires a special kind of flier.

Fortunately for Navy fliers, achieving that mastery is not laid solely on their shoulders. They don't have to do it alone. Since naval aviation is only a fraction of the size of the U.S. Air Force, everyone knows everyone else—and pays attention to everyone else. It's a lot like being part of a college fraternity (for good and for bad). Or—to put it more precisely—U.S. Naval aviation is a collection of small communities (F-14, F/A-18, EA-6B, etc.) in which an aviator spends his or her life for upwards of two decades. The good news here is that there's lots of support. The bad news is that aviators are hugely competitive. Your peers are always keeping score.

Such a world creates larger-than-life personalities—powerfully evolved human beings at the top of the food chain. To succeed you need a cast-iron ego, a lightning intellect, an excess of ambition, and fluent social skills. And the most successful have the ability to spread all this to others in their profession.

> *A Navy pilot (in legend, at any rate) began shouting, "I've got a MiG at zero! A MiG at zero!"—meaning that it had maneuvered in behind him and was locked in on his tail. An irritated voice cut in and said, "Shut up and die like an aviator." One had to be a Navy pilot to appreciate the final nuance. A good Navy pilot was a real aviator; in the Air Force they merely had pilots and not precisely the proper stuff.*
>
> *The Right Stuff* (Tom Wolfe, 1979)

The Navy likes to train its air crews hard. Frankly, they train the hell out of them. While other services emphasize providing officers with a "well-rounded" career, naval aviators in front-line units focus on getting ready for battle. This is not to say that Navy fliers are liberated from down-to-earth duties. They do paperwork like anybody else. Rather, the forward-deployed focus of the Navy requires more emphasis on combat training than usually is provided for the "garrison" units of the Army and USAF. An average naval aviator will spend fully half of his time getting ready to fight and staying proficient. While naval aviators fly about the same number of hours every month as their USAF counterparts, how and when they fly is vastly different. More of their flying is focused on actual combat and tactical training. And there is an almost manic devotion to flight safety, requiring extraordinary amounts of study and practice.

When a carrier air wing (CVW) is preparing to deploy, the air crews spend fully six months training and qualifying to prove their readiness for the job. This is concentrated training, with the entire CVW deploying to a special air warfare training center at Naval Air Station (NAS) Fallon, Nevada, for several weeks to learn composite strike warfare. Just before their deployment, they fly in a series of joint war games, which normally have higher operations tempos (Optempos) than actual warfare. Thus, by the time a naval aviator heads out to the carrier to begin his six-month overseas deployment, he is one of the best-prepared combat aviators in the world. That is not bragging. Consider, for instance, that no U.S. naval aviator has been shot down in air-to-air combat since 1972, and that in a generation of combat from Vietnam to Desert Storm, naval aviators have accumulated an average kill-loss ratio in the neighborhood of 17:1.

Along with the dangerous flying, the life of a naval aviator brings with it the expectation of long overseas deployments, usually lasting six months or more. A "normal" twenty-year career might send an officer on eight or ten of these "cruises." Once a carrier group is forward-deployed, even in relatively "friendly" waters like the western Pacific or the Mediterranean, the aircraft always (even when training) fly with live ordnance loaded. This means that when you are on cruise, the only difference between peacetime and combat flying is the position of the Master Arm switch on the control panel in front of you. As a result, national leaders have to put a lot of trust in individual naval aviators. With only the judgment of a young pilot between the President and a potential act of war, you can understand why they are trained so hard, and held to such exacting standards.

Naval Aviation Culture

Though I've met fighter pilots that enjoy getting shot at and being missed (they love living at that high pitch of excitement), by any true measure, no war is a good war. War is in no way "fun." Still, for the young men who served in it, World War II was the best of wars. They had good airplanes to fly, enemies to fight who were *real* enemies, and a just victory to win. American industry produced splendid aircraft (like the F-6F Hellcat and TBF/TBM Avenger) in which a young man with a couple of years of college and five hundred hours of flight training could expect to fly safely into combat, return to base, and go up to fight again. All kinds of young men flew into combat off carrier decks, from movie actors and Kansas farm boys to future U.S. Presidents. The string of victories that they achieved—Midway, Coral Sea, Leyte Gulf, and many others—testifies to the Navy's skill and wisdom in deploying and fighting naval aviation.

The key to this success was the vast array of training bases, which turned out naval aviators and crews by the tens of thousands. By comparison, as the war went more and more against them, the Japanese and Germans turned out air crews with ever fewer and fewer flying hours of training. American naval aviation leaders considered it a crime to let a young "nugget"[16] into the fleet with less than five hundred hours of flight time. Instead of leaving combat veterans in the fight until they died, as the Axis nations did, American naval aviators (often against their wishes) were sent home after a combat tour to rest and train new pilots before returning to combat. In that way, the veterans got a chance to recharge their batteries while the rookies got the benefit of their experience.

This meant practically that late in the war (the Battles of the Philippine Sea and Leyte Gulf in 1944, for example), American carrier air groups were being led by second- and third-tour commanders (O-5's). The Japanese units were lucky to have lieutenants (O-2's) with a few hundred flying hours. The results were predictable. In repeated one-sided victories, the Americans shot their opponents out of the air at a ratio of over ten to one.[17] So effective was the American juggernaut that the Japanese had to resort to Kamikaze suicide planes to try to stop the onslaught. But this too failed. Naval aviation had won the Great Pacific War, making the island assaults by Marine and Army units possible, as well as helping sweep the seas of enemy naval units. When surrender finally came, following the atomic bombs on Hiroshima and Nagasaki, the bombs were more an excuse than a reason.

How, you might ask, did the war impact on the culture of naval aviation in the U.S.? It gave it a tradition of success and confidence—success and confidence built on intense training. This tradition would hold, even in the dark days of Vietnam and the years following that horror.

Corrosion: The Vietnam Years

Even before the end of the Korean War, new carriers had been laid down, and a new generation of supersonic jets began to appear on their decks. Every month seemed to bring a new carrier aircraft, weapon, or innovation. This was a very good time for Naval aviation. Out of it came, for example, many of the astronauts who would take America into space and to the moon. There was a downside, however. The new jets were unreliable—their new engines being both underpowered and prone to fires and explosions. The practical consequence: Naval aviation, always a dangerous profession, became truly deadly. Naval aviators, always high-spirited and daring both in the air and their personal lives, began to take on a fatalistic attitude about their chances of reaching retirement age. The result was a "live for today" mentality, which they took with them into the 1960's and Vietnam.

16 Navy jargon for a rookie flier on their first cruise or deployment.

17 The most extreme of these engagements occurred early in the Battle of Leyte Gulf, when the commander of Air Group Nine aboard the USS *Essex* (CV-9), Commander David McCampbell, and a single wingman, Lieutenant Roy Rushing, engaged an incoming Japanese fighter force of over fifty enemy aircraft. McCampbell shot down at least nine, while Rushing killed six. No other American fighter mission—in any war—shot down so many. For this performance, McCampbell was awarded the Medal of Honor, and Rushing the Navy Cross.

This fatalism grew exponentially with the start of the Vietnam conflict, when losses to naval aviators who flew missions over Southeast Asia were staggering (due to enemy ground fire, surface-to-air missiles (SAMs), and MiG interceptors), and the chances of surviving a twenty-year Navy flying career became almost nil. Desperate for combat-ready air crews, and unable to send veteran Naval aviators on more than two "war" cruises because of personnel policies, the Navy suffered a severe pilot "crunch" during the conflict. Worse than just a shortage of fliers were the corrosive effects of the conflict itself on the culture of the community as a whole. Atlantic Fleet air crews, whose carriers rarely rotated to Southeast Asia, became almost second-class citizens next to the combat-hardened veterans from the Pacific Fleet. Even worse was the effect on the morale and morals of the aviators who went to Vietnam and came home.

> *I doubt that Mister McNamara and his crew have a morale setting on their computers.*
>
> Rear Admiral Daniel V. Gallery, 1965

Vietnam was a winless war for naval aviators. They lost their first comrades months prior to the Gulf of Tonkin Incident in 1964, and were the last Americans "feet dry" during the evacuation from Saigon in 1975. During the intervening dozen or so years, the Navy kept two or three aircraft carriers continually on "Yankee Station" (the U.S. code name for the carrier operating area in the northern Tonkin Gulf) as part of the bombing campaigns against North Vietnamese forces. It was a new kind of war for the Airedales,[18] most of who had grown up in the "Doomsday" mentality of the Cold War. Now they were saddled by absurd ROE ("rules of engagement"), guidance on targets, tactics, and weapons use. The brilliant but ultimately wrongheaded Secretary of Defense, Robert S. McNamara, and his crew of "whiz kids" devised this absurd situation. In one of the greatest military blunders in a century full of military misfortune, they failed to listen to on-scene commanders about how the air war should be fought. Instead, they tried to "micro-manage" the war from afar, and turned it into one of the worst military fiascos in America's history.

Denied the means to victory, the pilots on the carriers flew daily from Yankee Station, getting shot down, captured, and killed in numbers that still numb modern-day historians.[19] Their mission: not to take effective military action that could lead to victory, but to deliver to an enemy "political mes-

18 A pun. For the Navy, Airedales are not a breed of English terrier but the nickname used by ship's personnel to describe the Naval aviators of the embarked air wing.

19 The worst of these losses occurred on the 1967/68 cruise of the USS *Oriskany* (CVA-34) and CVW-16. During 122 days of action on "the line" in the Tonkin Gulf, thirty-nine CVW-16 aircraft were lost to combat and accidents, with twenty air crew killed, and another seven taken prisoner—over half the embarked aircraft, and something over 10% of the aircrew personnel. Vietnam combat cruises with losses of over twenty aircraft were not unusual.

sages'' from leaders in Washington who did not understand that the enemy did not care to listen to those messages. To say that air crews suffered a great deal of job-related stress is an understatement.

> *A fighter pilot soon found he wanted to associate only with other fighter pilots. Who else could understand the nature of the little proposition (right stuff/death) they were all dealing with? And what other subject could compare with it? It was riveting!*
>
> *The Right Stuff* (Tom Wolfe, 1979)

In any group that regularly undergoes stress, tragedy, and the insanity of a ''limited'' war, the survivors bond in unique ways. Thus it was with Vietnam-era naval aviators. They had faced off with death, and won (never forget that fighter pilots are incredibly competitive). They were the possessors of ''the Right Stuff,'' the keepers of the magic combination of courage, ego, and skills that allowed them to accomplish with fiendish precision actions that no machine could reliably repeat day after day. They were true warriors who—after the day's fighting was over—could imagine nothing better than to spend their off-duty time only with each other.

Soon, the entire naval aviation community had isolated itself, not only from American society in general, but even from the Navy that took them into battle. The result was a subculture that lived in the air wing spaces aboard ship and in the officers' clubs of the liberty ports (like Cubi Point in the Philippines) and home bases. Quite simply, naval aviators fresh from combat were permitted almost *any* behavior short of murder. This included drinking parties in the air wing berthing spaces on Yankee Station and wild sexual antics back at base, as long as they could get up the next day and fly again. Ships' captains and squadron commanders were not simply turning a blind eye on this madness of youth. The wild behavior of naval aviators was actually sanctioned and tolerated by senior Navy leaders all the way up to the Pentagon. The rationale was that the ugly nature of the Vietnam war entitled naval aviators to ''blow off steam'' in an equally ugly fashion. The fallout was a dozen years of drunken antics, womanizing, and wild partying anytime the air crews were not actually flying or in combat.

> *A law of nature holds that alcohol fuels all wars. And the lads at Cubi never suffered a fuel crisis. They got knee-walking, commode-hugging drunk the first couple of days, then recuperated with golf, swimming, or deep breathing.*
>
> *On Yankee Station* (Commander John B. Nichols and Barrett Tillman, 1987)

The effects of the Vietnam-inspired debauchery remained an integral part of naval aviation culture for a generation. Even though the end of the war restored a modicum of peacetime decorum to life aboard ship (alcohol under way became a *major* no-no!), it left a lasting mark on the souls of naval aviators. They now saw themselves as the keepers of a special tribal knowledge—the deep and esoteric knowledge only they possessed, that told them how wartime carrier operations had to be run. As tribal elders, they saw it as an imperative of their calling to pass their tribal knowledge on to the next generation of naval aviation leaders. Thus, when the junior officers who came of age during Vietnam became squadron commanders and carrier captains, they passed on to the new aviators they commanded the hard-drinking, hard-living, womanizing, daredevil culture that they grew up with. It would become a ticking time bomb.

The remaining years of the Cold War saw naval aviation and its personnel safely insulated from the great social changes that were taking place in American society. While the air crews went out on their regular rotations and cruises, thanks to the protection of their senior leaders, they lived in a virtual stasis, immune to outside forces, totally disconnected from the civilian culture. A disaster was waiting to happen. The storm hit in 1991 at the Las Vegas Hilton.

Dry Rot: The End of the Cold War

During the two decades following Vietnam, the civil rights and women's movements transformed American society. During those same two decades, those revolutions barely touched the military in general, the sea services in particular, and naval aviation least of all. In spite of reformers like Admiral Elmo Zumwalt (Chief of Naval Operations in the early 1970's), the culture of naval aviation remained unchanged.[20] As ever, it was a professional haven for middle-class white males, with strong second- and third-generation family associations. But a funny thing happened on the way to Desert Storm: Naval aviation found itself—slowly, reluctantly—setting off on the same road the rest of America was traveling.

The desegregation of the military began as far back as the late 1940s, when President Harry S. Truman issued an executive order to that effect. However the order had very little immediate effect on Naval aviation, for few Americans of color chose to make that a profession. Still, a tiny cadre of brave young men took the plunge; and the first of these, Jesse Brown, gave his life in combat while flying during the Korean War. Sacrifices like Brown's and others' went a long way toward validating minority naval aviators. The admission of women into naval aviation took much longer. Un-

20 Elmo Zumwalt was an early leader in improving conditions for enlisted personnel in the Navy. He provided much of the impetus for the necessary changes required for the all-volunteer military force that followed Vietnam. He also helped redefine the relationship between officers and enlisted personnel, greatly increasing respect and courtesy between the two groups.

fortunately, their acceptance there, with anything like real equality, remains to be achieved. All the same, the feminist revolution changed the U.S. military—even naval aviation—forever.

Broader questions still remain: Does humankind *need* women to be warriors? Does human nature demand it? Do equal rights before the law demand it? I'm not going to hazard an answer to these questions. But there's a much easier one I can safely field: Will women serve in combat in United States military services? The answer to that one, of course, is "yes." They already have and do. In principle, at least, there is no combat action that qualified women cannot handle. Meanwhile, fueled by the new all-volunteer military of the 1970s, the military began to recruit large numbers of women into the ranks. Initially they were limited to non-combatant and support jobs. But before long, the understanding of "non-combatant" and "support" began to change, and with those changes came an expansion of women's roles. By the early 1980s, they were flying transport aircraft and helicopters, as well as training and support aircraft.

But female naval aviators still remained landlocked, due to restrictions on women serving aboard ships. These restrictions, I should point out, were legal, *not* naval. That is to say, the legislation that restricted the role of women aboard ships—and still restricts the roles of women in combat—is contained in Title 10 of the U.S. Federal Code, which must be amended and approved by Congress. Professional military officers may have opinions about the rights and wrongs of these restrictions (which they are obligated to keep to themselves), but the ultimate responsibility for them goes higher up the ladder of government than the rungs they occupy.

In any case, the lot of women in naval aviation during the late stages of the Cold War was anything but pleasant. Since they were effectively barred from front-line fighter, support, and attack units, they would never have the command and promotion opportunities of their male counterparts, which went to "combat" air crews, thus making women second-class citizens in the military. The end of the Cold War in 1989 changed all that. Twice during the Bush years, American forces were committed to combat, in Operations Just Cause (Panama) and Desert Shield/Storm (Persian Gulf). During both operations (notwithstanding Title 10 and other limitations), women were prominently involved in combat operations. Several women commanded units in actual combat, though in "support" roles (military police, Patriot SAM batteries, transport helicopters, etc.). Some became prisoners of war (POWs), and a few died. After women performed in both conflicts with professionalism and bravery, Americans back home could not help but question the restrictions that kept them out of combat units.

Soon after the Gulf War, Congress rapidly amended Title 10, and opened up to women a variety of combat positions that had previously been reserved for male personnel. Women could now fill combat air crew slots and serve aboard warships. By the fall of 1997, only ground combat units (infantry, artillery, armor, etc.), special operations, and submarines remain barred to women. In fact, less than two years after the end of Desert Storm, the services were racing each other to put the first women into the cockpits of combat aircraft. Unfortunately, the change did not come smoothly.

The Air Force's first female bomber pilot, for example, was forced to resign over an adultery charge, all played before a noisy media circus. The Navy's first female fighter pilot died trying to eject from an F-14 Tomcat during a failed approach to the USS *Abraham Lincoln* (CVN-74). These lapses and failures, whatever you care to call them, didn't make life easier for other women flying in potential combat slots. But the greater failure remained in the cultural bias against female aviators. Also, male pilots had a legitimate beef. For in the force drawdown following the Cold War, many male naval aviators were "laid off" and forcibly sent into the civilian job market. Longtime naval aviators couldn't help but resent the invasion of women (in the name of perceived "political correctness") to replace their longtime male buddies. The cultural bias of these men, dating back to Vietnam, condemned such "social engineering" changes. So, predictably, early female naval aviators suffered harassment and hostility from the males they flew with.

But then, on the Labor Day weekend of 1991, some very ugly events happened at the Las Vegas Hilton Hotel, which blew up into the scandal called the "Tailhook Incident." Tailhook soon turned into an international indictment of the sea services' treatment of women in uniform. A yearly convention of naval aviators and their supporters in Las Vegas, Nevada, Tailhook had long had a reputation for drinking and wild behavior.[21] But Tailhook 1991 went over the top, when several female naval officers and other women were allegedly molested by drunk and out-of-control naval aviators. After one woman officer reported what had happened to her commanding officer and he refused to take action (other officers then and later lied about and tried to cover up the Tailhook events), she went to the Navy's criminal investigators. An official investigation was started, and the scandal hit the media.

Meanwhile, the Navy so badly botched the investigation that no convictions were obtained against the officers accused of assaulting women. And then Navy leaders lost control of the situation, resulting in the forced resignations of several high-ranking civilian and military leaders. In the process, thousands of naval officers, most of whom were not even there, had their careers harmed by the political fallout. Yet the botched investigation and the Navy's political folly were hardly the problem. Much less was it that naval officers had gotten drunk, molested women, and then lied about it (though this was bad enough). The problem was the hard-drinking, hard-living, womanizing, daredevil, isolated tribal culture of naval aviation. Naval aviators, a

---

21 The name derives from the Tailhook Association, a civilian organization that promotes and supports Naval aviation. The Association, which actually sponsors the Las Vegas conferences, had nothing at all to do with the Tailhook scandal (and was officially exonerated during the Department of Defense investigation). The Association is a fine organization, which publishes a superb magazine, *The Hook*.

Naval aviators finishing a day's flying in the "Dirty Shirt" pilot's wardroom aboard the USS *George Washington* (CVN-73). Naval aviators treasure such moments, and the comradeship that goes along with them.

JOHN D. GRESHAM

bastion of male exclusivity, had made it painfully clear that they did not want women in their combat flying units, and they had made their displeasure widely known. There would be further problems. But—slowly—progress was coming.

Naval Aviators in the Post-Tailhook Era

Though it has come at a high price, and with many fits and starts, much has changed in the culture of Navy flying since "Tailhook." Women in ever-greater numbers are serving aboard combat vessels. Every carrier group that deploys today has female air crews, along with a growing population of women aboard the ships that they fly from. From helicopter pilots flying off the back of escort vessels to fighter pilots flying patrols in no-fly zones, women have arrived and are in to stay. In the process, many longtime Navy traditions have gone by the wayside. Some of the changes have been as simple as the new rule that every person aboard ship sleep with (at least) a T-shirt and underwear on, to avoid "exposures" in a passageway at night; and sailors have learned to knock and wait for permission to enter female quarters. More substantially, ships have been rebuilt with separate berthing areas and heads (sleeping and shower areas). The result has been the greatest single change in Navy culture since the arrival of the all-volunteer force in the mid-1970's. Along the way, the Navy has learned important lessons about the effective integration of women into units and cultures that they previously have not been part of. These include:

- **Critical Mass**—Human beings are not built to handle difficult jobs alone. Without like-minded companions to share problems and solutions, emotions and trials, an individual can too easily give up, or bend under pressure. Thus women on board ships need other women to share their experiences with (just as men have other men). Armed with that realization, the Navy no longer drops women on their own into a squadron or wing, but puts a few women together—a concept the Navy calls "critical mass.'' Now that women have other women for support, the stresses of being ''new'' and ''different'' in the male-dominated world of naval aviation can be better managed. So now you'll find three or four women in each flying squadron where there are women, or none at all. This ''critical mass'' allows a young female ''nugget'' to survive the emotional rigors of her first fleet assignment.

- **Recruiting**—While ''critical mass'' helps integrate women into particular units, finding enough women to do the job is another matter. Recruiting qualified women is not easy. Because corporate America is already working hard to hire those few women (and minority) college graduates who master ''hard'' subjects like math, sciences, engineering, and computers, the pool available to join the military is quite limited. Many of the women attracted to the military choose to join the Army and Air Force, where the culture is less difficult for them to adapt to. Quite simply, the sea services have done a poor job of selling themselves to women (and minority) candidates, and will need to do a better job in the future.
- **Standards**—Since flying is unforgiving, strict standards of performance and proficiency among *all* aviators must be observed, a lesson the Navy has learned painfully. Cutting corners only produces failure, the loss of $50 million aircraft, and grieving families. The female naval aviators that are making it in today's squadrons are not cutting corners, nor have corners been cut in order to put them in a cockpit. They are doing it *right*! This means that they are doing *everything* that their male counterparts are expected to do in the cockpit, to the *same* standards; and this, more than anything else, has brought the acceptance of female naval aviators at the unit level.
- **Training**—Our society does little to prepare men and women for living and working in the kinds of conditions that a modern Navy imposes upon personnel. After the failures exemplified by ''Tailhook'' and the tribal culture of naval aviation, the Navy has started a series of *mandatory* leadership seminars for officers spaced at various points in their careers. At the same time, *all* Navy personnel have been given sensitivity training to improve their understanding of how professional relationships between officers and sailors of the opposite sex are supposed to work in the modern military. The Navy's justification for these educational efforts is not ''political correctness.'' Rather, since families and schools train ever fewer young people today in civics, manners, and social skills, the sea services feel that it is up to them to make sure their people know these skills and can act accordingly. Manners do count!

All of these initiatives have started to ''level'' the naval aviation playing field for women, and allowed them to gain a foothold in fleet aviation units. Still, some things cannot be mandated or trained into professional warriors. You can't teach a young ''nugget'' how to become ''one of the boys'' in his or her first squadron, for instance. Doing that is especially tough, even if you are equipped with a "Y" chromosome. All naval aviators, no matter what their sex, must be "bonded" into their squadron if they are to survive the emotional and character-building strains that they will face on their first real "cruise." First-tour naval aviators are traditionally "pushed" by the members of their squadrons, and for good reason. The pressure dished out in the ready rooms is designed to separate the winners from the "also-rans."

Lots of male naval aviators fail to survive their first squadron assignments due to the pressure, and so have many of the women who have tried. Frankly, some of these women have shown every bit as much personal courage as civil rights pioneers like James Meredith and Rosa Parks. They have gone where no other women have been before, and the survivors are frequently among the best in their class groups upon graduation. They have to be.

Meanwhile, future squadron and air wing commanders will have to show greater sensitivity and leadership to the conditions of all "nugget" aviators, women included. This may help the entire naval aviation community, since keeping more junior officers after their first tours means fewer personnel will have to be trained. At over a million dollars per trainee, that quickly adds up to real money.

Raw Material: Recruiting

How exactly does one go about becoming a naval aviator? Let's take a quick tour of a hypothetical naval aviation career. Though this may seem like a bit of ego puffery, it's not: Young people choose to try out for naval aviation because they want to be among the "best of the best." If you can launch and land a modern aircraft from the flight deck of an aircraft carrier, cruiser, destroyer, frigate, or amphibious ship, you will never have to justify your flying skills to anyone. Nobody else—not the Israelis, British, not even our own U.S. Air Force—makes pilots better than the USN. Much like Marine Corps basic training, which produces the world's finest combat riflemen, the Navy trains fliers with basic flying and combat skills that are unsurpassed. Of course the USAF and others train excellent combat aviators. That goes without saying. However, when you want superb combat skills, *and* the ability to fly off of a rolling and pitching deck at night in rough weather, you'd better plan on calling the Navy for the air crews.

What kind of person does the Navy want to fly its airplanes? For starters, he or she has to be a college graduate from an accredited four-year university.[22] Prior to World War II, the Naval Academy supplied the majority of naval aviation cadets. But when the war demanded a vastly expanded pool

22 There are still a few enlisted billets in naval aviation, but these are limited to personnel in charge of cargo loading, para-rescue, and some sensor operations. In general, any position of responsibility is going to have an officer in it.

of air crews, the requirements for naval aviation cadets were lowered to completion of just two years of college. Today, the sea services feel that the responsibility for flying a fifty-million-dollar aircraft (with more computing and sensor power than a whole fleet just a generation ago) should go to someone with a university education. For a modern pilot will have to be a systems operator, tactician, and athlete, as well as a naval officer with duties to lead and manage.

Once you have the college degree, and assuming that you want to fly over the water for your country, that your eyesight and physical condition are good, and that you can pass the required batteries of mental and coordination tests, what else do you need? First, you need to be an officer in the U.S. Navy or U.S. Marine Corps.[23] If you are a graduate of the Naval Academy (or, for that matter, West Point or Colorado Springs), then you have automatically earned a reserve officer's commission as an ensign or 2nd lieutenant.[24] The same is true if you have completed an accredited Reserve Officers Training Corps (ROTC) program at a university. However, if you are a simple college graduate with an ambition to fly for the sea services, then there are several Officer Candidate Schools (OCSs) that can give you the basic skills as a Navy or Marine Corps officer, as well as the commission. Though there were once a number of these schools around the country, today there are just two, one at Quantico, Virginia, for the Marines, and the Navy school at Pensacola, Florida. However you get the commission to ensign/2nd lieutenant (O-1), the path to the cockpit of an aircraft in the sea services starts at Naval Air Station (NAS) Pensacola.

NAS Pensacola: Cradle of Naval Aviation

NAS Pensacola, on the shore of the bay whose name it borrows, was originally founded as a Naval Aeronautical Station in 1914. But the region's relationship with the Navy goes back much further. The bay itself, discovered in the 16th century by the Spanish explorer Don Tristan de Luna, attracted official U.S. Navy interest in the early 1800s because of its proximity to high-quality timber reserves, a staple of 19th century shipbuilding. Starting in 1825, the Navy built yard facilities near the site of the present-day NAS. From this Naval station came patrols that suppressed the slave trade and piracy in the mid-1800s. Destroyed by retreating Confederate forces during the Civil War, the base was rebuilt shortly after the end of that conflict. Severely damaged again by hurricane and tidal events in 1906, the excellent

23 The naval aviation program also trains air crews for the Coast Guard, which is technically a part of the Department of Transportation. These include graduates of the Coast Guard Academy in New London, Connecticut, as well as the Coast Guard Officer Candidate School at Yorktown, Virginia. Other nations also send their naval aviation candidates to take their training in the U.S.

24 Just a few years ago, service academy graduates automatically received a regular commission upon graduation. However, in an attempt to even the playing field for non-academy graduates, all new officer commissions are now reserve commissions. Once officers have risen to the rank of lieutenant, they can apply for what is called "augmentation" to a regular status.

The Flightline at Naval Air Station Pensacola, Florida. Every Navy, Marine Corps, and Coast Guard aviator starts his or her career at this base.

Official U.S. Navy Photo

location and facilities proved too valuable to surrender to the elements, and the base was not only rebuilt, but also expanded.

Pensacola's association with naval aviation began in 1913, when recommendations were made to establish an aviation training station in a location with a year-round climate that was favorable to the needs of early aviators. Opened in 1914, it was the home to a rapidly expanding aviation force that by the end of World War I included fixed-wing aircraft, seaplanes, dirigibles, and even kites and balloons! But the lean years following the war meant that only about a hundred new aviators per year were being trained. That time ended in the 1930s with the creation of Naval Aviation Cadet Training Program, which was designed to expand the air crew population in anticipation of the coming world war. To support the growth in the training program, several other training bases were constructed, including NAS Corpus Christi, Texas, and NAS Jacksonville, Florida. Eventually, the combined U.S. naval flight training facilities were turning out over 1,100 new naval aviators a month, though this was reduced following the end of World War II. On average, during the Korean and Vietnam Wars, about two thousand naval aviators a year were trained to meet wartime requirements, while more peaceful times saw that number drop to around 1,500. Today, NAS Pensacola is the home of a still-robust naval air crew training capability.

Training: Into the Pipeline

Soon after an aviation cadet arrives at Pensacola, he or she has to make a major decision: whether to train to become a Naval Aviator (NA—pilot) or Naval Flight Officer (NFO—airborne systems operator). Or rather, just about everybody starts out wanting to be pilots, but then the decision about which way to go is often made for them when the vision test results come in. Eyesight is the first great pass/fail point among fliers. In general, the services look for good distance vision, though excellent night vision is also desired. Many of those who wind up as NFOs do so because they fail the initial eyesight cut for pilots. As it happens, though, life as an NFO very rarely proves disappointing. More often than you might believe, squadron and air wing commands are won by NFOs, many of whom have been noted for their superior leadership and management skills.

Whichever career path beckons the incoming cadets, they all start training in the same classroom. Specifically, there's a six-week course known as Aviation Preflight Indoctrination (API), which comprises a syllabus designed to bring all of the Student Naval Aviators (SNAs) and Student Naval Flight Officers (SNFOs) up to a common knowledge and skill base. API covers aerodynamics, engineering, navigation, and physiology. Along with the classroom work, the students receive physical training in water survival, physical conditioning, and emergency escape procedures. API "levels" the skill base of the cadets, and provides a fighting chance to those who did not (for example) study physics or computer science in college. When API is completed, the training pipeline splits into two separate conduits. One of these is the Primary Flight Training (PFT) pipeline for SNFOs, while the other is for SNAs wanting to pilot Naval aircraft.

Pilot Training: The SNA Pipeline

SNA PFT is designed to teach pretty much the same basic flight skills that a civilian would need to obtain a private pilot's license. It consists of some sixty-six hours of flight training, as well as a syllabus of ground classroom and simulator training. The actual flight training includes basic aerobatics, formation flying, and military flight procedures. This is quite similar to that of the Army and USAF. However, the way that training is conducted has recently changed a great deal for all U.S. military air personnel. These changes have resulted from the 1986 Goldwater-Nichols defense reform legislation. Specifically, Goldwater-Nichols encouraged the Navy, Air Force, and Marine Corps to find ways to combine common tasks into "joint" (i.e., multi-service) programs and units. The consequence for pilot training has been to combine primary/undergraduate flight training, as well as training for a number of different missions and airframes. To that end, the services have established joint training squadrons around the country. They have further teamed up to build a new common primary/undergraduate trainer, the T-6A Texan II, which will enter service in 1999. Based upon the Swiss Pilatus PC-9 turboprop trainer, it will provide a truly economical joint training solution for primary/undergraduate flight training.

Thus a young SNA going through PFT in 1997 might be found at Vance AFB near Enid, Oklahoma. Assigned to the 8th Flying Training Squadron (FTS), he will have done his PFT flight training in an Air Force T-37B, in a joint unit commanded by a naval officer, Commander Mark S. Laughton. Similar squadrons are located at NAS Pensacola, Randolph AFB and NAS Corpus Christi in Texas, as well as other bases. Since the joint training squadrons have proved successful, plans are under way to provide joint training at the airframe level where it is appropriate. For example, since all the services with fixed-wing aircraft fly variants of the venerable C-130 Hercules, there will soon be a single C-130 pipeline unit for training the air crews.

At the end of the PFT phase of training, cadets find out what "community" they will be headed for at the completion of their training. Though just a fraction the size of the USAF, the air forces of the sea services are even more diverse in their roles and missions. Therefore, following the basic phase of PFT, cadets move onto one of five training pipelines (all of which have intermediate and advanced phases). These include:

- **Strike (Tactical Jets)**—This course of training provides student trainees for the F-14 Tomcat, F/A-18 Hornet, AV-8B Harrier II, EA-6B Prowler, S-3 Viking, and ES-3 Shadow aircraft. Normally, strike pipeline SNAs train at the same base where they did their PFT work. Along with further classroom work in aerodynamics, engineering, meteorology, communications, and navigation, there is flying. A *lot* of flying! All told, the intermediate and advanced phases of the strike pipeline PFT provide for around 150 flight hours, covering a great range of required skills and knowledge. These include flight instruction in visual and instrument flying, precision aerobatics, gunnery/weapons delivery, high- and low-altitude flight, air combat maneuvering (ACM), and formation flying. Night flying is also taught, along with flying in a variety of weather conditions, and radar approaches/landings. During this time also comes the dreaded carrier qualification, where the SNA meets up with the deck of an actual aircraft carrier for the first time. To help the students along, extensive use is made of part-task trainers based upon personal computers (PCs), as well as high-end full-motion simulators. However, no amount of simulation and preparation can insure that everyone completes the roughly sixteen-month course.

  For years, this phase of training had the SNAs flying either the T-2C Buckeye or TA-4J Skyhawk, both classic two-seat training aircraft. But a long-overdue replacement is finally coming into service after a series of problems and delays. Known as the T-45 Goshawk training system, it is based upon a heavily modified British Aerospace Hawk trainer, and is designed to provide a beginning-to-end training for the Strike pipeline. This means that the contractor (Boeing, through the acquisition of McDonnell Douglas) provides everything required—simulators, computer-based-trainers, the T-45 training aircraft, and all the maintenance personnel. In order to make the training system work for PFT students, the sea services only need to provide personnel (instructors and students), a base, and fuel. The newest version, the T-45C, incorporates a fully functional "glass" cockpit, similar to the F/A-18's and that of other modern tactical aircraft that the students will eventually fly.[25] The T-45C can

25 Most new aircraft have replaced traditional dial and "strip" instruments with computer-driven Multi-Functional Displays (MFDs). These have the advantage of better presenting data to the air crews, and they can be reconfigured in flight. This means that during takeoff, for example, the air crew can pick the instruments most important to them at that time. So-called "glass cockpits" have between five and a dozen such MFDs, and have become quite popular.

A flight of Boeing T-45 Goshawk trainers. Based on the British Aerospace Hawk-series trainers, they provide the Navy with an economical jet trainer that is replacing the aging T-2 Buckeye and TA-4 Skyhawk.
Boeing Military Aircraft

be used for a much more varied curriculum than the two aircraft it replaces; and thanks to a fuel-efficient engine and all the new avionics systems, the T-45 training system will actually not only save money, but also improve the quality and fidelity of the various training curriculums.

- **E-2/C2**—This training course supplies air crews to fly the E-2C Hawkeye airborne early-warning aircraft and its transport cousin, the C-2 Greyhound, both of which are powered by twin-engine turboprops. Because the airframes that it supplies air crews for are among the most heavily loaded and difficult to fly on and off carriers, the E-2/C-2 pipeline is unique. Thus, for example, the E-2/C-2 pipeline deletes some of the combat/weapons-oriented portions of the Strike PFT course work. Utilizing the T-44A Pegasus (essentially a twin-engine Raytheon/Beech King Air), the intermediate training is carried out by Naval Training Squadron 31 (VT-31), and is run at NAS Corpus Christi, Texas. The advanced phase is handled by VT-4 at NAS Pensacola, Florida, flying T-45's.
- **Maritime**—Since the sea services fly several types of four-engine turboprop aircraft (the P-3/EP-3 Orion and C-130/KC-130/HC-130 Hercules), a separate pipeline (Maritime) supports these communities. The Maritime syllabus begins with six additional weeks of flying at the primary PFT base. For the remaining twenty weeks of the course (intermediate and advanced), the students fly the T-44A Pegasus with VT-31 at NAS Corpus Christi for an additional eighty-four flight hours of instruction. Since these aircraft never land on carriers, the syllabus concentrates on multi-engine aircraft operating procedures, especially in emergency and all-weather operations.
- **E-6**—One of the more chilling missions flown by naval aviators (a mission unique to the Navy) involves flying the E-6 Mercury—the TACMO

(Take Charge and Move Out) aircraft. TACMO was originally the control function for the Navy's Trident Fleet Ballistic Missile (FBM) submarines, but its mission has grown. Based on a Boeing 707 airframe, the E-6 Mercury is packed with secure communications and battle-management equipment. Along with the gear for the TACMO mission, the E-6 carries a fully equipped battle staff from the U.S. Strategic Command (STRATCOM—based at Offut AFB near Omaha, Nebraska). This allows the E-6's to control the launch and weapons release of *all* U.S. nuclear forces (bombers, land-based missiles, and sub-launched missiles) from a (relatively!) secure airborne command post (this job was previously handled by the USAF fleet of EC-135 Looking Glass aircraft). In the event that a nuclear strike were to destroy the National Command Authorities in Washington, D.C., and other land-based locations, the TACMO aircraft would still be able to order a counterstrike.

To support this highly specialized mission, the Navy has a specific pipeline to supply air crews for this single type of airframe. While generally like the Maritime pipeline, the multi-engine-trainer time is carried out on the new T-1A Jayhawk Tanker/Transport Trainer System (TTTS—based on the Raytheon/Beech 400A business jet). Like the T-45 training system, the Jayhawk training curriculum makes extensive use of computerized task trainers and simulators. Overall, the E-6 pipeline emphasizes all-weather flight techniques and cockpit resource management.

- **Helicopter**—Since about half of sea service aircraft are helicopters, the rotorcraft course of study is second only to the strike pipeline in numbers of aviators trained. The Helicopter intermediate-phase PFT is composed of six additional weeks at the primary training base, with an emphasis on instrument flying. This is followed by the twenty-one-week advanced phase of the Helicopter pipeline, which is composed of 116 hours of flight training in the TH-57B/C Sea Ranger helicopter (the Navy's trainer version of the famous Bell Jet Ranger business/utility helicopter). Along with the flying, the classroom work includes helicopter aerodynamics and engineering, night and cross-country flying, as well as combat search-and-rescue techniques. Finally, the Helicopter pipeline SNAs actually take off and land from the Helicopter Landing Trainer (HLT), a specially configured barge at NAS Pensacola.

The decision about where an individual goes is based on several factors, most importantly where he or she finishes in the first part of their PFT class. Normally, high-scoring students are funneled into the "glamor" Naval aviation assignments, like the fighter/attack communities. Since air wing and carrier skippers have traditionally come from the "fast movers," assignment to one of these communities carries great weight, status, and self-esteem. Still, more than a few young aviators choose other specialties, such as helicopters or support aircraft. Though one reason is that the skills of flying transport and cargo aircraft have greater value in the civilian job market, sometimes trainees just want to fly a particular kind of aircraft, or a specific mission. Whatever community the trainees want, the personnel detailers do their best to match these desires with the needs of the Navy and Marine Corps.

While every SNA undergoes a rigorous training regime, those in the Strike and E-2/C-2 pipelines clearly have the toughest challenge—learning to make arrested landings aboard aircraft carriers. You cannot overemphasize how *this* one skill, more than any other, sets Naval aviators apart from their land-based counterparts. Landing on a moving ship at sea is insanely difficult, and it must be done with absolute precision *every* time. In fact, no other phase of SNA training "washes out" so many young fliers. The defining moment for every naval aviator occurs when they come out of the break and line up into the "groove" for their first carrier qualification. Terrifying. Heart-stopping. Insane. That's what they all think when they first look down and out at a carrier and realize they'll have to land on *that* in just about fifteen seconds!

To survive your first set of carrier qualifications (naval aviators have to requalifiy literally dozens of times in the course of a career), the key is to make "good" landings as early as possible during qualifications. This is because your final score is an average of *all* your landing attempts. If you start out poorly, then you've dug yourself a hole that is almost impossible to get out of. The Navy likes SNAs who are "comfortable" and "natural" with the carrier landing process (as if this is *ever* possible!), and pilots who have to "learn" or "force" it are considered potentially dangerous, and not suited for the trade.

NFO Training: The Guys in Back

Pilots and NFOs need each other just to survive. And it's not just part of the job. The men and women who fly for the sea services have a special bond; they look out for each other in the air and on the ground. This comradeship, added to the many other rewarding aspects of Navy flying, helps keep naval aviators coming back to reenlist. Just as with pilots, the path to becoming an NFO begins at NAS Pensacola, with the same six-week API course taken by SNAs. But then the SNFOs are assigned to their own PFT, run by VT-10. Here they spend fourteen weeks learning basic airmanship, including twenty-two hours of flying time in a PFT trainer. Though they spend eight of these in the pilot's seat, they are not allowed to solo. The SNFOs then undergo an extensive PC-based training course in aircraft systems, which includes training on radio and navigation procedures, and classroom work in aerodynamics, emergency procedures, flight rules and regulations, and cockpit resource management. Once the basic PFT course is completed, the SNFOs continue onto their intermediate PFT courses via one of two pipelines: Navigator and Tactical Navigator Intermediate Training:

- **Navigator**—The Navigator pipeline supplies personnel for the P-3 and EP-3 Orion; C-130, KC-130, and HC-130 Hercules; and E-6 Mercury TACMO communities. Twenty-two weeks long, the Navigator course is run by the Air Force's 562nd FTS at Randolph AFB, Texas. There, SNFOs in the Navigator pipeline complete eighty hours of airborne flight training in the T-43A trainer (a modified Boeing 737), learning the difficult trade of long-range and over-water navigation. These include use of celestial, radio, and satellite navigation equipment, as well as secure voice and data transmission systems.
- **Tactical Navigator Intermediate Training**—Every SNFO who is not assigned to the Navigator course at Randolph AFB goes into the Tactical Navigator Intermediate Training (TNIT) pipeline. This course is designed to provide NFOs for all the "tactical" (i.e., combat) aircraft communities in the sea services—such as the F-14 Tomcat, the S-3B Viking, the E-2C Hawkeye, and the EA-6B Prowler. The TNIT SNFOs take their training with VT-10 at NAS Pensacola, and the course lasts fourteen weeks. The flight training for TNIT SNFOs primarily provides experience in low-level navigation and air-traffic-control procedures and is currently accomplished in contractor-operated T-39Ns (modified Sabreliner business jets); but this will change shortly, as the services begin transitioning over to jointly operated T-1A Jayhawks. Already, the T-1As are augmenting the T-39Ns for navigational training hops. Upon completion of TNIT, SNFOs are then assigned to one of three advanced training courses:

    - **Strike SNFO:** The Strike SNFO course provides advanced training for NFOs heading into the S-3 Viking, ES-3 Shadow, and EA-6B communities. This course is run by VT-86 at NAS Pensacola. Flying in the T-2C (soon to be replaced by the T-45), T-39N, and T-1A. Strike pipeline SNFOs spend sixty flight hours over eighteen weeks learning over-water and low-level navigational procedures. The key course objective is to build crew coordination skills, so that in the heat of a combat or emergency situation, they will be ready to act to survive and complete their assigned missions. Once they complete the Strike course, the SNFOs destined for the EA-6B and ES-3 communities go to a special electronic warfare course at Corry Station on NAS Pensacola. S-3 SNFOs go straight into the S-3 community once they finish their training.

    - **Strike/Fighter SNFO:** The Strike/Fighter SNFO pipeline provides NFOs for the small community of two-seat strike fighters in service in the Navy (F-14 Tomcats) and Marine Corps (F/A-18D Hornets). While similar to the Strike syllabus, the Strike/Fighter SNFO course is longer (twenty-five weeks) to allow the teaching of airborne intercept and radar skills, air combat maneuvering, and air-to-ground weapons deliveries.

- **Aviation Tactical Data System (ATDS) SNFO:** The ATDS SNFO course provides airborne controllers for the E-2C Hawkeye community. This pipeline is unique in that it is run by an actual fleet unit, Carrier Airborne Early Warning Squadron 120 (VAW-120) at NAS Norfolk, Virginia. The thirty-two hours of ATDS flight training (spread over twenty-two weeks) take place aboard actual fleet E-2C aircraft (also unique in SNFO training).

Now the new aviators can savor their achievements. They have reached the crowning moment when they are issued their naval aviator number and their "Wings of Gold." Since the earliest days of naval aviation, this small pin has been the symbol that has set them apart from other officers in the sea services. It is now time for them to join the communities and aircraft that will be at the center of their naval careers for the next two decades.

But before they head out to their first fleet assignment, there is one more school for some of the new naval aviators. This is the notorious SERE (Survival, Evasion, Resistance, and Escape) training course, one of the toughest courses any military officer can take. Though its exact details are classified, I do know that it is designed to take "at risk" pilots who will be entrusted with "special" knowledge or responsibilities, and place them into a "real-world" prisoner-of-war (POW) situation. SERE training faces the student with physical and mental stresses similar to those they might expect to experience if they are captured by one of our more unpleasant enemies (North Korea, Iran, Iraq, etc.). As of 1996, there was a single joint SERE school, located at Fairchild AFB near Spokane, Washington. Normally a student attends prior to arriving at his first squadron assignment.

Into the Fleet

By now, officers intending to fly for the sea services have been in the military for something over two years and are ready to pass the final hurdle before they begin to repay the million-dollar investment the taxpayers have so far put into their careers. This is their final certification in a Fleet Readiness Squadron (FRS), which teaches the specific skills necessary to operate each type of Navy or Marine Corps aircraft. During the FRS rotation the Navy teaches its Naval aviation professionals the skills that will make them dangerous out in fleet units. Under the supervision of the FRS instructor pilots (IPs), the new NAs and NFOs learn the tactically correct methods for employing the weapons, systems, and sensors of their community's aircraft. The IPs themselves, normally very skilled airmen who have completed a tour or

The moment of truth. A U.S. naval aviator prepares to launch in an F/A-18C on the deck of the USS *George Washington* (CVN-73).
OFFICIAL U.S. NAVY PHOTO

two at sea, are the final quality check that determines whether a new aviator is allowed to go out to sea. In general, the FRS is the vessel where a particular community's "tribal knowledge" is kept to be passed along to the nextgeneration of air crews. And at FRSs, many of the new concepts for weapons and systems are born.[26]

For the new NAs or NFOs, the FRS phase of their career can go quickly, or last a while. Exactly how long depends on how fast they learn to operate a fleet aircraft to the exacting standards of the FRS IPs and how soon jobs become available in one of the fleet squadrons. The more difficult aircraft like the F-14 Tomcat or EA-6B Prowler might require a young aviator to be held back so that certain skills can be reinforced; and some are "washed out" of one aircraft type and moved to another that's less demanding.

Second Home—Squadron Life

Once the FRS IPs have concluded that a "nugget" (rookie aviator) is ready, a call goes out to the detailing office to look for a spot in one of the fleet squadrons. Squadrons are the basic fighting unit and building block of CVWs (and of all naval aviation); and for the next ten years or so, squadron life will dominate the new nugget's career. But before we get to that, let's take a quick look at some Navy jargon and designations. Though the Navy is notorious for its clumsy and awkward-sounding acronyms and conjunctive designations, these batches of alphabet soup do actually serve a purpose. Consider the following table:

26 In the 1960s when air-to-air kill ratios against North Vietnamese MiG fighters began to fall off, the dedicated efforts of a couple of F-8 Crusader FRS IPs (James "Ruff" Ruliffson and J.R. "Hot Dog" Brown) created the famous Topgun school. More recently, the F-14 FRS at NAS Oceana, Virginia, managed to hang a modified LANTIRN laser targeting pod onto a Tomcat, so that it could deliver laser-guided bombs. This little trick increased the number of aircraft that could deliver precision weapons in every CVW by about 25%, which is not shabby for an *ad hoc* effort!

Naval Squadron Designations

Squadron Designator	Squadron Type	Acronym
HC	Helicopter Combat Support Squadron	HELSUPPRON
HCS	Helicopter Combat Support Squadron—Special	HELLSUPPRONSPEC
HM	Helicopter Mine Countermeasures Squadron	HELMINERON
HS	Helicopter Anti-Submarine Squadron	HELANTISUBRON
HSL	Helicopter Anti-Submarine Squadron—Light	HELANTISUBRONLIGHT
HT	Helicopter Training Squadron	HELTRARON
VA	Attack Squadron	ATKRON
VAQ	Tactical Electronic Warfare Squadron	TECELRON
VAW	Carrier Airborne Early Warning Squadron	CARAEWRON
VC	Fleet Composite Squadron	FLECOMPRON
VF	Fighter Squadron	FITRON
VFA	Strike Fighter Squadron	STRIKEFITRON
VFC	Fighter Composite Squadron	FITCOMPRON
VP	Patrol Squadron	PATRON
VPU	Patrol Squadron—Special Projects Unit	PATRONSPECPROJUNIT
VQ	Fleet Air Reconnaissance Squadron	FAIRECONRON
VQ	Strategic Communications Squadron	STRATCOMMROM
VR	Fleet Logistics Support Squadron	FLELOGSUPPRON
VRC	Fleet Logistics Support Squadron	FLELOGSUPPRON
VS	Sea Control Squadron	SEACONRON
VT	Training Squadron	TRARON
VX	Air Test and Evaluation Squadron	AIREVRON
VXE	Antarctic Development Squadron	ANTARCTICDEVRON
TACRON	Tactical Air Control Squadron	TACRON

If you understand the squadron designation, and add the squadron's number behind it, you know what *kind* of unit you are talking about. For example, VF-14 is a fighter squadron, which just happens to fly F-14 Tomcats. They are known as the "Tophatters," and their heritage dates back to the 1920s, when they were originally designated VF-2, flying aboard the old *Lexington* (CV-2). The system is actually quite logical and simple, if you take the time to understand it.

Other facts about Navy squadrons are not quite so obvious; the number of aircraft and personnel within a particular kind of unit, for example. An F/A-18 Hornet squadron usually deploys with a dozen aircraft, eighteen air crew, and a support/maintenance base of several hundred personnel. Conversely, each EA-6B Prowler squadron has only four airplanes, but more air crew (about two dozen) and maintenance personnel than the Hornet unit. For each Prowler carries four air crew (compared with the F/A-18's single pilot), and the jamming aircraft require much more maintenance than the Hornets. The squadrons themselves are structured pretty much alike. A full commander (O-5) generally commands, with a lieutenant commander as the executive officer. Backing them up are department heads for maintenance, intelligence, training, operations, and even public affairs. Watching over the enlisted troops will be a master chief petty officer, who is the senior enlisted advisor to the commander. Under normal peacetime conditions, the squadron personnel will spend about three to four years in the unit, about enough time for two overseas deployments.

The new nuggets, meanwhile, are getting ready for their first overseas deployment. But before that happens, they are assigned a "call sign" (frequently "hung" on the new aviator during a squadron meeting). Call signs are nicknames used around the squadron to differentiate all the Toms, Dicks, Jacks, and Harrys that clutter up a ready room and make identification over a crowded radio circuit difficult. Most call signs get "hung" on a pilot because of some unique characteristic. Sometimes they are inevitable. Thus, every pilot named Rhodes is going to be named "Dusty," just as any Davidson will be "Harley." Others are more unique. One F-4 RIO (Radar Intercept Officer) who lost several fingers during an ejection over North Vietnam became "Fingers." Another pilot became "Hoser" because of his tendency to rapidly fire 20mm cannon ammunition like water out of a fire hose. Most call signs last for life, and become a part of each naval aviator's personality.

New pilots and NFOs normally arrive in a squadron during the first few months after it comes home from its last deployment. There they will be expected to get up to speed in the squadron's aircraft, weapons, and other systems, as well as in the proper tactics for employing all of these. Thus by the time the squadron deploys, it is hoped the nuggets will be more dangerous to a potential enemy than to themselves or their squadron mates. To help them get started, new aviators are usually teamed with an older and more experienced member of the squadron. For example, in F-14 squadrons you normally see a nugget pilot teamed with a senior (second or third tour) RIO, who is probably a lieutenant commander. If the squadron flies single-seat aircraft like the F/A-18 Hornet, then the nugget pilot will be made the wingman to a more senior section leader. The final six months prior to the nugget's first deployment are spent "working up" with the rest of the squadron, air wing, and carrier as they mold into a working team.

During the cruise, nuggets are expected to fly their share of missions in the flight rotation, stand watches as duty officers, and generally avoid killing themselves or anyone else without permission. If the nugget does these tasks well on his or her first overseas cruise (normally lasting six months), it is likely he or she has a future in the Naval aviation trade. It is further hoped that the rookie will have become proficient in flying all the various missions assigned to the squadron, and qualified to lead flights of the squadron's aircraft. When the squadron returns from the cruise, the nuggets will (hopefully) have enough experience and enthusiasm to do it again the following year.

Most naval aviators have by this time been promoted to lieutenant (O-3), and have been entrusted with minor squadron jobs like public affairs, welfare, or morale duty. It is also the time that the Navy begins to notice those young officers who have promise. One sign you've been noticed is to be sent to school. If you are a good "stick" in an F-14 or F/A-18 squadron, for example, you may get a chance to head west to NAS Fallon near Reno, Nevada, to attend what the service calls the Naval Fighter Weapons School, which you probably know better as Topgun). Topgun is a deadly serious post-graduate-level school designed to create squadron-level experts on tactics and weapons employment. The E-2C community also has its own school co-resident at NAS Fallon, called Topdome, after the large rotating radar domes on their aircraft. Graduates of these schools have an automatic "leg up" on other aviators at their level, and will likely get choice assignments if they continue to shine. More than a few Topgun graduates have gone on to the Navy's Test Pilot School at Patuxtents River, Maryland, or even to fly the Space Shuttle.

All too soon however, the second cruise arrives. Though second-cruise aviators are expected to show some leadership and help the new nugget air crews with their first cruises, most of what they do is fly. They fly a *lot*! Now is the time when taxpayers begin to get back the million-dollar-plus investments made in these young officers. Most naval aviators find life good at this stage. With a cruise of seniority over the nuggets, and none of the command responsibilities that will burden them later in their career, it is a nice time to be a naval aviation professional.

The Good Years—The Second and Third Tours

The Navy, wisely, is well aware that after two cruises, young naval aviators tend to be burned out and need shore duty to recharge their batteries. During this first shore tour (which lasts about three years), a young man or woman can earn a master's degree (a necessity for higher promotion these days), start a family, and perhaps build a "real" home.

An officer who shows special promise for higher command may also be offered graduate work at one of the service universities (such as the Naval Post-Graduate School, the Naval War College, the National War College at Fort McNair, in Washington, D.C., or the Air University at Maxwell AFB, Alabama). Staff schools like these are designed to teach officers the skills needed for high-level jobs like running a squadron, planning for an air wing or battle group staff, or working for a regional commander in chief. There may also be an opportunity for the young officer to get some time as an IP at one of the FRSs. They might also serve in a staff job for an admiral or other major commander.

By the end of this three-year period, they will probably be ready to go back to a flying unit at sea. Our aviator is by now around thirty years old, with over eight years of service in the Navy, meaning that this flying tour represents a halfway point in his or her flying career. Here they will do some of their most demanding work. The second sea tour (of three to four years) puts the aviator out on a carrier for another two cruises—either as a member of a squadron, or perhaps as an officer on an air wing staff. Whatever the case, the aviator will get another heavy dose of flying, though this time there'll be a great deal more responsibility. For it is during this time that officer enters the Navy equivalent of middle management. Specifically, this means that officers now have to provide more flight and strike leadership on missions, as well as expertise in the various planning cells that support flight operations.

Once this tour is completed, the aviator is almost guaranteed a two-year shore tour as an IP at either a training squadron or a FRS. There will also probably be a significant raise in pay, since promotion to lieutenant commander (O-4) normally occurs during this time. After the IP shore tour comes a department head tour, which is the start of their rise to command.

Command—The Top of the Heap

For naval aviators, the path to combat command starts when they arrive at their squadron for their third flying tour (another three-to-four-year, two-cruise sea tour) and are assigned a major squadron department (maintenance, training, operations, safety, supply, etc.) to run. How well they do here will ultimately determine how far they will go in the Navy. After the department head tour, officers who prove to be ''only'' average will go back to another shore tour, perhaps on a staff or to a project office at the Naval Air Systems Command, and will probably be allowed to serve their twenty years and retire. But if the Navy feels an officer has command potential, then things begin to happen quickly, starting with a two-year ''joint'' staff tour, which is designed to ''round out'' the officer's career and provide the ''vision'' for working effectively with officers and personnel from other services and countries. Following this, the officer heads back to what will probably be his or her final flying tour, as the executive officer (XO) of a squadron. If the first cruise as XO goes well, the second cruise comes with a bonus—promotion to full commander (O-5) and the job of commanding officer (CO) of a squadron of naval aircraft.

It also is the beginning of the end of the officer's squadron life. In less than eighteen months, our aviator will be handing over command of the unit to *his or her* XO, and the cycle moves on. From here on, aviators take one

Captain Lindell "Yank" Rutheford, commanding officer of the aircraft carrier USS *George Washington* (CVN-73). Rutheford is a longtime F-14 Tomcat pilot who has risen to the top of his profession.
JOHN D. GRESHAM

of two paths. They can take another staff tour, followed by "fleeting up" to take over their own air wing (with a promotion to captain, O-6). The other option is that they can take the path to command of an aircraft carrier. This includes nuclear power school, an O-6 promotion, and a two-year tour as a carrier XO. Following this comes a command tour of a "deep draft" ship (like a tanker, amphibious or logistics ship), and eventually command of their own carrier. Beyond that comes possible promotion to rear admiral and higher command. However, it is the "flying" years that make a naval aviator's career worth the effort. Years later when they have retired or moved on to other pursuits, the aviators will likely look back and think about the "good years," when they were young and free to burn holes in the sky, before heading back to the "boat."

Building the Boats

Officially, the Navy calls it a ''CV'' or ''CVN.'' Sailors on the escorts call it a ''bird farm.'' Submariners wryly call it a target. But naval aviators call it—with something like reverence and religious awe—''the boat.'' It is the central icon of *their* naval careers. In addition to being their home and air base, aircraft carriers hold an almost mystical place in the world of naval aviators. As we've already seen, young naval aviators' skills (and future chances of promotion) are judged mainly on their ability to take off and land safely on ''the boat.'' Later, as they gain seniority, they'll strive to command one of the giant supercarriers. Finally, at the sunset of their naval careers, they will be expected to lead the fight to obtain authorization and funding for construction of the new carriers that will serve several future generations of naval aviators.

Why this community obsession about ''the boat''? The answers are both simple and complex. In the first chapter, I pointed out some of the reasons why sea-based aviation is a valuable national asset. However, for the Navy there is a practical, institutional answer aimed at preserving naval aviation as a community: ''If you build it, they will come!'' That is to say, as long as America is committed to building more aircraft carriers, the nation will also continue to design and build new aircraft and weapons to launch from them, and train air crews to man the planes. In other words, the operation of aircraft carriers and the building of new ones represent a commitment by the Navy and the nation to all of the other areas of naval aviation. New carriers mean that the profession has a future, and that young men and women have a rationale for making naval aviation a career. The continued designing and building of new carriers gives the brand-new ''nugget'' pilot or Naval Flight Officer (NFO), a star to steer for—a goal to justify a twenty-year career of danger, family separation, and sometimes thankless work.

This is fine, as far as it goes. And yet, as we head toward the end of a century in which aircraft carriers have been the dominant naval weapon, it is worth assessing their value for the century ahead. More than a few serious naval analysts have asked whether the kind of carriers being built today have a future, while everyone from Air Force generals to Navy submariners would

The USS *George Washington* in the Atlantic during JTFEX 97-3 in 1997. Once "worked up," carrier groups are the "big sticks" of American foreign policy.
John D. Gresham

like the funds spent on carrier construction to be reprogrammed for their pet weapons systems. Two hard facts remain. First, big-deck aircraft carriers are still the most flexible and efficient way to deploy sea-based airpower, and will remain so for the foreseeable future. Second, sea-based airpower gives national leaders unequaled options in a time of international crisis.

With this in mind, let's take a quick tour of the "boats" that America has been building for the past half century. In that way, you'll get an idea not only of the design, development, and building of aircraft carriers, but also of the size, scope, and sophistication of the industrial effort all that takes.

American Supercarriers: A History

The atomic bombs that forced Japan to capitulate in 1945 almost sank the U.S. Navy's force of carriers. With the end of the war, as a cost-saving measure, most U.S. carriers were either scrapped or mothballed. And by 1947, the wartime fleet of over one hundred carriers had shrunk to less than two dozen vessels. Meanwhile, President Harry S Truman had decreed a moratorium on new weapons development, except for nuclear weapons and bombers to carry them. The Navy, desperate for a mission in the atomic age, began to design a carrier and aircraft that could deliver the new weapons.[27] The USS *United States* (CVA-58—the "A" stood for "Atomic" combat), would have been the biggest carrier ever built from the keel up (65,000 tons displacement). The Navy argued that immobile overseas Air Force bases were vulnerable to political pressure and Soviet preemptive attack, while carriers, secure in the vast spaces of the Norwegian Sea, the Barents Sea, or the Mediterranean, could launch nuclear strikes on Soviet Naval bases or deep into the Russian heartland.

Claiming that the newly created Air Force could better deliver the new

27 The atomic combat requirement was outlined in a famous 1947 memorandum prepared by Rear Admiral Dan Gallery. He was a legendary Naval aviation figure (he commanded the escort carrier group that captured the German U-505 in 1944), and his paper would eventually start a virtual war between the Navy and the newly created Air Force.

The USS *Forrestal* (CV-59), the first of America's supercarriers. She is cruising here in the Gulf of Tonkin during combat operations in 1967.
OFFICIAL U.S. NAVY PHOTO FROM THE COLLECTION OF A. D. BAKER

atomic weapons with their huge new B-36 bombers, Air Force leaders like General Carl "Tooey" Spaatz lobbied intensively to kill the new carrier program. By persuading the Truman Administration that they could deliver nuclear weapons more cheaply than the Navy, the Air Force succeeded in having the *United States* broken up on the building ways just days after her keel was laid (April 23rd, 1949). Soon afterward, the Secretary of the Navy, John L. Sullivan, resigned in protest, leading to the "Revolt of the Admirals" (discussed in the first chapter), which allowed the Navy to make a public case for conventional naval forces. Once the Truman Administration realized the political cost of killing the *United States*, the cuts in naval forces were stopped. It was just in time, as events turned out. For the carriers recently judged obsolete in an age of atomic warfare held the line in the conventional war that erupted in Korea on the morning of June 25th, 1950.

Even before the end of the Korean War, the Truman Administration recognized the need for new, bigger, more modern aircraft carriers. Though he was never a friend of the Navy, President Truman nevertheless belatedly authorized construction of a new class of "supercarriers" similar to the *United States,* canceled just three years earlier. The first of the new flattops was USS *Forrestal* (CVA-59—the "A" now reflecting the new "Attack" carrier designation), which was followed by three sister ships: *Saratoga* (CVA-60), *Ranger* (CVA-61), and *Independence* (CVA-62). These were huge vessels, at 1,039 feet/316 meters in length and almost sixty thousand tons displacement. The *Forrestal* class incorporated a number of innovations, almost all of British origin. A 14° angled deck enabled planes to land safely on the angled section, while other planes were catapulting off the bow. Steam catapults allowed larger aircraft to be launched. Also, a stabilized landing light system guided pilots aboard more reliably than the old system of hand-held signal paddles. Along with the new carriers came the first-generation naval jet aircraft. Meanwhile, the Navy initiated a huge Fleet Rebuilding and

The USS *Enterprise* (CVN-65), the world's first nuclear-powered aircraft carrier. Here she is cruising in the Mediterranean Sea with the nuclear cruisers *Long Beach* (CGN-9) and *Bainbridge* (CGN-26) during Operation Sea Orbit in 1964.
OFFICIAL U.S. NAVY PHOTO FROM THE COLLECTION OF A. D. BAKER

Modernization (FRAM) program for older carriers and other ships, both to give them another twenty years or so of service life and to delay the need to buy so many expensive new ships like *Forrestal.*

The first Cold War confrontation in which aircraft carriers played a major role was the Suez Crisis in 1956; carrier groups assigned to the U.S. Sixth Fleet spent the next year supporting operations by U.S. Marines and other forces trying to restore stability in Lebanon following the Arab-Israeli war. In 1958, Task Force 77 got a workout in the Far East when it interposed between the forces of Taiwan and Communist China during the crisis over the islands of Quemoy and Matsu. Meanwhile, two new follow-on supercarriers were ordered—*Kitty Hawk* (CVA-63) in 1956 and *Constellation* (CVA-64) in 1957. Essentially improved and enlarged *Forrestal*-class vessels, they approached the upper limits of size and capability for oil-fueled carriers. The time had come for a break with fossil-fueled power plants, and the carrier that followed was truly revolutionary.

The successful development of nuclear reactors to propel submarines encouraged the Navy to put them in surface ships. Backed by the mercurial Director of Naval Reactors, Vice Admiral Hyman Rickover, an improved *Kitty Hawk* design was developed to accommodate a nuclear propulsion plant. Ever eager to maximize the influence of nuclear power in the Navy, Admiral Rickover dictated that the new carrier should have just as many nuclear reactors (eight!) as there were oil-fired boilers in each *Kitty Hawk*-class carrier. When the new carrier, designated USS *Enterprise* (CVAN-65), was commissioned in the early 1960's, she was so overpowered that the structure of the ship could not stand the pounding of a full-power run. There are stories of speed runs off the Virginia capes in which the *Enterprise* went so fast (some say over forty knots; the actual numbers are still classified), that she left her destroyer escorts far behind, without tapping her full power.

Though *Enterprise* more than lived up to the heritage of her proud name, she was to be a one-of-a-kind ship.[28] Then-Secretary of Defense Robert S.

28 The original carrier USS *Enterprise* (CV-6) was arguably the U.S. Navy's greatest warship, with a combat record second to none. She fought in five of the six great carrier-versus-carrier clashes, surviving serious

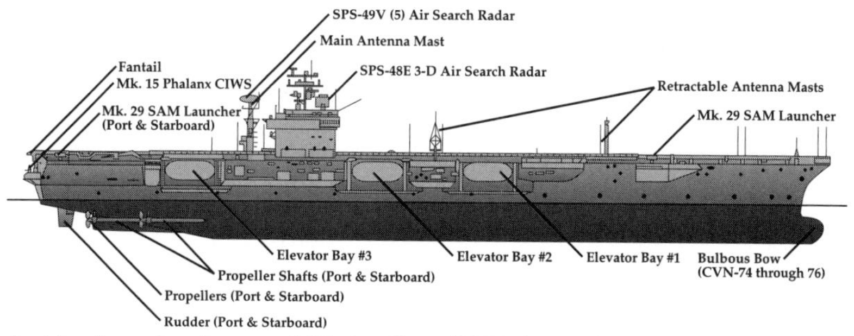

A side view of an improved *Nimitz*-Class (CVN-68) nuclear-powered aircraft carrier.

JACK RYAN ENTERPRISES, LTD., BY LAURA DENINNO

MacNamara, no friend of the Navy, blocked construction of more nuclear-powered carriers. Over the next decade, only two new carriers, *America* (CVA-66) and *John F. Kennedy* (CVA-67), would be constructed. These flattops, essentially repeats of the earlier *Kitty Hawk*-class, were powered by oil-fired boilers. After MacNamara's resignation in 1968, the ban on nuclear carrier construction lifted, and the Navy received authorization for a new class of three nuclear-powered attack carriers. This would become the mighty *Nimitz*-class (CVN-68) program.

The *Nimitz*-Class (CVN-68) Supercarriers

Because of the vast base of experience developed over the previous four decades, even before design of the *Nimitz*-class carriers began in the late 1960's, the Naval Sea Systems Command (NAVSEA) had a number of good ideas about what they wanted from their next generation of flattops. Frankly, they wanted a lot! The largest warships (in dimensions and displacement) ever planned at the time, the *Nimitz*-class carriers were to be the ultimate expression of sea-based airpower. Some of the "fighting" qualities of the *Nimitz*-class included:

- **Aircraft Capacity**—For over seventy-five years the value of a flattop has been measured by the number and types of aircraft it can carry. Ever since the Navy learned that the original USS *Ranger* was too small to carry a credible air wing, U.S. carrier designs have emphasized big flight and hangar decks to park, stow, and operate aircraft.[29] In addition, growth in

combat damage many times. The *Enterprise* was so hated by the Japanese that they claimed to have sunk her by name on a number of occasions.

29 USS *Ranger* (CV-4), was the first American carrier built from the keel up. At only about fourteen thousand tons displacement, *Ranger* was tiny compared to *Lexington* and *Saratoga*, and it showed when she went into service. With less than half of the aircraft capacity of the two larger ships, *Ranger* was simply too small to support a powerful air group, and was never considered a front-line vessel. Despite this, the Navy learned valuable lessons from building *Ranger*, and it showed in the next class of aircraft carriers.

the size and weight of combat aircraft has driven the design of carriers. For example, an F4F Wildcat fighter of 1941 left the deck at a maximum weight of 7,952 lb/3,607 kg, but today's F-14 Tomcat fighter has a maximum takeoff weight of 74,348 lb/33,724 kg! The *Nimitz*-class carriers were designed to handle ninety or more aircraft (though they currently operate with air groups of about seventy-five), depending on "spot factor" (the amount of deck space each aircraft type requires).

- **Armament**—Experience with heavy guns and long-range surface-to-air missile (SAM) batteries on earlier classes of aircraft carriers proved that the deck space, interior volume, manpower demands, and blast effects of such weapons interfered with air operations, the carrier's true reason for existence. Therefore, weapons on newer carriers would be limited to point defense (i.e., "last ditch" self-defense) systems like the RIM-7 Sea Sparrow surface-to-air missile (SAM) and Mk. 15 Phalanx/CIWS 20mm automatic cannon. A few .50-caliber machine guns would also be mounted for defense against suicide motor boats or terrorist swimmers.
- **Crew Size**—For centuries, experience has shown that the more sailors you cram aboard a warship, the better her fighting qualities, especially when you need to repair battle damage. On the other hand, sailors take up a lot of space, and generate large "hotel" loads on the ship's power plant (for electricity, water, heating, and cooling) that have nothing to do with fighting. Modern sailors are volunteers, who expect a minimum level of comfort. The Royal Navy's eighteen-inch spacing between hammocks aboard warships two centuries ago may have worked for impressed seamen, but would hardly do for today's sailors. Therefore, naval designers are constantly balancing the advantages of larger crews with the costs of personnel on ship size and capability. The *Nimitz*-class carriers would be designed to sail with about six thousand personnel on board: 155 officers and 2,980 sailors for the ship; 365 officers and 2,500 enlisted personnel for the air wing. Now add an admiral's staff, a few dozen civilian contractors to maintain the high-tech equipment, and a constant trickle of distinguished visitors and media representatives, and a carrier can get really crowded!
- **Deployability**—Since a crisis may be halfway around the world, a carrier needs to go fast. On the other hand, high speed is worthless if the carrier does not carry sufficient fuel to get where it has to go without frequent refueling. The interior space consumed by a large power plant and its fuel is not available for aircraft, crew berthing, ammunition, jet fuel, and other useful stowage. In the final analysis, the choice of a nuclear power plant was a no-brainer. The *Nimitz*-class carriers were designed to carry two General Electric A4W/A1G nuclear reactors, and were expected to operate for fifteen years between refuelings.[30] That's up to one million nautical miles of steaming on just one set of reactor cores.

30 A new reactor design under consideration for future carriers will *never* need refueling. This is a tremendous advantage, since refueling is a complex overhaul that takes three years in a shipyard.

The carrier USS *George Washington* (CVN-73) conducting an underway replenishment (UNREP) from the fleet logistics ship USS *Seattle* (AOE-3). UNREP is a vital capability in keeping battle groups forward-deployed, and utilizes both ''high lines'' and helicopters to transfer cargo and fuel.

OFFICIAL U.S. NAVY PHOTO

- **Sustainability**—Once a carrier has reached an operating area, it must conduct operations for as long as possible without resupply since it may take weeks for fleet supply vessels to catch up with the carrier battle group. The enemy may not wait while you replenish at sea, so the amount of fuel, food, ammunition, and spare parts carried on board has a direct effect on how long a carrier can stay in action. It is also essential when fleet supply vessels reach the carriers; for when carriers are conducting Underway Replenishment (UNREP), basic safety rules dictate that they cannot operate aircraft or maneuver freely. Thus, the less often they take aboard fuel and supplies, the more time they can spend ''on the line'' conducting combat operations. The *Nimitz*-class carriers were designed to store up to nine thousand tons of jet fuel and almost two thousand tons of bombs, ammunition, and missiles. This is a vast improvement over earlier designs.
- **Survivability**—All of the above are worthless if the carrier is a blazing hulk about to turn turtle and sink. *Nimitz*-class carriers were designed in an era when the threat of Soviet cruise missiles and torpedoes armed with 1,000-kg/2,200-lb warheads was quite real. These weapons could blow a cruiser or destroyer in half, and do considerable harm to an aircraft carrier. The Navy was especially conscious of these dangers after three deadly fires aboard USN carriers during the Vietnam War had taken a high toll of lives, aircraft, and equipment. Remember that these ships are basically big boxes filled with explosives, jet fuel, and people, all packed tightly together. With all this in mind, the NAVSEA designers went to extreme lengths to make the new carriers both durable and survivable. The flight and hangar decks, as well as the hull, would be built from high-tensile steel, with a vast scheme of compartmentation and built-up structure. In addition, the new flattop would make only minimal use of light metals like aluminum, which are flammable under some easily reached fire conditions.

By the late 1960's the characteristics of what was initially known as SCB-102 (Ship Control Board Design 102) were firming up, with the following providing some idea of what the Navy desired:

- **Displacement**—Approximately 95,000 tons fully loaded.
- **Size**—A length of 1,092 feet/332.9 meters, beam of 134 feet/40.85 meters, a flight deck width of 250 feet/76.5 meters, and a maximum loaded draft of no more than 39 feet/11.9 meters.
- **Power Plant**—Two Westinghouse A4W nuclear reactors driving four General Electric steam turbines, turning four screws for a total of 280,000 shp. While the top speed is still classified, it is *well* over thirty-three knots.
- **Manning**—SCB-102 provided for a ship's company of 2,900 enlisted personnel and 160 officers. Room was additionally provided for two thousand air wing personnel, thirty Marines, and seventy members of the flag staff. This added up to almost 5,200 embarked personnel.
- **Aircraft Complement**—Approximately ninety aircraft. These would include improved models of aircraft like the F-4 Phantom II, A-6 Intruder, A-7 Corsair II, and E-2 Hawkeye, as well as newer and larger planes like the F-14 Tomcat, S-3 Viking, and EA-6B Prowler.
- **Defensive Armament**—Three eight-round RIM-7 Sea Sparrow SAM point-defense missile systems.

All of these features added up to the biggest class of warships ever built. Only the *Enterprise* had dimensions, displacement, and performance anything like the proposed SCB-102 design, and "the Big E" was lugging around eight nuclear reactors, the power of which could not be fully used. SCB-102 would be a much better balanced design—a fully integrated warship that would grow and modernize as the Cold War moved into the post-Vietnam era.

On the other hand, this very impressive package was going to be expensive and difficult to build. Because of foreign competition, America's private shipbuilding industry was in decline during the late 1960's. At the same time, government-owned yards run by the Navy were getting out of the ship construction business altogether to concentrate on overhauls and modernization work. This meant that only one shipyard in America was large enough to build the ships of the SCB-102 design—Newport News Shipbuilding (NNS) in Virginia. By 1967, NNS had been awarded a sole-source contract for the initial units of the new *Nimitz* class (CVN- 68). These eventually included the lead ship, which was named for the World War II Commander in Chief of the Pacific Fleet (CINCPAC), Admiral Chester Nimitz, and two other ships would be the *Dwight D. Eisenhower* (CVN-69—named for the former President) and the *Carl Vinson* (CVN-70—named for the Georgia

Secretary of the Navy John Lehman, who headed the Navy Department from 1981 to 1986 during the Administration of President Ronald W. Reagan.
OFFICIAL U.S. NAVY PHOTO FROM THE COLLECTION OF A. D. BAKER

senator and political architect of America's World War II "Two Ocean Navy").

It would, however, be years until all three of the new ships were completed. Labor strikes and management problems plagued the construction of *Nimitz*, which took over seven years to complete (compared with four years for *Enterprise*). All three ships wound up costing hundreds of millions of dollars more than planned, making them fat targets for Congressional critics of Pentagon "fraud, waste, and abuse." The multi-billion-dollar price tag of the new ships meant that new carriers were going to be hard to sell to a nation that increasingly saw the military as a liability. In fact, not one new carrier was authorized by the Administration of President Jimmy Carter. However, a fourth unit of the *Nimitz* class, *Theodore Roosevelt* (CVN-71—after the late President and father of the "Great White Fleet"), was forced upon President Carter by Congress, who funded the unit in Fiscal Year 1980 (FY-80). Others would follow.

The election of President Ronald Reagan launched a period of rebirth for the Navy. This rebirth, directed at the perceived threat of a growing and aggressive Soviet "Evil Empire," was the personal achievement of one man: then-Secretary of the Navy John Lehman. Lehman, himself a Naval aviator and heir to the wealth of a great Wall Street investment firm, called for a "600 Ship Navy," with fifteen aircraft carriers at its core.[31] Fiscal Year 1983 (FY-83) saw the authorization of two *Nimitz*-class nuclear-powered aircraft carriers, *Abraham Lincoln* (CVN-72) and *George Washington* (CVN-73). Navy leaders dubbed this program the "Presidential Mountain," because three of the presidents honored are carved on the Mount Rushmore monu-

31 Secretary Lehman also authorized the reactivation of the four World War II-era *Iowa*-class (BB-61) battleships armed with antiship and long-range cruise missiles.

ment, and were strong supporters of the Navy.[32] Along with the three new carriers, over a hundred new nuclear submarines, guided-missile cruisers, destroyers, frigates, and support ships were authorized by the end of the 1980's. It was the biggest Naval building program since the Second World War.

Before the ''Presidential Mountain'' was completed, the global oceanic conflict they were designed to fight (or deter, if you thought that way) evaporated. With the end of the Cold War in 1991, the supercarriers acquired new roles and missions. In operations like Desert Shield/Desert Storm (Persian Gulf—1990/1991) and Uphold Democracy (Haiti—1994), they showed their great staying power and flexibility. Meanwhile, two more *Nimitz*-class carriers had been authorized in FY-88 to replace the last two units of the *Midway* class. This was just enough to keep the NNS shipyard alive. By the early 1990's it was time to plan on replacing the fossil-fueled carriers like *Forrestal* (CV-59) and *America* (CV-66), which were due to retire. Though at one point the Clinton Administration cut the number of carriers to eleven, the number was eventually stabilized at an even dozen (considered the minimum needed to sustain two or three forward-deployed carrier battle groups). In addition, in FY-95, another *Nimitz*-class ship was authorized, rounding out the third group of three. These three ships, *John C. Stennis* (CVN-74), *Harry S. Truman* (CVN-75), and *Ronald Reagan* (CVN-76), will hold the force level at twelve.[33]

In many ways, the *Nimitz*-class ships represent a ''worst-case'' design, able to accommodate the most difficult conditions and threats. Designed against a Cold War expectation of immense Soviet conventional and nuclear firepower, they are almost too much warship for an age where there is no credible threat against them. Whether America needs so much capability right now and in the near future is a matter I'll take up shortly. Meanwhile, let's look at how these great ships are put together.

Newport News Shipbuilding: Home of the Supercarriers

The Virginia Tidewater has been a cradle of American maritime tradition for almost four centuries. The first English colony in North America was established in 1607 on the south bank of the York Peninsula at Jamestown. Later, Hampton Roads was the scene of the world's first fight between ironclad ships, when the USS *Monitor* and CSS *Virginia* dueled in 1862.[34] Across the

32 Thomas Jefferson also appears on Mount Rushmore, but he was always skeptical about sea power, and in the Navy's eyes he did not merit the naming of a carrier.

33 Originally, CVN-75 was to have been named the USS *United States*, after the original supercarrier (CVA-58) broken up on the building ways in 1949. In fact, there exist photos of her keel being laid under that name. However, for political reasons, the Clinton Administration decided to rename her *Harry S. Truman*. So for the second time, Harry Truman ''sank'' the USS *United States*!

34 The *Virginia* is frequently and incorrectly referred to as the *Merrimac*, which was previously a steam frigate in the Federal Navy. Incompletely burned and scuttled when the Gosport Naval Yard (near the present-day Norfolk Naval Base) was abandoned in 1861 by Federal forces, it was raised and then used to build the Confederate ironclad.

The nuclear-powered aircraft carrier *Harry S. Truman* (CVN-75) being constructed at Dry Dock 12 in the Newport News Shipbuilding (NNS) yard. The large bridge crane in the foreground is used to place superlifts and other components into the dock.
OFFICIAL U.S. NAVY PHOTO

James River is the port of Norfolk, the most important naval base in the United States. And along the north bank of the James River is the town of Newport News, a twenty-mile-long snake-shaped community that is the birthplace of American aircraft carriers.

As you drive from Interstate 64 south onto Interstate 664, the yard makes its first appearance in the form of the huge pea-green-painted construction cranes that dominate the skyline of the city. And then as you turn off onto Washington Avenue, you will see the name on those cranes: Newport News Shipbuilding. Founded in 1886 by Collis P. Huntington, Newport News Shipbuilding (NNS) is the largest and most prosperous survivor of the American shipbuilding industry.[35] Seven of the battleships in "Teddy" Roosevelt's "Great White Fleet" were built here. Now one of just five U.S. yards still building deep-draft warships, NNS is the largest private employer in the state of Virginia, with some eighteen thousand workers (about half of the Cold War peak). The builder of the *Ranger* (CV-4—America's first carrier built from the keel up), NNS is the last U.S. shipyard capable of building big-deck nuclear carriers. Like most shipyards, NNS was originally built along a deep-channel river with inclined construction ways. Many of the original machine shops and dry docks are still in use after over a century of service. However, the facility has gradually been rebuilt into one of the most technically advanced and efficient shipyards in the world.

On the northern end of the yard you find the building area for aircraft carriers and other large ships. The centerpiece of this area is Dry Dock 12, where deep-draft ships are constructed. Almost 2,200 feet/670.6 meters long and over five stories deep, it is the largest construction dock in the Western Hemisphere. The entire area is built on landfill, with a concrete foundation supported on pilings driven through the James River silt into bedrock several hundred feet below. The concrete floor of Dry Dock 12 is particularly thick, to bear the immense weight of the ships built there. The end of the dock

35 After years of being a part of Tennaco Corporation, Newport News Shipbuilding separated in 1996 and is now a full-time shipbuilding concern.

extends into the deep channel of the river, and is sealed off by a removable caisson (a hollow steel box). Running on tracks the length of Dry Dock 12 is a huge bridge crane, capable of lifting up to 900 tons/816.2 metric tons, while a number of smaller cranes run along the edge of the building dock. Dry Dock 12 can be split into two watertight sections by the movable caisson, so that one carrier and one or more smaller ships can be constructed at the same time.

Only a decade ago NNS could expect to start a new *Nimitz*-class aircraft carrier every two years or so. NNS also had a share of the twenty-nine planned *Seawolf*-class (SSN-21) submarines on order. There were also new classes of maritime prepositioning ships, as well as massive overhaul and modification contracts to support John Lehman's "600 Ship Navy." But today the outlook is dramatically different, and the number of projects under way has been scaled back radically:

- With the carrier force set at twelve flattops instead of fifteen, the U.S. only needs to build a carrier about every four years.
- The *Seawolf* program was terminated at just three boats, and the work on all three went to the General Dynamics Electric Boat Division. Thus the massive investment in specialized facilities and tooling for submarine construction will lie unused at NNS until the start of the New Attack Submarine (NSSN) program in the early 21st century.
- Now that several hundred U.S. Naval vessels are being retired because of cost and manpower, the massive overhaul and modification program is only a fraction of what was originally planned.

NNS nevertheless remains the only American shipyard capable of building nuclear-powered surface warships. If future carriers or any of their escorts are to be nuclear-powered, then NNS will build them. Since at least one more *Nimitz*-class carrier is planned (the as-yet-unnamed CVN-77), the yard will stay fat in flattop construction for another decade. Meanwhile, Congress has guaranteed NNS a share of the NSSN production with Electric Boat, allowing the company to utilize its investment in submarine construction facilities built for the *Seawolf* program years ago. There has also been a steady flow of Navy and commercial refit and modernization work, and this is proving to be highly lucrative. In fact, NNS is preparing for one of the biggest refits ever, when USS *Nimitz* (CVN-68) comes back into the yard for its first nuclear refueling.

Building the Boat

Before we actually go on board a *Nimitz*-class carrier, let's take a look at how the ship is built. A *Nimitz*-class CVN is among the largest man-made moving structures. And with a price tag around $4.2 billion, it is also among the most expensive. Only the biggest commercial supertankers are larger. Such vessels are mostly hollow space, and they aren't built to take anything like the punishment a warship must be able to absorb. On top of that, carriers must hold six thousand personnel and operate over ninety aircraft. And finally, no supertanker has a power plant of such impressive capability as the nuclear power plants on *Nimitz*-class—or one that requires such obsessive care. Every component of the nuclear power plant comes under the meticulous scrutiny of the Office of Naval Reactors. Very early in the history of U.S. Navy nuclear propulsion, it was realized that the first nuclear accident would mean the end of the program. Therefore, rigid inspection standards and elaborate safeguards were applied to every step of design, construction, and testing. For example, every welded pipe joint (there are thousands of them!) is X-rayed, to ensure that it has no flaws, cracks, or voids.

Strange as it may sound, building a 95,000-ton aircraft carrier is a precision operation, which requires immensely detailed planning. For example, the maximum draft of a ship being built at NNS is limited both by the size of Dry Dock 12 and by local tidal conditions. Even at an unusually high tide, Dry Dock 12 can be flooded only to a depth of about thirty-three feet/ten meters, meaning that construction of a carrier can be taken only so far before it must emerge out of the dock into the James River. Once that's done, the hull is moored to a dock on the eastern end of the yard for final construction and outfitting. Because of the quick-moving tidal conditions near the mouth of the Chesapeake Bay, the launching is normally timed to the minute, and there are never more than a few inches to spare.

A *Nimitz*-class CVN gets its start in Washington, D.C., about a decade before its launching, when admirals at the headquarters of the Naval Sea Systems Command (NAVSEA, formerly known as the Bureau of Ships, the agency that manages ship construction) fix the retirement date of an aging carrier. This determines the time line for budgeting a new flattop. The time line, almost a decade long, starts at the point when money begins to be committed to the building of the new ship. Soon after that, contracts are signed for "long-lead items"—those components that can take years to order, design, manufacture, and deliver. These include nuclear reactors, turbines, shafts, elevators, and other key items that must be installed early in the construction of the ship.

Budgeting must also take into account changes and new items that go into each new carrier, for each has literally thousands of changes and improvements over earlier ships of the class. To lower the drag of the hull, the most recent *Nimitz*-class carriers have bulbous bow extensions below the waterline. Lowering the hull drag extends the life of the reactor cores and allows power to be diverted from propulsion to the "hotel" systems like air-conditioning and freshwater production. Most design changes are not so significant, and usually involve nothing more than a material or component change, like a new kind of steam valve, electrical switch, or hydraulic pump. Even so, every change involves written change orders, as well as stacks of engineering drawings. Back in the 1960's and 1970's, a small army of draftsmen, engineers, and accountants was required to produce the mountain of paper documenting the changes on a new carrier. Today, a much smaller force manages a computerized drawing and change-management system custom-programmed for NNS. In fact, in the interest of efficiency and competitiveness, the entire NNS operation has become heavily computerized.

A prime example of computerization is the ordering-and-materials-control system. NNS cannot afford a huge inventory of steel plate and other materials sitting around rusting in the humid Tidewater climate. There is only limited space for storage and construction, and every bit must stay busy for NNS to turn a profit. To minimize this potential waste, NNS has installed a computerized "just-in-time" ordering-and-materials-control system. The many components and raw materials (steel plate, coatings, etc.) that go into a *Nimitz*-class carrier arrive exactly when they are needed. No earlier, and no later. In this way NNS's investment capital is not needlessly tied up, and the final cost to taxpayers is reduced by millions of dollars. The NNS workforce has also become more efficient, since fewer items need to be stored, protected, hauled from place to place, and inventoried.

The actual start of construction begins some months prior to the official date of the ceremonial keel-laying. At that time, the Dry Dock 12 cofferdam is placed so that about 1,100 feet/335.3 meters of room are opened at the rear of the dock. This leaves 900 feet/274.3 meters at the river-gate end of the dock for construction of tankers or other projects. NNS workers then begin to lay out the wooden and concrete structural blocks that the carrier will be built upon. Building a ship that displaces over 95,000 tons/86,100 metric tons on wood and concrete blocks may sound like building a skyscraper on a foundation of paper, but NNS uses *lots* of these blocks to spread the load around. This very old technique is also used when ships are brought into dry dock for deep maintenance. Some things just work, and cannot be improved upon.

The close tolerances in the construction of a *Nimitz*-class carrier demand absolute precision from the start. Exact placement of the first keel blocks is critical, as they represent the three-dimensional "zero" points upon which everything else is built. This preliminary work goes on for four to six months, until the keel-laying ceremony draws near. At the same time, some initial assemblies are welded together and stored on the floor of the dry dock, since storage space in the main construction yard is tight. At the ceremonial laying of the keel on a *Nimitz*-class vessel, the guests include the Secretary of the Navy, the Chief of Naval Operations, and hundreds of other dignitaries. By tradition, the ship's "sponsor" (a sort of nautical godmother) is appointed—usually the wife of a high-ranking Administration official or politician whose favor is being sought by the Navy. Then a ceremonial weld is made in the first "keel" member (a steel box girder built up along the centerline of the lowest part of the hull), and the carrier's construction is officially under way.

Now a thirty-three-month countdown clock starts. From this day forward

Automated flame-cutting of steel plates at NNS.
John D. Gresham

to the launch date, the construction process is a race to determine the milestone bonuses and resulting profits for NNS stockholders. Meanwhile, Navy officials plan dates for commissioning and first deployments, select the "plankowner" officers and crew who will first man the new carrier, and assemble the "pre-commissioning unit" (PCU). These are the sailors who will report on board the ship while it is still under construction, in order to learn every detail of maintenance and operation.

Back at Dry Dock 12, the thirty-three-month construction moves forward rapidly. The secret to staying on schedule is "modular construction," a technique originally pioneered by Litton-Ingalls Shipbuilding in Mississippi. Rather than constructing a ship like a building, from the bottom up, the ship's designers break the design down into a series of modules. Each module is completed alongside the construction dock, with piping, fixtures, and heavy equipment already installed. Then it is lifted into place and "stacked" with other modules to form the hull. When that is done, the modules are "joined" (welded together). Pipes, ducts, and electric wiring bundles are connected into a mostly finished configuration, and the ship is "floated" out of the dock (or launched), with final work done alongside a "fitting-out" dock elsewhere in the yard. This mode of construction has many advantages. For one thing, the ship can be launched at a more advanced stage of construction than used to be the custom, which reduces costs considerably. Work that takes an hour to do in an NNS workshop usually takes three hours out in the yard, or eight hours in the ship once it is floating in the water. So anything that can be built in the shops or installed in the yard before it is assembled reduces costs; it is money in the bank.

Though modular military shipbuilding was pioneered by Litton-Ingalls, the scale at NNS is far greater. At NNS, they call this the "Superlift" concept. By way of comparison, Litton's largest module weighs around 500 tons/453.6 metric tons, while NNS utilizes modules up to 900 tons/816.6 metric tons lugged in place by the huge bridge crane. NNS can build a *Nimitz*-class carrier with about a hundred "Superlift" modules. Two dozen "Superlifts" make up a *Nimitz*-class carrier's flight deck, while the bow bulb and island structure are individual Superlifts.

A Superlift starts as a small mountain of steel plates, brought by rail and truck to NNS. Flame-cut to exact tolerances in the shops just south of Dry Dock 12, the plates are tack welded together by spot welds, then permanently joined by robotic welders along a pair of side-by-side production lines. These are then linked into the structural assemblies that form each Superlift. Once the basic structure is completed, cranes move it to the large assembly area next to Dry Dock 12. Then NNS yard workers crawl over and inside it to ''stuff'' electrical, steam, fuel, sewage, and other lines, fittings, and gear into place. Sometimes Superlifts are turned upside down, to make ''stuffing'' easier. When a Superlift is ready for joining, the nine-hundred-ton bridge crane is moved into position overhead, the lift cables are fastened, and the assembly in Dry Dock 12 made ready. Despite a Superlift's gigantic size and weight, this is a precision operation, with tolerances frequently dictated by the relative temperatures of the ship assembly and the Superlift. Depending on temperature, the metal structure of a Superlift can easily expand or contract over an inch during a given day on the Tidewater.

Around the assembly yard, several dozen Superlifts are in various stages of preparation at any given time. Some interior and exterior painting is done on Superlifts, to make this nasty and environmentally sensitive job a little safer. Because power, water, and air-conditioning can be installed in a Superlift while it is being assembled, the construction process is considerably facilitated. This is particularly helpful in the hot, muggy summers and cold, wet winters of the Tidewater region. There is a particular order to how Superlifts are stacked. The initial Superlifts—including the double bottom, reactors, steam power plants, ammunition magazines, and heavy machinery—are laid around the keel structure. In general, these items (making up the bottom of the middle third of the carrier) are the heaviest and most deeply buried components, and cannot be accessed or installed easily later on. They take some four months to assemble.

At twenty-two months to launch, everything aft to the fantail and up to the main/hangar deck is in place. Many of the living and habitation spaces are also included in this phase, as well as the majority of the carrier's protection systems (double bottoms, heavy plating, and voids—hollow spaces like fuel tanks, etc.). Now the assembly is beginning to look like a ship. At eighteen months to launch, the hangar deck is taking shape, along with the great overhanging ''sponson'' structures that extend out to port and starboard. Assembly of the bow is beginning. The flag (admiral's staff) and air wing spaces are fitted out, as well the offices for the various ship's departments. By fourteen months to launch, the hangar deck, sponson, and bow structures are in place, and the first parts of the flight deck are filling in amidships. After four more months, the hangar and flight decks are almost finished. Meanwhile, the lower bow has been completed, as well as the entire fantail structure. At two months before launch, the entire island structure—an eight-story building—is lifted onto the deck of the ship. This final Superlift represents the completion of major construction.

While the NNS yard workers seal up the hull and make it watertight, the managers and planners get ready for the actual launching of the ship. The launching ceremony is similar in many ways to the keel-laying just over two-and-a-half years earlier. Again, the Secretary of the Navy and the Chief of Naval Operations are present, as is the carrier's sponsor. She gets to break the traditional bottle of champagne over the new carrier's bow. A hint, though: Scratch the bottle first with a diamond-tipped scribe to ensure a clean break. Long-winded speeches, prayers, and benedictions complete the launching ritual. Then things get deadly serious and precise.

Since Dry Dock 12 is not deep enough to float off a finished *Nimitz*-class carrier, as soon as the hull structure is complete, it must be quickly floated out of the dock. Then the uncompleted carrier can be moved to a deeper part of the James River channel, where it can be moored to a fitting-out wharf for completion. The depth of the dock and the tidal conditions of the Tidewater region allow very little margin for error—meaning that the launching of a carrier is synchronized with the highest tide in a given month, to provide maximum clearance over the end of the dry-dock gate.

Before this can begin, any other ships in Dry Dock 12 are floated out and the movable cofferdam is removed. Then the dock is carefully flooded, with hundreds of NNS and Navy personnel monitoring tidal conditions and the watertight integrity of the carrier. When the dock is fully flooded and the ship has lifted off the keel blocks, the gate is opened. Now things happen fast. As a small tugboat pulls the carrier out of the dry dock, other tugboats wait just outside in the river to take control of the massive hulk. When the carrier is finally clear of the gate and safely into the deep channel of the river, it is turned and towed downstream to the fitting-out wharf on the southern end of the NNS property. Here it will be moored until it is turned over to the Navy, approximately two years later.

While it is an impressive sight sitting at the fitting-out dock, the mass of metal floating there is hardly a ship of war. It is still, in naval terminology, just a "hulk." Making it into a habitable vessel is the job of almost 2,600 NNS yard workers—everything from nuclear-reactor engineers to diesel-engine mechanics, computer specialists to roughneck welders. Building a modern warship takes almost every technology and tradecraft known. Imagine a skyscraper with offices, restaurants, workshops, stores, and apartments that can steam at more than thirty knots, with a four-and-a-half-acre airfield on the roof. That is a fair description of a *Nimitz*-class aircraft carrier.

During a visit to NNS in the fall of 1997, I spent some time aboard the USS *Harry S. Truman* (CVN-75) while she was about nine months from commissioning and delivery. I'd like to share with you some of my experiences there. My first stop, after NNS and Navy officials led me aboard, was the massive hangar deck. At 684 feet/208.5 meters long, 108 feet/33 meters wide, and 25 feet/7.6 meters tall, it is designed to provide a dry, safe place to store and maintain the aircraft of the embarked wing. As we walked forward, I passed several large access holes that led into the two nuclear reactor

The hulk of the USS *Harry S. Truman* (CVN-75) at the NNS fitting-out wharf in the fall of 1997. By mid-1998, the *Truman* was conducting sea trials off the Atlantic Coast.
JOHN D. GRESHAM

Catapult-testing deadweights aboard the *Harry Truman* (CVN-75).
JOHN D. GRESHAM

compartments below. These would be buttoned up shortly, my guides told me. The nuclear fuel packages would then be installed, followed by testing and certification of the twin A4W reactor plants. All around the hangar deck, workers were busy welding and installing pieces of equipment.

After climbing several ladders, we emerged on the flight deck, where hundreds more NNS workers were hustling about at their tasks, and then moved forward to the catapults, which were in the process of testing and certification. They are installed in pairs on the bow and the deck angle port-side, and each of the four 302-foot/92.1-meter-long C13 Mod. 1 catapults is capable of launching an aircraft every few minutes (the cycle time depends largely on the skill of the deck crew). Each catapult is powered by a pair of steam cylinders, which are built into the flight deck, and normally use high-pressure saturated steam from the reactor plant; but since the reactors were not yet powered up, *Truman* drew her power, water, and steam from plants dockside.

Testing such powerful machines is a dramatic procedure. Scattered

The island structure of the *Harry S. Truman* (CVN-75) being finished at NNS.
John D. Gresham

around the deck were a number of orange-painted, water-filled, wheeled trolleys called deadweights. Each deadweight simulates a fully loaded aircraft, with attachment points that allow it to be hitched to the shuttle of a catapult. After the bow has been pointed into the James River channel, and the Coast Guard and local boaters have been suitably warned, each catapult fires the entire range of deadweights. The tests are noisy and the sight of the weights flying hundreds of yards/meters into the channel is bizarre. Nevertheless, this is a highly effective way to prove that the machinery is ready. After leaving the catapults, we headed aft to inspect the catapult control station between Catapults 1 and 2.[36] Set on a hydraulically raised platform under an armored steel door, the control station is a pod where the catapult officer—or "shooter"—can control the catapults in safety and comfort. Another identical station is located on the port side, controlling Catapults 3 and 4.

Next we walked over to the island structure, where our guides showed us how the many systems on the flag and navigation bridges, the primary flight control, and the meteorology office were installed. Although the basic *Nimitz* design is over thirty years old, the many changes bringing it into the 21st century are quite visible. Up on the *Truman's* navigation bridge, for example, are many of the "Smart Ship" systems (mentioned in the second chapter) that make it possible for three people to steer the ship from auto-

36 The four catapults on every carrier are numbered 1 through 4, from the starboard bow (Catapult 1) to the port angle (Catapult 4).

The cluttered flight deck of the *Harry S. Truman* (CVN-75) while being fitted out at NNS.
JOHN D. GRESHAM

mated control stations (before, almost two dozen people were required to do the same job). Similar systems will be scattered throughout the *Truman*, and will be tested when she goes to sea in 1998.

As we moved farther aft, we passed by the kinds of tool sheds and other temporary storage buildings that you find at any construction site. Then we dropped down a ladder back to the hangar deck and down another into the bowels of the ship. At this point, the primary work on *Truman* involved preparing some eight hundred (out of a total 2,700) compartments for turnover to the Navy. Those compartments contain crew berthing, medical facilities, galley and mess areas, office spaces, the ship's store, the post office, and storage rooms. Everything needed to finish these spaces must be carried up and down ladders and through narrow passageways by hand. Sprained knees and ankles are the price paid to haul paint cans, power cables, and tools into the ship.

Shortly after this job was completed, just after New Year's of 1998, the first of the Navy's crew of "plankowners" arrived. Several of the ship's spaces that had already been turned over proved to be spotless when we visited them; and the quality and workmanship are very impressive. In particular, the communications spaces, which were just being brought to life by a Navy crew, had the look and smell of a new automobile. As the final stop on my visit, I was allowed to visit the magazines and the pump room in the very bottom of the ship.

It was close to quitting time when we made our way back to the hangar deck, aft to the fantail, and down the access ramps to the dock. As we sat waiting for our tired leg muscles to loosen, the shift alarm went off, and we watched 2,600 NNS workers come off shift and head for home—an impressive sight. As they passed by us on the dock, I was reminded of the builders of the Egyptian pharaoh's pyramids. Both groups labored to build a wonder of the world. Unlike the pharaoh's slaves who hauled and stacked the stones in the desert, these people *have chosen* to labor at their "wonder of the world." They *want* these jobs, take pride in what they do, and make good livings. For those who think that Americans don't build anything worthwhile

The ''NNS Miracle'': Some of the 2,600 Newport News Shipbuilding workers leave the *Harry S. Truman* (CVN-75) at the end of an afternoon shift.

John D. Gresham

these days, I say go down to NNS and watch these great men and women build metal mountains that float, move, and fly airplanes off the top. It truly is the ''NNS'' miracle.

When the initial crew cadre came aboard *Truman* in early 1998, they began to help the NNS yard workers bring the ship's various systems to life. This process (ongoing until the ship is handed over to the Navy) is designed to make her ready for her ''final exams,'' when the carrier will become truly seaworthy, with her reactors powered up and most of her ''plankowner'' crew aboard. Combat systems tests occur when the ship is about 98% complete, with evaluations of the radar and radio electronics, defensive weapons, and all the vast network of internal communications and alarms. After these tests, it is time for sea trials off the Virginia capes, including speed runs to evaluate the power plant. After these trials are completed, the Navy conducts one last series of inspections prior to the most important ceremony of the entire building process (at least for NNS). This is the signing of the Federal Form DD-250, which indicates that the Navy has taken possession of the vessel and NNS can now be paid!

The next six to eight months are filled with training and readiness exercises, including the traditional ''shakedown'' cruise. Following this is a short period of yard maintenance (known as ''Post Shakedown Availability'') to fix any problems that have cropped up. The new carrier will then spend much of her time over at the Norfolk Naval Station, moored to one of the long carrier docks, where she will get ready for commissioning. At the commissioning ceremony, the high officials, the dignitaries, and the ship's sponsor once again gather. Again there are speeches and presentations. And almost a decade after the decision was made to build this mighty warship, a signal is given, the commissioning pennant is raised, the crew rushes aboard to man the sides, and she is finally a warship in the U.S. Navy.

The *Nimitz* Class: A Guided Tour

Let's now take a short walking tour of a *Nimitz*-class carrier. We'll start the way most guests come aboard, at the officers' accommodation brow on the

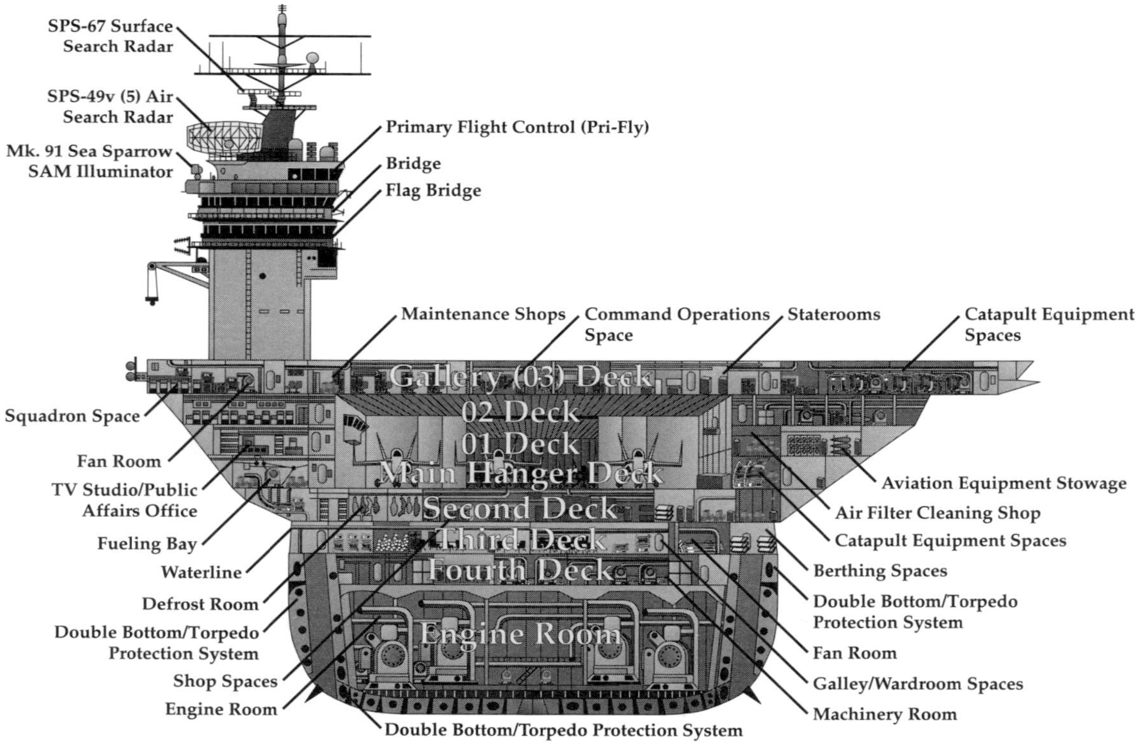

SPS-67 Surface Search Radar
SPS-49v (5) Air Search Radar
Mk. 91 Sea Sparrow SAM Illuminator
Primary Flight Control (Pri-Fly)
Bridge
Flag Bridge
Maintenance Shops
Command Operations Space
Staterooms
Catapult Equipment Spaces
Gallery (03) Deck
02 Deck
01 Deck
Main Hanger Deck
Second Deck
Third Deck
Fourth Deck
Engine Room
Squadron Space
Fan Room
TV Studio/Public Affairs Office
Fueling Bay
Waterline
Defrost Room
Double Bottom/Torpedo Protection System
Shop Spaces
Engine Room
Double Bottom/Torpedo Protection System
Aviation Equipment Stowage
Air Filter Cleaning Shop
Catapult Equipment Spaces
Berthing Spaces
Double Bottom/Torpedo Protection System
Fan Room
Galley/Wardroom Spaces
Machinery Room

A front cutaway view of an improved *Nimitz*-class (CVN-68) nuclear-powered aircraft carrier.

JACK RYAN ENTERPRISES, LTD., BY LAURA DENINNO

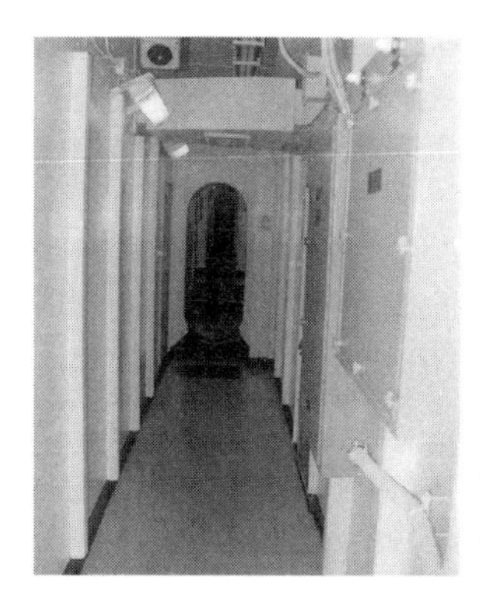

The main starboard corridor on the O-2 level of a *Nimitz*-class (CVN-68) carrier.
JOHN D. GRESHAM

starboard side just under the island. One of the first things you notice is the thickness of the hull, which is composed of high-strength steel several inches thick. It is that thick to protect against battle damage and fires. The same material makes up the flight and hangar decks, providing them with a similar resistance to damage and fires. Everywhere, there are redundant water and firefighting mains, with damage control stations in every passageway. The Navy is deadly serious about firefighting, and there even is a water deluge system, which can flood the deck, or wash it down in the event of a nuclear or chemical attack.

Past the entryway hatch, you take the first of many tall steps over structural members the crew calls ''knee knockers.'' Though they are a constant nuisance to movement throughout the ship, these steel thresholds provide structural strength to the entire vessel. A *Nimitz* has miles of virtually indistinguishable passageways. And there are dozens of places in them where just standing around watching can be hazardous—due to noise, fumes, moving machinery, or simply wet, slippery decks. These passageways are considerably narrower than those in other combat vessels, particularly amphibious ships which have room for combat-loaded Marines to move around. Despite their huge size, carriers are volume-limited, and space for people to live, work, and walk takes away capacity for fuel, bombs, and fighting power. So getting around with any sort of load can be a genuine chore. You often see ''bucket brigades'' of sailors moving loads of food and other supplies from one place to another.

The narrow corridors are one important reason for the Navy's constant emphasis on simple courtesy. A senior officer or chief headed in the opposite direction always gets a respectful greeting and the right of way in these narrow passages. I learned a valuable lesson sometime ago from a civilian analyst who had spent many years on board Navy ships: ''If you're standing anywhere and you're not touching metal, you're probably in somebody's way.''

Moving inboard through several hatches, you emerge into the vast hangar deck; 684 feet/208.5 meters long, 108 feet/33 meters wide, and 25 feet/

7.6 meters tall—about two-thirds the total length of the ship. Three immense sets of power-driven sliding armored doors divide the hangar bay into zones, to limit the spread of a fire or damage from explosions. In good weather, daylight floods in from four huge oval openings in the sidewalls where the elevators are located. In bad weather sliding barriers seal off the elevator openings to keep the interiors safe and dry. The elevators themselves are the largest aluminum structures on the ship (to save weight). Each of these mammoth lifts (one on each side aft, with two others forward on the starboard side) can raise two fully loaded F-14 Tomcats (the heaviest carrier aircraft) to the flight deck at one time. This is one of the few places on the ship where you can actually see the sea and sky, and remind yourself of the outside world. The flight deck, by contrast, is a highly restricted area. Since there are no portholes, most of the crew rarely sees the light of day. You often find crew members who go days and weeks at a time without either a breath of fresh air or a view of the outside world.

The hangar deck is one of the three main horizontal structures on a carrier (the flight deck and keel/double bottom are the other two), and it provides much of the stiffness and protection for the rest of the ship. Any damage from hits on a carrier should be contained outside the armored boxes that surround the hangar deck and engineering/living spaces below. When it's empty, you would have room to play two games of American football in the hanger bay. But when it's filled with fifty or sixty aircraft only inches apart, there is barely room to worm your way through the mass of landing gear, pylons, and maintenance equipment. The hangar deck is always packed with airplanes and equipment, though there is not enough room to strike down all of the air wing's birds at one time. This means that some of the birds must always be parked on the flight deck. Fortunately, Naval aircraft are designed to withstand the corrosive effects of salt water, and can take the punishment fairly well.

Just aft of the elevator bay is a large stowage area where the ship's boats are stacked, along with bulky items like forklifts, spare arresting cable reels, and spare engines. Moving aft from this holding area, you find the engine and maintenance shops, which completely fill the stern of the ship. Here the ship's Aircraft Intermediate Maintenance Division (AIMD) repairs, overhauls, and tests engines, hydraulic pumps, electronics boxes, and countless other mechanical components that keep planes flyable and combat-ready. The maintenance shops are divided up into small spaces where work is done that normally takes acres of workshops and hangars back ashore.

Farther aft of the AIMD shops, you again break out into daylight on the stern, or fantail, of the ship, an open area the full width of the hull, roofed by the flight deck, with projecting platforms and catwalks on either side. Mounted on the fantail are massive test stands, where aircraft engines can be strapped down and run at full power. Because no bit of open space goes to waste on a carrier, you'll only rarely find a time when you can just stand back here and watch the ocean go by. This is especially true during flight

The hangar bay of the USS *George Washington* (CVN-73), a *Nimitz*-class (CVN-68) carrier.

JOHN D. GRESHAM

operations. If an aircraft should hit the stern (in what aviators dryly call a "ramp strike"), the fantail is going to be showered with flaming jet fuel and debris. Such accidents are very rare, but they *do* happen, which means that unless you work there, you aren't permitted on the fantail. So if you get to see this spot while under way, count yourself lucky.

Here also are one of the four (three on the *Nimitz* (CVN-68), *Dwight D. Eisenhower* (CVN-69), and *Carl Vinson* (CVN-70)) Mk. 15 Phalanx Close-In Weapons Systems (CIWS). A pedestal-mounted 20mm Gatling gun with its own tracking radar, the Mk. 15 is designed to knock down incoming missiles and aircraft. Phalanx has now been in service for almost twenty years, and is considered marginal against the latest threat systems (like the sea-skimming, Mach 2 Russian Kh-41/SS-N-22 Sunburn missile). The Mk. 15's will eventually be replaced by twenty-one-round launchers for the Rolling Airframe Missile (RIM-116A RAM). RAM is based on the classic AIM-9 Sidewinder air-to-air missile, with a modified seeker from a Stinger (FIM-92) man- portable SAM. RAM—much more capable than the Mk. 15—can actually destroy an incoming Mach 2 missile before it hits (or showers the ship with supersonic fragments).

Located below the Phalanx mount are the twin ports for the ships SLQ-25A "Nixie" torpedo countermeasures system. Nixie is a towed noise-maker streamed behind the ship when there is a threat of incoming torpedoes. The idea is that the "fish" will chase the towed decoy, and detonate against it instead of the ship. Since each decoy can be used only once, two Nixie decoys are kept at the ready, each at the end of a spooled tether in the stern. Finally, on a platform at the stern next to the Mk. 15 stands the instrument landing system. This is a stabilized "T"-shaped bar of vertical and horizontal lights, which helps a pilot on final approach judge the roll and motion of the ship.

Heading back forward into the hangar bay, you will probably notice the "spongy" feel of the deck, which comes from the grayish-black non-skid coating that is applied to seemingly every horizontal surface exposed to the weather. Non-skid—a mix of abrasive grit and synthetic rubber applied in a rippled pattern—keeps you from slipping on a wet, oily, or tilted deck, an all too common occurrence on a naval vessel. Up on the flight deck, the constant pounding and scraping of landing gear and tailhooks quickly erode the coating and expose bare steel. When this happens, maintenance crews

Looking down into the well of one of the hundreds of ladders aboard a *Nimitz*-class (CVN-68) carrier. These are tall and narrow, and are quite grueling to climb.

JOHN D. GRESHAM

mix up a batch and ''touch up'' worn spots. Also notable is the hangar deck's elaborate fire-suppression system, which can put enough foam into the hangar bay to drown the unwary. Fire hoses and mains sprout from every corner of the hangar bay, and damage control gear is also in evidence.

In the overhead are storage racks for everything from aircraft drop tanks to spare engines. You can even see a spare catapult piston—a steel forging as long as a bus—racked high on the wall of the hangar bay. In the forward part of the hangar bay on the starboard side are two more aircraft elevators, as well as the passageways that lead into the forecastle. Here you find more AIMD offices and shops, as well as most of the berthing spaces for enlisted personnel from the embarked air wing. Cramming almost six thousand personnel into a ship, even though it's close to a quarter mile long, makes for tight quarters. Even so, the enlisted and chiefs' berthing spaces on a *Nimitz* are still more comfortable than those aboard a submarine or older Navy surface warship.

For a young person coming aboard a warship for the first time, the cramped personal space may seem harsh. In fact, while personal space *is* spartan, it is nevertheless quite functional. Enlisted personnel get a stowage bin under their bunks, and a single upright locker about the size of the one you had back in high school. They can also stow some personal items in their workspaces, but they still must always plan ahead when packing to go aboard ship. For sleeping, crew will normally be assigned to a bunk (called a ''rack''), which will be one in a stack of three. You will find around sixty racks in a berthing space, with an attached rest room/shower facility (what the Navy calls a ''head''), and a small common area with a table, chairs, and television connected to the ship's cable system. Television monitors can be found in almost every space on board, displaying everything from the ship's Plan of the Day (called the ''POD''), to movies, CNN Headline News, and the ''plat cams''—a series of television cameras that monitor activities on the flight deck.

The racks themselves are narrow single beds, with a comfortable foam-rubber mattress, and basic bedding. There are also privacy curtains, a small reading lamp, and usually a fresh-air vent—often a vital necessity. While most of the interior spaces of a *Nimitz* are air-conditioned, even nuclear-powered chillers sometimes have a hard time keeping up with the hot and humid conditions in the Persian Gulf or the Atlantic Gulf Stream in summer. That stream of cool air on your face is sometimes all that lets you sleep. Other distractions on board can also keep you from getting rest, such as the launching and landing of high-performance combat aircraft on the roof. Crew members with quarters just below the catapults and arresting gear have a hard time sleeping when night flight operations are running, which is why the air wing personnel are berthed here. When the wing is flying, they would not be in their racks anyway.

Forward of the living spaces, in the very bow of the ship, is the forecastle. Here the anchors, handling gear, and their huge chains are located. It is also the domain of the most traditional jobs in the Navy: the Deck Division. In an era of computers and guided weapons, these are the sailors who can still tie every kind of knot, rig mooring lines, and handle small boats in foul weather. You need these people to operate anything bigger than a rowboat, and aboard a carrier they are indispensable. On the port side of the forecastle you find the first of a set of "stairs," which we'll use to climb up several levels. These are not conventional stairways, but very nearly vertical ladders, and they are quite narrow. You learn to move up and down ships' ladders carefully, and finding a handy stanchion to grasp when you're on them becomes instinctive.

Opening another hatch, you find yourself on a small platform adjacent to the bow. From here, you can climb a few steps and move out onto the four and a half acres that is the carrier's flight deck. Again, the spongy feel of the deck tells you that there is non-skid under your feet. Around the deck, two or three dozen aircraft are packed in tight clusters, to free as much deck space as possible. During flight operations, the noise is incredible. It is so loud that you must wear earplugs just to watch from up on the island, while flight deck personnel who must work among the aircraft wear special "cranial" helmets with thickly padded ear protectors to preserve their hearing. Only Landing Signals Officers (LSOs, the people who guide aircraft during landings) are allowed on deck during flight operations without a cranial, since they have to clearly hear and see aircraft as they approach the stern for landing.

There are other hazards as well. In fact, the flight deck of a modern aircraft carrier is arguably the most dangerous workplace in the world. Aircraft are constantly threatening to either suck flight deck personnel into their engines, or blow them off of the deck into the ocean. For this reason, the entire perimeter of the flight deck and the elevators is rigged with safety nets. In addition, everyone on the flight deck also wears a "float coat," which is an inflatable life jacket with water-activated flashing strobe light, and a whistle to call for help—just in case the safety nets don't catch you. Standard flight deck apparel also includes steel-toed boots, thick insulated fabric

Flight deck personnel aboard the USS *George Washington* (CVN-73).

JOHN D. GRESHAM

gloves, and goggles (in case a fragment of non-skid or some foreign object/debris—FOD—is blown into your face).

Each float coat and cranial is color-coded by job. Under the float coats, deck crews also wear jerseys—heavy, long sleeved T-shirts—of the same color as the float coat (though they may be a different color from the cranials). These color-code combinations are universal aboard Navy ships. Here is what they mean:

DECK PERSONNEL IDENTIFICATION GUIDE

Personnel Function	Jersey/ Coat Color	Cranial Color
Aircraft Handlers/Tractor Drivers	Blue	Blue
Elevator Operators/Messengers	Blue	White
Aircraft Directors	Blue	Yellow
Handling Officers/Plane Directors	Yellow	Yellow
Catapult/Arresting Gear Officers	Yellow	Green
Arresting Gear/Catapult/Maintenance Crew	Green	Green
Arresting Gear/Catapult Officers	Green	Yellow
Cargo Handlers/Replenishment Officers	Green	White
Quality Assurance Personnel	Green	Brown
Liquid Oxygen (LOX)/Medical/Safety/Transfer Crew	White	White
Landing Signals Officer (LSO)	White	None
Plane Inspector	White	Green
Plane Captains	Brown	Brown
Helicopter Plane Captains	Brown	Red
Aviation Fuels Crew	Purple	Purple
Ordnance/Crash and Salvage/Explosives Disposal Crew	Red	Red

For example, only sailors wearing purple coats, jerseys, and cranials are allowed to handle fuel and other flammable fluids on deck (they are nicknamed the "grapes").

Keeping an eye on flight-deck operations is a vital task. Up on the island, observers constantly watch the position and flow of planes, personnel, and equipment around the deck. Any deviation from standard procedures or safety rules calls down a sharp and angry rebuke over the flight deck loudspeaker (loud enough to hear through your cranial—and that is really LOUD) telling you *exactly* what you must do *RIGHT NOW!* To help these commands make sense, there is a standard set of coordinates and definitions for the various parts of the flight deck. For example, the catapults are numbered from 1 through 4 in order, starboard to port, bow to stern. The elevators are numbered, with 1 and 2 ahead of the island on the starboard side, number 3 just aft, and number four on the port side aft. The jet blast deflectors (JBDs) are matched to the catapults, 1 through 4. The arresting wires are also numbered, running from number 1 farthest aft, to number 4 up forward. Areas of the deck also have specific names, so that when an observer or lookout yells out a warning, he can direct other eyes to it without delay. Some examples include:

- **The "Crotch"**—The point where the roughly 14° landing deck "Angle" ends and the port bow begins.
- **The "Junkyard"**—The area at the base of the island aft. Here tractors, forklifts, a wrecking crane, and the world's smallest fire truck (collectively known as "yellow gear" even though some are now painted white) are parked, always ready to move when needed.
- **The "Hummer Hole"**—The area just forward of the Junkyard. Here the E-2C Hawkeyes (nicknamed "Hummers") and their cargo-carrying cousins, the C-2 Greyhounds, are parked.
- **The "Street"**—The " Street" is up on the bow in the area between Catapults 1 and 2; the forward catapult control pod is located there.
- **The "Rows"**—Also on the bow are the "1 Row" and "2 Row." These are the zones outboard of Catapults 1 and 2 and are normally used as parking areas for the F/A-18 Hornets when a landing event is active.
- **The "Finger"**—A narrow strip of deck just aft of Elevator 4, with parking space for a single plane.

Working in this noisy, hot, and dangerous world is the job of some of the bravest young men and women you will ever meet. Most are under twenty-five; and some look so naive (or so scary), you might not trust them to valet park your car at a restaurant. Yet the Navy trusts them to safely handle aircraft worth several *billion* dollars, not to mention the infinitely precious lives of air crews, each representing millions of dollars in training and experience.

Theirs is a world of extremes. For up to eighteen hours a day, they're subjected to noise that would deafen if not muffled; heat and cold that would

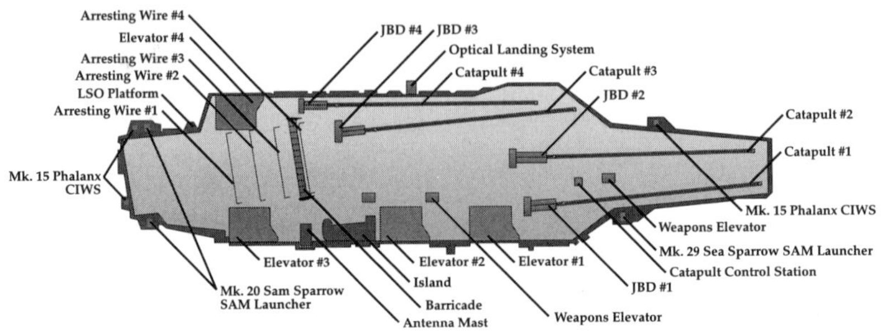

A top view of an improved *Nimitz*-class (CVN-68) nuclear-powered aircraft carrier.
JACK RYAN ENTERPRISES, LTD., BY LAURA DENINNO

kill if not insulated. They are surrounded by explosives, fuel, and other dangerous substances,[37] and are frequently buffeted by winds of over sixty knots. For this, they receive a special kind of respect and a "hazardous duty" bonus (in 1998, about $130 per month) in addition to their sea pay. These young men and women know their work makes flying aircraft on and off the boat possible, and they take quiet pride in this dirty, dangerous job up on "the roof." Because of the extreme noise, a richly expressive sign language is used to direct operations on the flight deck. Using a series of common and easily understood hand signals, the deck crew personnel tell each other how to move aircraft and load bombs and equipment, and warn each other of emergencies. They constantly watch out for each other, for only the brother or sister sailor looking out for you keeps you safe. All of these efforts are dedicated to just two basic tasks: the launching and landing of aircraft. Let's now look at how it is done in somewhat greater detail.

If you move aft from the bow down the "Street," you walk between the two bow catapults, each as long as an American football field. And there is a similar catapult arrangement on the landing "angle" on the port side. Most of the machinery for each C13 Mod. 1 catapult is concealed under the flight deck: two slotted cylinders in a long steel trough, each with a narrow gap along the top. Overlapping synthetic rubber flangles cover and seal the gaps. In each cylinder is a piston, with a lug projecting through the sealing strips on top. Each of these lugs leads to a small crablike fixture called a "shuttle," which is up on the flight deck.

When an aircraft is ready to launch, it is maneuvered into position under the guidance of a plane handler. When the nosewheel is just behind the shuttle, a metal attachment on the gear strut, called a towbar, is lowered into a slot on the shuttle. Meanwhile, the Jet Blast Deflector (JBD) just aft of the plane is raised, and another mechanical arm is attached to the rear of the

37 For example, the tiny "LOX crew" cares for a tank of immensely hazardous liquid oxygen, which is used to refill the breathing air systems of some aircraft. This tank sits on an inclined ramp on the deck edge. A quick-release fitting allows it to be sent into the sea in the event of a fire, to prevent a catastrophic explosion.

The nose gear of an F/A-18C Hornet on the #1 Catapult of the USS *George Washington* (CVN-73). The forward towbar is linked to the catapult shuttle, and the holdback device is in position.

John D. Gresham

nose gear strut with a device called a "holdback."[38] This allows the aircraft to run its engines up to full power, far beyond the ability of the plane's brakes to keep it on the deck. In this way, the bird will have a considerable forward thrust even before it starts moving. Each aircraft type in the wing has its own special color-coded holdback, to prevent them from being used mistakenly on the wrong bird. The exceptions are the F-14 Tomcat and F/A-18 Hornet, which have permanent holdback devices built into their nosewheel gear struts.

Once the aircraft is properly hooked up by one of the green-shirted catapult crewmen, another "green shirt" holds up a chalkboard with the plane's expected takeoff weight written on it for the pilot and catapult officer (down in the catapult control pod) to see. If both agree that the number is correct (confirmed by hand signals), then the catapult officer (known as the "shooter") begins to fill the twin pistons with a pressurized charge of saturated steam from the ship's reactor plant.[39] The steam pressure is carefully regulated to match the takeoff weight of the aircraft, the speed of the wind over the deck (this is the natural wind speed plus the speed of the ship), and other factors like heat, air pressure/density, and humidity. This has to be *very* precise. Too much pressure will rip the nosewheel gear out of the plane, while too little will cause a "cold shot." In a cold shot, the aircraft runs down the deck and never reaches takeoff speed; the catapult then hurls it into the water ahead of the onrushing carrier.[40] At best, the crew will eject and

38 Because of the high temperatures generated by the engine afterburners of aircraft like the F-14 Tomcat and F/A-18 Hornet, the JBDs contain a system of cooling channels, through which are pumped seawater. This system keeps the hydraulically erected JBDs from melting under the thermal pounding.

39 The Navy does *not* use radioactive steam to power its catapults. The steam that powers everything on the ship is actually heated in the secondary (non-radioactive) loop of the reactor plant. All of the radioactive components of the reactor plant are contained in either the reactor vessels or the primary cooling loop of the system.

40 Some people get lucky. In 1983, during an attempted launch on board the USS *John F. Kennedy* (CV-67), the crew of an A-6E Intruder suffered a "cold shot," and ejected just before the aircraft pitched over the end of the bow into the water. The pilot's ejection seat fired him up, and his parachute let him down gently, unhurt, onto the deck just in front of the JBD of the catapult that had misfired his aircraft! The bombardier/navigator was not quite so lucky. Because his seat fired an instant earlier, he was thrown farther aft and to the side, and his parachute caught the overhanging tail of an EA-6B Prowler before he hit the ocean. The emergency crews searched for over a half hour before they found the crewman hanging over the side aft of the island, bruised from banging heavily against the hull, but alive.

the aircraft will be lost. At worst, both the aircraft and flight crew will be lost. As might be imagined, catapult officers (who are themselves veteran carrier aviators) take this highly responsible job quite seriously.

Once the pressure is at the desired level, there is a final check of the aircraft by the green shirts. If all appears to be at readiness, the catapult officer signals this to the pilot. The pilot selects the proper engine setting (usually maximum power or afterburner), snaps a salute back to the catapult officer in the pod, and braces for what is about to come. At that point, the catapult "shooter" hits a button in the control pod, and the twin cylinders are released. This snaps the holdback and throws the aircraft down the catapult track. The pilot/crew is hit with several times the force of gravity (what pilots call "G" forces), and their eyes are driven back into their sockets. Approximately one hundred yards/ninety meters and two seconds later, the towbar pops out of the shuttle, and the aircraft is on its own. Having achieved flying speed (usually around 150 knots), the pilot has now gained control of the airplane (that is, he or she can actually fly it).

Back on deck, a cable and pulley system retracts the shuttle to its start position, and the cycle repeats. A well-trained crew can complete this process in less than two minutes. A normal launch sequence using all four catapults can put an airplane into the air every twenty to thirty seconds. This means that launch events for several dozen aircraft can take less than fifteen minutes from start to finish. However, since the aircraft just launched will be back to land in only a couple of hours, the timing of what gets done next can be critical.

Configuring the flight deck for a landing "event" requires that the deck be "respotted," with as many aircraft as possible moved forward. In most cases, these are parked on Rows 1 and 2, so that the "angle" will be clear for returning aircraft; and this means that Catapults 1 and 2 are now blocked and unavailable for use. While it is theoretically possible to launch aircraft during landing operations, this is rarely done. To do so would require much of the air wing to be struck below to the hangar deck, a time-consuming and tiring exercise for the deck crews. In fact, carrier captains like to use the aircraft elevators as little as possible, since these constitute part of the flight deck and parking area for aircraft when they are in the "up" position. It's hard to find anything more precious to a carrier skipper than flight deck space, and even the four and a half acres on a *Nimitz*-class flattop seems small when filled with airplanes, ordnance, equipment, and people.

The flight deck can not only get crowded, it can easily become dangerous. For this reason aircraft that are not actually taking off or landing are parked and chained down as quickly as possible. Chaining down is also necessary because a slight list on a slick deck can send an aircraft sliding around like a rogue hockey puck on an ice rink. In fact, almost everything on deck is chained down when it is not in use, including the low-rise fire-fighting and aircraft tractor vehicles. Normally, as soon as an aircraft is shut down and parked, a crew of strong-backed young blue shirts moves in to attach tie-down chains to some of the thousands of tie-down points imbedded in the plating of the flight deck.

On the port side aft is a sponson holding what is called the "Lens." This is a stabilized (against the motion of the ship) system of lights and directional lenses, designed to provide approaching pilots with a visual glide path down to the deck. If an approaching aircraft has the proper attitude and sink rate, then the pilot sees an amber light—or "meatball"—from the system. If the pilot can keep the "ball" centered (with a row of green lights) all the way down (any offset from the proper attitude shows the pilot a row of "red" lights), then it should put him down in the perfect spot for a landing on the deck aft.

Once the flight deck has been respotted for the coming landing event, and the ship has once again come into the wind, things again get exciting. Modern carrier aircraft are too heavy and their stall speeds are too high to possibly land in the roughly 500 feet/152 meters of space on the flight deck. In fact, the only way to get a high-performance airplane onto a carrier deck is to literally fly it to a "controlled" crash, and stop it forcibly before it falls into the sea. The lens system and other special landing instruments (some aircraft even have an automatic landing system) are useful aids, but pilots usually need additional help. This formidable task is the job of a lot of very special equipment and is overseen by the Landing Signals Officers (LSOs). Back in the old days of propeller-driven planes and the early jets, LSOs were the *only* landing aid for pilots. They did their job with nothing more than a pair of lighted paddles (to show the pilots their landing attitude) and a few hand signals. LSOs today do their job from a small platform on the port side aft, and it is there that we now will go to get a perspective on the fine art of a carrier landing.

Landing a carrier aircraft starts in the aircraft cockpit, when the pilot makes the break into the ship's landing pattern. The pattern itself is controlled by the Carrier Air Traffic Control Center (CATCC) located one level down from the flight deck. The CATCC is a miniature of what you would find at any major airport, and it functions in exactly the same way. The controller's job is to "stack" the aircraft, prioritize them into an oval-shaped pattern about a mile wide and four miles long around the port side of the carrier, and "stagger" them, so the LSO has the necessary time to bring each aboard. (They can land an aircraft about every thirty seconds under good conditions.) The aircraft in the pattern are prioritized by their "fuel state," a polite way of saying that the first planes to be brought aboard are the ones that are about to fall into the ocean from fuel starvation. Just to be sure this does not happen, the carrier usually has an airborne tanker overhead during flight operations to refuel airplanes too close to the Empty point on their fuel gauges.

When the landing event has been properly organized, the "Lens" is turned on, and the first pilot in the pattern makes the "break" out of the pattern to line up on the stern of the carrier. During the "downwind" leg of the pattern, the pilot drops the plane's landing gear, tailhook, and flaps, makes

Detail of a landing wire and capstan on the USS *George Washington* (CVN-73).

John D. Gresham

sure the radio is set up on the LSO frequency, and turns left toward the boat. Assuming all this has been done properly, the aircraft should start its final approach at eight hundred feet altitude, about three-quarters of a mile from the stern of the carrier, and just fifteen seconds from touchdown.

As the aircraft finishes its break, the LSO orders the pilot over the radio to "Call the ball!" This tells the pilot to let the LSO know that he has spotted the amber "meatball" of the landing system. If the pilot does see it, he or she calls "Roger ball!" back to the LSO to confirm that. At this point, the final ten-second dash to the deck is on. On the LSO platform, the LSO and an assistant are watching and judging the aircraft's attitude. Highly experienced pilots themselves, LSOs are expert judges of all this. In his or her hand, the LSO holds what is known as the "pickle." This controls a series of lights near the LSO platform, which are visible to aircraft approaching the stern. As long as the aircraft continues properly on course, the pilot gets a green "OK" light. But the LSO can also activate "more power" and "wave off" lights with the "pickle." The LSO can also coach the pilot by radio, but this is not normally done. Since an enemy could intercept radio signals in wartime conditions, "emissions control" procedures (called EMCOM Alpha in its most extreme form) dictate that combat landing operations be done only with lights. If the aircraft is set up properly, it should now be about thirty feet over the fantail, with airspeed of around 130 knots/240 kph, and a decided nose-up attitude. At this point, the pilot and LSO have done their part of the job, and it is the turn of machinery to finish it.

Handling this task is the ship's arresting gear system, located in the middle of the 14° angle aft. Stretched across the deck are four braided steel cables (called "wires" by the crew), numbered 1 through 4, from rear to front. The wires are spaced about fifty feet apart, and each is hooked to a pair of hydraulic cylinders located one deck below. If the pilot and LSO have set the landing up properly, the aircraft should hit the deck in the roughly two-hundred-foot/sixty-one-meter-by-fifty-foot/fifteen-meter rectangle formed by the wire system. If this happens, the tailhook hanging from the rear of the aircraft should snag one of the wires. If a successful "trap" occurs, the aircraft and hook pull the wire out of its spools belowdecks, and the hydraulic cylinders slow the aircraft to a stop in about 300 feet/91.4 meters, in just two seconds. The crew is then thrown forward in their straps, and lots of negative (forward) "Gs" nearly push their eyeballs out of their sockets.

Once the aircraft is safely aboard, a green-shirted deck crew member called a "hook runner" clears the landing wire from the hook, while a "blue shirt" plane handler starts directing the pilot to taxi forward out of the landing area. When the aircraft is clear of the angle, the arresting cable is retracted and made ready for the next landing. While all this is happening, the LSO is writing down a "score" for each pilot's landing. They grade two factors. First, the general way the pilot actually flew the approach and landing. An "OK" means that this was done safely and to accepted standards. Second, the wire the pilot "snagged." As we saw earlier in the first chapter, the favored target is wire number 3, which provides the safest landing conditions and the least strain on the aircraft. Landings on wires 2 and 4, while acceptable, merit a lower score; but hitting wire number 1 is considered dangerous and usually brings the pilot counseling from the LSO.

Each pilot's landing scores are posted on what is known as the "greenie" board down in the squadron ready room for all to see. These scores are accumulated, and by the end of an entire cruise, a "Top Hook" award is given to the pilot with the best landing record. The scores also frequently affect the ratings of the pilot's airmanship, which affects their future promotion hopes. Great "Hooks" may go to test pilot school or become instructors, while those with lower scores may never fly off a ship again.

In the first chapter, I had occasion to mention one of the rules that every Naval aviator learns early: As soon as the aircraft hits the deck, push the throttles to full power. In this way, if the tailhook fails to snag a wire (called a "bolter"), he has the necessary speed to fly off the end of the angle, and get back into the landing pattern for another try. Bolters happen fairly rarely these days, though every Naval aviator still experiences them now and again. Sometimes the tailhook skips off of the deck, or just fails to connect. Whatever the reason, the 14° angled deck makes it possible for the pilot to go around again, and get aboard another time. Angled decks have saved more aircraft and aviators' lives than any invention since the development of tailhooks. The pilot just climbs out into the traffic pattern and sets up for another try. There also is an emergency net or "barrier" that can be rigged to catch an aircraft that cannot be otherwise snagged by an arresting wire. This, however, is something that no Naval aviator cares to try out if it can be avoided.

Continuing the tour of the flight deck, you can see scattered around the perimeter of the deck many different fittings and nozzles. These provide everything from jet fuel to AFFF (Aqueous Film-Forming Foam). There is also a seawater deluge system, for nuclear/chemical washdowns and fighting *really* bad fires, as well as "chutes" where deck personnel can drop ordinance in danger of "cooking off," should they get too hot from a deck fire. This is another of the many risks faced by flight deck personnel, though they would tell you that not doing the "dangerous" things on "the roof" is a good way to get everyone aboard killed. These are brave people, who do heroic things every time a flight evolution takes place. I defy any nation to effectively operate sea-based aircraft without such folks.

Moving on to the island, you open another hatch, head inside, and climb up six ladders to the 010 level and the Primary Flight Control, or "Pri-Fly," as it is called. Here, some six stories above the flight deck, is the control tower for the carrier, where all the operations of the flight deck and the local airspace are handled by the Air Boss and the "Mini" Boss, his (or her) assistant. They are surrounded by computer displays showing everything they need to help them control the air action around the ship.

Climb down another ladder, and you arrive on the bridge, where the captain spends most of his time. On the port side is a comfortable elevated leather chair, which belongs to the commanding officer, and from which he normally cons the ship (flanked by computer screens). Over on the starboard side of the bridge are the actual conning stations, including the wheel, chart table, and positions for several lookouts. Even though the bridge is equipped with a GPS receiver, advanced radars, and all manner of electronic aids, human eyes and binoculars are still important to the safe conning of a carrier.

Just aft of Pri-Fly is arguably the most popular spot on board, "Vultures Row"—an open-air balcony overlooking the flight deck (and a good place to take in some sun). There anyone can safely watch the comings and goings below (bring your camera and earplugs!). It also offers a wide view of the whole ship, especially the defensive and sensor system.

From there you can see the sponson mounts for the eight-round Mk. 29 Sea Sparrow SAM launchers. The *Nimitz*-class carriers each have three of these systems, one forward on the starboard side, with the other two aft (port and starboard). The RIM-7M Sea Sparrow is a short-range SAM, designed to support the Mk. 15 CIWS mounts in defending the ship against any "leaker" aircraft or missile that makes it past the screen of Aegis missile cruisers and destroyers supporting the carrier group. Based upon the venerable AIM-7 Sparrow air-to-air missile (AAM), Sea Sparrow was originally developed to provide small ships like frigates and destroyers with a short-range point-defense SAM at a reasonable cost. NATO adopted the system as the standard short-range SAM system for small escorts. Like its AAM cousin, Sea Sparrow utilizes a guidance system known as "semi-active" homing. This means that a Mk. 91 fire-control radar (each *Nimitz*-class carrier has three of these) "illuminates" an incoming missile or aircraft, much as a flashlight is aimed at an object in a dark room. The seeker head of the missile "sees" the targets reflected radar energy from the Mk. 91 radar. The guidance system of the missile then automatically provides it tracking to the target.[41]

41 The only known "live" service firing of Sea Sparrow occurred in 1992, when the USS *Saratoga* (CV-60) accidentally launched a pair of the SAMs, one of which struck the Turkish destroyer *Mauvenet*. Five Turkish sailors were killed by the detonation of the warhead, including the ship's captain.

An eight-round Mk. 29 RIM-7M Sea Sparrow launcher aboard the USS *George Washington* (CVN-73).

John D. Gresham

Sea Sparrow is an excellent point-defense system that gives the ship good protection out to a range of up to 10 nm/18.5 km. Back in the 1980's, it was enhanced through the addition of a Mk. 23 Target Acquisition System (TAS) radar. This fast-rotating system can detect low-flying and high-angle targets, and then pass them along automatically to the Sea Sparrow system for engagement. The system's only drawback is that once the eight ready rounds have been fired from the Mk. 29, the launcher must be manually reloaded. Sea Sparrow is being improved through the development of the Enhanced Sea Sparrow Missile (ESSM) System, which marries the basic seeker system with a new airframe. This will give ESSM more range and performance than RIM-7M, as well as the ability to be fired from both Mk. 29's and the Mk. 41 vertical launch system (VLS) launchers found on newer warships.

Unlike surface ships, flattops do not have many convenient spots for placing antennas for radios and sensors. This has to do partly with maintaining appropriate separation between emitting antennas, and partly with the need to avoid clutter on the flight deck during flight operations. For this reason, the island structures of American carriers have always been antenna farms. You'll also find a number of UHF/VHF radio antennas on the edge of the flight deck, placed on special mounts that rotate horizontally during flight operations. On *Nimitz*-class carriers there is additionally a large antenna mast just aft of the island, to hold those radar and communications antennas that need to be as high as possible. These masts and mounts hold a variety of sensors including:

- **SPS-48E**—A 3-D air-search radar that provides air traffic control and battle management functions. This high-resolution radar has a reported range out to approximately 60 nm/110 km.
- **SPS-49(V)5**—This is the best current Naval 2-D air-search radar. Extremely reliable, with a detection range of up to several hundred miles/kilometers, SPS-49's are found on most major combatants in the U.S. Navy, as well as many foreign vessels.
- **SPS-64(V)9**—This is primarily a surface-search/navigation radar for keeping formation and operating close to shore. It is a development of the classic Litton LN-66 navigation radar.

The array of antennas on the island structure of the USS *George Washington* (CVN-73). This is representative of the configuration on late-production *Nimitz*-class (CVN-68) carriers.
JOHN D. GRESHAM

- **SPS-67**—The SPS-67 is a general-purpose surface-search radar, designed to provide precise targeting data against surface targets.
- **Mk. 23 Target Acquisition System (TAS)**—This is a small, fast-rotating radar for detecting sea-skimming or high-angle missile attacks. It feeds data directly into the SYS-2 (V)3 weapons-control system, which can automatically activate the RIM-7/Mk. 29 Sea Sparrow SAM systems.
- **Mk. 91 Fire Control System (FCS)**—The three Mk. 91 FCSs provide guidance for the RIM-7M Sea Sparrow SAM launched by the three Mk. 29 launchers.
- **SLQ-32 (V)4**—The SLQ-32 is a family of electronic-warfare systems, which can be tailored to the protection requirements of a particular ship. The (V)4 version has a wide-band radar-warning receiver, a wide-band radar jammer, and a bank of Mk. 137 Super Rapid Blooming Chaff (SRBOC) launchers. These six-barreled mortars throw up a cloud of chaff (metal-coated Mylar strips) and infrared decoys to blind or confuse an incoming missile at the last moment prior to an attack.
- **WRL-1H**—The WRL-1H is a general-purpose wide-band radio/radar-warning/intercept receiver, designed to provide a basic intercept capability for everything from radio traffic to bearings on radar sets.

These systems give the carrier's commanding officer and battle group staff good situational awareness of the battle space surrounding their ship and the ARG. Along with the supporting sensor systems, the island also provides mounts for many of the ship's communications systems. While many of these are classified, they cover the full range of the electromagnetic spectrum and functions. The most interesting of these are the domed antennas for the satellite communications systems, which provide much of the high-reliability secure communications for the battle group.

Since they were originally designed primarily to transmit encoded text messages, even these systems have limits. Today, carriers need a lot more than just a relatively slow, secure means of receiving words. This problem surfaced with particular impact during Desert Storm, when none of the U.S. Navy carriers had the ability to receive the daily Air Tasking Order (ATO) from CENTCOM's air command in Riyadh. Every other air unit in the theater, including those of our allies, could get the ATO (which ran to hundreds of pages of densely formatted text), even if only by high-speed FAX machines over secure phone lines. But the Navy, having always planned on fighting on their own in the open ocean, was ill prepared for the communications required for joint operations with other services. As a result, the Navy did not receive its daily delivery of the ATO by high-tech satellite or data link, but by hand-delivered paper copies flown in by an S-3 Viking. As might be imagined, this was quite an embarrassment for the Navy, and as a result it began to put together systems to relieve this lack of joint connectivity.

The first try at a solution to the problem was known as the "Challenge Athena" experiment. Challenge Athena I—initially an experimental system on board the USS *George Washington* (CVN-73)—is a two-way, low-speed (around 768 kilobytes per second—kps) satellite link based upon commercial antenna technology. Originally developed for use in delivering intelligence photos and conducting video teleconferences, it has grown into a much broader communications system, and in the process has become incredibly popular with everyone in the fleet. Along with the obvious benefits to top planners and commanders, Challenge Athena provides the crew not only with two-way E-mail contact home, but also with direct live access to commercial television channels like CNN and ESPN. A new high-speed version of the system, Challenge Athena III, is about to be installed throughout the carrier force, as well as on fleet flagships, big-deck amphibious ships, and perhaps even major combatants like the Aegis cruisers and destroyers. A comparable system is being developed for use by submarines, to support Tomahawk cruise-missile targeting, special operations, and unmanned aerial vehicle (UAV) missions. The domed Challenge Athena antennas are located on the flight deck level, outboard of the island and the crotch.

Now it is time to go below. After a drop down a stack of six ladders from the bridge, we find ourselves on the 03 or "Gallery" level, directly under the flight deck. Heading inboard, we find two central passageways running the length of the full ship. Almost a quarter-mile long, these passageways seem to go on forever, with only an occasional cross-passageway to break the monotony of "knee knockers" and watertight hatches. Most of what we see here are doors, lots of them, behind some of which are the real "brains" of the ship—the various command, air wing, and squadron spaces. In addition, most of the air wing officers and flag staff personnel live here. If you turn left and head aft down the main starboard passageway, you pass compartments filled with the hydraulic cylinders for the arresting-gear system. These are gigantic, filling the space between the two main corridors. The compartments here are also even more spotlessly clean than the rest of the ship, since one of the first signs of trouble in a hydraulic system is telltale leaks of fluid.

Farther aft are many of the squadron ready rooms. These large spaces are the headquarters for the various flying squadrons and detachments attached to the carrier's embarked air wing. The ready room is the inner sanctum of a flying squadron, a combination of clubhouse, rest area, and meeting/briefing/planning center. Since the rules of naval aviation allow a freedom of speech and expression that would not be tolerated in other areas aboard ship, ready rooms are extremely private places (where life as a naval aviator is seen at its most raw and splendid). This means that they are for aviators and *only* aviators, and permission is required before *anyone* else is allowed inside.

Ready rooms are wondrous places, filled with historic photos, trophies, and plaques from the unit's past. At the front of the ready room is the desk for the squadron duty officer and a large white board for briefings and discussions. There also are rows of the most comfortable chairs you will ever sit in. Based on a design that predates the Second World War, they are soft but firm, with thick leather covers embossed with the squadron's colors and logo. They can also recline for a short nap between sorties, and have fold-down writing tables for scribbling notes.

At the rear of the room is a small enclosed area where the terminal for the Tactical Aircrew Mission Planning System (TAMPS) is located. TAMPS is an automated system that allows air crews to perform route and mission planning. Since it can take into account effects like terrain masking and enemy air defense weapons envelopes, TAMPS is a major improvement over the old system of paper maps, photos, and air crew intuition. After each squadron does their planning over the networked TAMPS system, the staff of the air wing can review an entire strike/mission plan before the mission is flown.

After leaving the ready room, we'll head forward. After we've passed through about a third of the ship, the tile changes from normal Navy gray to a bright blue, meaning that we have reached what the crew calls "blue tile country." This is the central command and control complex for both the ship and the carrier battle group. The deck in "blue tile country" is subdivided into a series of spaces, each dedicated to a different set of warfare tasks. These include:

- **Combat Information Center (CIC)**—This is the battle nerve center of the ship, with displays for all of the ship's sensors, as well as information acquired from data links and national sources (the DoD term for reconnaissance satellites, aircraft, and other systems). The CIC is specifically designed to present all the available data on the combat situation to the officers making the decisions about how to "fight" the ship. Filled with consoles, terminals, and big-screen displays, this space has separate zones for antisub, antiair, and antisurface warfare, communications, damage control, and other functions. Back in World War II a captain normally fought his ship from the bridge, but today's Arleigh Burke or Phillip Vian will normally be found at a glowing console within a dimly lit CIC. Aircraft carriers' CICs are somewhat different from those of other ships. On a carrier, not all of the terminals and personnel are in a single room, as they are on an Aegis cruiser or destroyer. This better hardens the ship against attack, and avoids a huge and overmanned space, which could be destroyed by a single hit. Thus, the various warfare specialties—antiair (AAW), antisubmarine (ASW), antisurface (ASUW), etc.—have their own small control centers, which forward their data into the main CIC.

- **Carrier Air Traffic Control Center (CATCC)**—The CATCC is a control center for handling airspace and traffic control around the battle group. This one is different from a local FAA control center, in that it moves with the ship and has the ability to data-link information from offboard sensor systems like Aegis ships and AEW aircraft (E-2Cs, E-3's, etc.).
- **Tactical Flag Command Center**—The TFCC is essentially a duplicate in miniature of the CIC. The difference is that the TFCC is specially configured to maximize access to data that flag officers (i.e., admirals/ battle group commanders) need. To support this requirement, the TFCC was developed with the same kinds of large-screen displays and workstations that you would find aboard the Aegis ships that screen the carrier. (The TFCC used to be called "Flag Plot," but that space now resides up on the island.)
- **Joint Intelligence Center (JIC)**—The Joint Intelligence Center is a clearinghouse for information required by the ship, the battle group, and embarked air units. Analysts in the JIC can draw from vast databases of National Imagery and Mapping Agency (NIMA) maps, satellite photography, and anything else the intelligence community provides. The JIC staff is a "rainbow" organization from every unit in the battle group, as well as from other services and intelligence organizations. Even better, they can *probably* tell you what it all means.
- **Ships Signals Exploitation Space (SSES)**—This small sealed space is for the *really* secret stuff: "exploitation" of enemy radio signals and electronic emissions. Equipped with data links to national and theater-level intelligence systems, the SSES can provide battle group leaders with up-to-date information on enemy intentions and activities. Only specially cleared intelligence and communications technicians are allowed inside.

Normally, these are all quiet places manned by a small staff working in shifts. But when an operation or exercise is under way, they resemble a darkened beehive without the buzz, everyone working around the clock until the exercise is finished. By the way, it's *really* cold there, due to the vast amounts of air-conditioning and chill water needed to keep all the electronics and computers from literally melting down. Even in the dog days of August, you often find console operators and other watch-standers wearing windbreakers and pullover sweaters to keep the chill out of their bones.

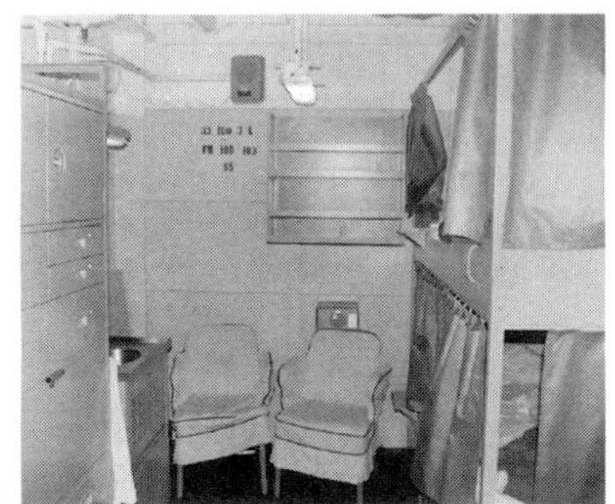

A two-person officer stateroom aboard a *Nimitz*-class (CVN-68) carrier.

JOHN D. GRESHAM

Forward of the command spaces are the flag quarters, where the battle group commander and his staff live. If any place on a carrier can be called luxurious, this is it. There is fine furniture and wood paneling, a large mess and briefing area, a private galley, and the admiral's stateroom, office, and head. Comfortable and functional, all of these spaces are within a few seconds walk of the TFCC. Its comfort notwithstanding, nobody I know likes working in the flag quarters. That is because the flag spaces are directly under the launch shuttle and JBD for Catapult Number 1. The noise during deck operations is deafening, and living and working here during round-the-clock flight operations is downright unpleasant. Such things in fact rarely bother the admiral and staff personnel, however, since they don't get that much sleep anyway. The demands of running a battle group mean that if they are getting more than six hours of sleep every day, they are probably not working hard enough! By a strange irony, the nicer the quarters, the less time an occupant gets to spend in them. While rank and responsibility bring physical rewards, most senior officers rarely have the free time while aboard to enjoy them.

Moving forward again, we find more ready rooms, as well as dozens of staterooms for the air wing personnel and ship's officers. Most of these are two-man units, and are actually quite pleasant to live in (as I did for several days). The racks are doubled-decked, and somewhat larger than those of the enlisted personnel. There is a fair amount of personal stowage space, as well as a small fold-down desk. Each officer has a safe for classified materials and personal items, as well as a small sink and mirror. Though a few staterooms have shared heads and shower facilities, most officers use one of the many community head/shower spaces around the ship. Roommates also usually go in together on electronic items like a ''boom box'' stereo, television, and VCR; and there is a box for plugging these into the ship's cable television and radio network, as well as the commercial feeds from the Challenge Athena system.

Forward of the living spaces, there is a truly wonderful place, called the ''Dirty Shirt'' galley and wardroom area. This is the only officers' wardroom aboard where wearing flight suits and flight deck work gear is ''acceptable.'' While the other wardrooms belong to the ship, the ''Dirty Shirt'' wardroom ''belongs'' to the air wing, which means that aviator traditions apply here. ''Dirty Shirt'' menus tend to be more informal, and talking ''shop'' is allowable. Each squadron has its own table, and etiquette dictates that you ask permission to join anyone who is already there. Still, more often than not, you will find a warm smile and an invitation to join the conversation. In the ''Dirty Shirt'' mess there is also is a neat, little-known secret: the ''dog'' machine—the nickname for the soft-serve ice cream dispenser, which is kept going around the clock.[42] It is a wonderful diversion from the sometimes-spartan life aboard ship; and the ''Dirty Shirt's'' dog machine is usually the best on the ship.

Heading aft, about two-thirds of the way back, we come to a cross-corridor intersection with what looks like a small store on each corner. These are the various squadron ''shops'' for the flying units of the air wing, with one such space for every squadron in the air wing. Here all the data on the readiness, flying and maintenance status, and ordinance/stores loadouts of every squadron's aircraft is managed. Here also is where the Command Master Chief (CMC) for each squadron works. The CMC is the senior enlisted sailor in each squadron, and functions as the shop foreman who keeps the aircraft ready to fly and fight. The CMC also functions as an advisor and advocate for the enlisted personnel of the squadron to the unit's officers. Along with the entire corps of petty officers, the CMCs are the institutional ''glue'' of the Navy, and a good officer rapidly learns this fact. Finally, they are the keepers of the ''Squadron Store.'' This sells coffee mugs, T-shirts, patches, and stickers of the squadron logo (called ''zaps''). If you get aboard a carrier, be sure to pick up a few of these, since the money always goes into the squadron relief fund. I always do.

Returning aft to the island ladder well, we head down four more levels to the Second Deck (deck levels above the hangar or main deck have numbers—01, 02, etc.—while decks below are spelled out). Here most of the crew (officers and enlisted personnel) take their meals. Both have galley and eating facilities here, and something like fifteen thousand meals a day are served on this deck alone. The enlisted personnel eat cafeteria-style in three large spaces amidships that can hold about five hundred personnel at a time. The officers' wardroom (called ''Number Three'') is farther aft, and is essentially a sit-down-style restaurant, though there's a buffet line if you desire. Always open, Wardroom Three is the social center of the ship. Here the officers can come together for a few minutes and share news of the day with their shipmates. Coffee, ''bug juice'' (the Navy version of ''Kool Aid''), and nacho machines are always powered up, and you can usually beg a meal from the mess stewards if you look as though you've worked hard enough. There even is what is jokingly known as the ''nuclear-powered cappuccino machine,'' which dispenses a passable cup of that delicious brew.

Surrounding the officers' wardroom on the Second Deck are the state-

42 The name is a particularly rude reference to a habit of man's best friend.

The main control panel of the pump room aboard the carrier *Harry S. Truman* (CVN-75). This panel controls the main pumps for the entire ship, and is located between the magazines at the bottom of the vessel.
John D. Gresham

rooms for most of the ship's senior officers and department heads. Like the flag quarters on the 02 level, these are very pleasant, with private offices and head/shower facilities. Also like flag quarters, they are used very little since there is very little time for sleep and relaxation while aboard a nuclear supercarrier. Aft of the wardroom are more enlisted quarters. These are much like the ones we've already visited, except that flight deck sounds are muffled by the mass of the ship; and you'll probably hear and feel instead the ship's engineering plant. At high speeds (over twenty-five knots), when the hull begins to resonate, the background buzz can be annoying. Another annoyance is the heat on the lower decks when the ship passes through warm water like the Gulf Stream or Persian Gulf. Things can get downright steamy under some conditions.

Dropping down another ladder, you come upon the machinery spaces on the Third Deck, where most of the systems that keep the ship "alive" are contained. Here and on the deck below are machine shops, electrical switchboards and emergency diesel generators, the ship's laundry, medical and dental facilities, and the air-conditioning plant. Also on the Third Deck is the ship's store, the post office (a surprisingly large facility), and the newly installed banks of satellite phones. These allow sailors to call home from anywhere in the world for about a dollar a minute, and make a real difference in the lives of the crew.

Below the Fourth Deck are the heavily protected and restricted spaces dedicated to the nuclear reactors, propulsion machinery, ammunition magazines, and pump rooms. Surrounded by a double hull with massive voids (specially designed buffer zones to absorb explosions) as protection against damage, these are the safest and most secure areas of the ship. Due to the security restrictions placed upon the Navy by the Department of Energy and the Director of Naval Reactors (NAVSEA 08), I'm not able to describe their layout or equipment.[43] I can say, however, that the two Westinghouse A4W reactors provide enough saturated steam to run the ship at thirty-plus knots while leaving enough electricity to power all the ship's other systems com-

43 For those of you with a desire to fully understand the workings of nuclear reactors in detail, see my book *Submarine: A Guided Tour of a Nuclear Warship* (Berkley Books, 1992).

fortably. The four General Electric steam turbines put out 280,000 shp to four shafts, and are highly agile at starting and stopping.

With the tour at an end, we drag our weary bones and joints up to the hangar deck, and walk over to the accommodation ladder back to the dock. By now you have a pretty good idea of the layout of today's *Nimitz*-class carriers. However, the four-decade production run of this design is starting to wind down, and new ideas are beginning to be put forth for a new generation of flattop. Read on, and I'll try and give you some ideas about what they will look like.

The Future: CVN-77 and CVX-78

The *Nimitz*-class carriers are as capable as their designers and builders could manage back in the late 1960's, representing an almost optimum mix of capabilities for operations during the Cold War. Yet SCB-102 is a design in its third decade of continuous production, the Cold War is now history, and it is time to think about a replacement after the *Ronald Reagan* is launched in a few years. That is exactly what the Navy is doing. The U.S. Navy will always have the mission of projecting forward presence with a regular cycle of carrier rotations. At the same time, the Navy also foresees dealing more frequently with irregular, unpredictable situations. And finally, there is the necessary requirement to keep costs of building, operating, and maintaining carriers reasonable.

Question: How can the Navy do all that?

Answer: Accept the fact that is it time for a new direction in flattop design and construction.

To do this, NNS founded a carrier ''Skunk Works'' called the Carrier Innovation Center, based a stone's throw from Dry Dock 12 at Newport News.[44] Here the NNS design engineers are studying ways to build carriers that will be more suited to the operations the post-Cold War will bring. Working in concert with a number of other corporate partners, as well as NAVSEA, NNS has helped the Navy form a two-step plan for taking carrier construction and sea-based Naval aviation into the 21st century.

Phase one of the plan involves the building of one additional *Nimitz*-class carrier after the USS *Ronald Reagan* (CVN-76), which is now under construction. This unnamed carrier, known today as CVN-77, will be a *Nimitz*

44 The term ''Skunk Works'' refers to the original Lockheed Advanced Projects Division in Burbank, California, which was headed by the legendary Kelly Johnson and Ben Rich, and was designed to produce ''out-of-the-box'' ideas that could be rapidly and economically produced. Examples of the Skunk Works concept in action include the F-80 Shooting Star, the U-2 and SR-71 reconnaissance aircraft, and the F-117A Nighthawk stealth fighter. A number of companies, including Newport News Shipbuilding and Boeing Military Aircraft, have set up similar organizations.

One of several proposed Newport News Shipbuilding designs for CVN-77. Based on a *Nimitz*-class (CVN-68) hull and power plant, the new carrier would incorporate stealth technology, as well as a number of improved operating features.
NEWPORT NEWS SHIPBUILDING

only under the skin. Current plans have CVN-77 utilizing a basic *Nimitz* power plant and hull structure up to the main deck level, but from there on up everything else will be new. CVN-77 will be used as a technological "bridge" ship where a number of new technologies and ideas will be tried out. While some of these technologies have yet to be fully defined, most have already been inserted into the mass of requirements documents being produced at NAVSEA. They include:

- **Signature Reduction**—This is stealth technology, or more accurately "low observables." Can anyone actually hide a quarter-mile-long monster from modern sensor systems? The answer is "yes," but with qualifications. You have to remember that an object's radar, thermal, electronic, and acoustic signature has very little to do with its actual size. Shaping, materials, and other engineering details have much more to do with these characteristics. By way of example, an expert I spoke with claimed that a 90% reduction in the radar cross section of a carrier could be achieved through relatively minor, though detailed, changes to the ship's island; sponsion, and deck structures. This would mean that a *Nimitz*-sized ship might be given a radar signature smaller than a guided-missile frigate's. Already, outstanding signature reduction work has been done on *Arleigh Burke*-class (DDG-51) Aegis destroyers, which are extremely tough to see on radar and infrared sensors.
- **Automation/Reduced Manning**—A key Navy initiative is to reduce manning aboard ships (primarily as a cost-saving measure). With over 70% of every defense dollar going to personnel costs, the Navy figures it can save over $50,000 per year for every sailor who can be eliminated or replaced by automation. According to current plans, CVN-77 will implement many of the "Smart Ship" systems that are being tried out on the USS *Yorktown* (CG-47). These systems have already reduced the size of the *Yorktown*'s crew by 15%. The Navy has even greater goals for CVN-77, and a cut of from 25% to 33% is considered possible. This could mean a reduction of up to one thousand personnel from the ship's com-

pany, and a savings of over $50 million a year over a ''standard'' *Nimitz*. That translates into some $2.5 billion during the fifty-year service life of CVN-77.

- **Adaptive Mission Features**—CVN-77 will be capable of rapid reconfiguration for missions other than those traditionally associated with ''big deck'' aircraft carriers. Operations ''short-of-war'' and disaster/humanitarian relief missions are becoming the rule rather than the exception. To this end, the Navy has decided to redesign the interior spaces of CVN-77 to provide more adaptability. The changes include air wing enlisted berthing areas with the kinds of personal stowage (weapons, ammunition, etc.) required by Marines or other ground personnel who might go into ground combat. Likewise, air wing planning, control, and unit spaces will be more capable for joint operations, so that units like Army helicopter battalions or special operations forces could use them with a minimum of modification. Finally, the hangar bays and elevators are being redesigned to increase aircraft options, so that tilt-rotor aircraft, UAVs, and even the planned new generation of unmanned combat aerial vehicles—UCAVs—can be carried and operated. One senior Naval analyst has even suggested the inclusion of a ''Roll-On, Roll-Off'' (Ro-Ro) ramp on the fantail for loading of vehicles and cargo. All of this adds up to a carrier with more capability and variety than any ever built.
- **Process/Work Flow Improvements**—NNS has made a formal review of the jobs done on board a carrier in order to identify key areas where ''process improvements'' can be implemented into the CVN-77 design. NNS is looking at what is called a flight deck ''pit stop.'' There the crews servicing aircraft or waiting to launch could do so under shelter from the elements. Performing more flight deck functions in the hangar deck (arming, fueling, etc.) would also reduce the wear and tear on both personnel and equipment. And several tasks like ordnance loading (for very strong backs) and critical movement paths through the ship for supplies and personnel will be automated. This would eliminate the many ''bucket brigades'' of sailors moving supplies through the corridors. There is even some consideration of putting a ''ski jump'' on the bow to enhance the launching of the new generation of carrier aircraft, which might eliminate the need for catapults.[45]
- **Materials Improvements**—A wide variety of new materials are being considered for inclusion in the CVN-77 design. Heat-resistant silica tiles should allow the jet blast deflectors to dispense with the traditional water-cooling system. A new lightweight blown fiber-optical local area network

45 ''Ski Jumps'' were developed by the Royal Navy in the 1970's to improve the takeoff and load-carrying characteristics of V/STOL aircraft like the FRS.1/2 Sea Harrier and AV-8B Harrier II. The addition of a slight incline to the end of a flight deck provides the aircraft an upward ''push'' at the critical point of takeoff. So effective are ski jumps at giving V/STOL aircraft ''something for nothing,'' that almost every nation with carriers, with the exception of the United States, utilizes them in their carrier designs.

(LAN) cabling will increase the speed and capacity of the ship's data network by up to 100,000 times. Composites for interior and topside structures (to reduce weight) and radar-absorbing materials (RAM—to assist in signature reduction) will also make their debut on CVN-77. Hull paints and non-skid coatings with vastly expanded service lives (measured in years instead of months) are also being developed, and all of these substances will be more environmentally "friendly." Finally, with an eye to the day in the middle of the 21st Century when CVN-77 will itself go to the scrap yard, a master material list will be prepared, so that whoever takes it apart will know what to be careful with. The Navy is still having nightmares removing asbestos lagging (insulation) aboard ships built before the EPA banned the stuff. The master materials list should put an end to such problems.

- **Weapons**—While the Mk. 29 Sea Sparrow launchers and Mk. 16 Phalanx have provided adequate point defense to past *Nimitz*-class carriers, it is likely that CVN-77 will be equipped with more potent armament. Following the lead forged by the new *San Antonio*-class (LPD-17) amphibious dock ships, CVN-77 will probably be equipped with several clusters of Mk. 41 VLS systems, suitable for launching the Evolved Sea Sparrow Missile (ESSM) that is being developed as a follow-on to the RIM-7M Sea Sparrow SAM. Each eight-cell Mk. 41 module (which can be clustered with up to seven additional modules to build a 64-cell missile launcher) can carry up to four ESSM rounds per cell. Since the Mk. 41 launcher can also launch other weapons (like the BGM-109 Tomahawk cruise missile), you might see quite a few VLS cells scattered about the deck edges of the CVN-77. Also expect that three or four 21-round Mk. 49 launchers for RIM-116 point defense SAMs will be there as well. RAM is rapidly replacing the old Mk. 15 20mm Phalanx CIWS aboard Navy warships, and it is likely that CVN-77 will be equipped with RAM from the start.
- **Data/Electronic Systems**—Though computer-based systems are used aboard warships for everything from propulsion control to sending E-mail home, warship designers did not actually take the digital revolution into account until fairly recently. The technology of personal computers, networks, and workstations has moved so quickly that equipment and technologies in NAVSEA ship specifications are usually obsolete before they go out for contract. NNS is therefore recommending that the Navy "open" the specification for the data, electronic, and electrical systems to include what is known as commercial, off-the-shelf (COTS) technologies, and to specify performance beyond anything currently in production. For example, the fiber optical LAN currently installed in the USS *George Washington* (CVN-73) is a 10-BaseT/T-1-style system, with data-transfer rates of around ten megabytes (MB) per second. For the CVN-77 design, NNS is thinking about a shipboard LAN with data-transfer rates in the

terabit (TB—that is, 1,000,000 MB)-per-second range. Though specifying a LAN with a capacity 100,000 times greater than the one aboard ships today may sound absurd, it makes perfect sense if you consider that computer and LAN technology is doubling in speed and capacity every eighteen months. By allowing commercial-style equipment and software aboard ship (such as using Windows NT as a shipboard-wide operating system), costs are reduced and the crew will be given equipment that is as up to date as government procurement can make it. Finally, NNS will try to use COTS systems in the future wherever a military-specification, custom-built electronic system might be used now.

- **Zonal Electrical Distribution Systems**—While the computer/electronics revolution is generally a good thing, you still have to power all this new stuff. Unbelievable as it may seem, all of the laptop computers, televisions, VCRs, and personal stereo equipment aboard ship are now causing significant electrical problems for carriers. Even though a nuclear power plant gives you enough electrical power to light a small city, you still have to effectively distribute all that power to where it is required, when it is needed, without overloading the power-distribution system. To do this, the Navy and NNS want to install what is known as Zonal Electrical Distribution Systems. Using this system, for example, the ship's systems involved with daytime operations (in offices and work spaces like laundry and galley facilities) can be powered when they are most active, and isolated when they are idle. Zonal Distribution will also improve damage-control capabilities because of increased system redundancy.
- **Communications Systems**—Ever since Desert Storm pointed out its relative isolation, the USN has been trying to catch up with the other services in communications technology. Although the Challenge Athena system is a good start, it lacks both the reliability and bandwidth (i.e., data-flow capacity) to handle the volume of data required in a major war. Further, the need for additional bandwidth, especially in the satellite frequencies, has been growing almost as fast as the speed and power of computer/LAN technology. Therefore, CVN-77 will have a communications capacity far beyond that of current ships. In particular, the new high-speed satellite systems preferred by the regional CinCs will be emphasized, as well as secure data-link systems for distribution to other ships in the battle group.

All of these features will make CVN-77 the most powerful and capable aircraft carrier ever built. Though it will be a *Nimitz* in the hull and propulsion systems, it will be totally new in almost every other way. Though the schedule for CVN-77 is based upon funding dates that will be controlled by a President and Congress that have not yet been elected, current plans have the ship funded in FY-2001, with delivery in Fiscal Year 2008 (it is planned to replace USS *Kitty Hawk* (CV-63)).

The second element in the Navy's carrier production plan is currently known as CVX (Aircraft Carrier—Experimental), which will be the lead ship of a new class of carriers, the first in almost a half century. The program, which will hopefully deliver its first ship in FY-2013, is designed to incorporate all of the ''bridge'' technologies from CVN-77, as well as some other improvements that will be possible because of the new hull and power plant that will be part of the design. Some of these new features will include:

- **Hull Design**—The hull form of the CVX is still under study, though it will probably be a traditional monohull design. It is likely that the CVX will displace something more than the 95,000 tons of the *Nimitz*-class carriers. What the ship will actually look like, however, is anyone's guess.
- **Propulsion/Power Plant**—If there is any sticking point in the design of the CVX-class carriers, it will be over the question of the power plant. Though powerful arguments against nuclear-powered warships remain, for all its vices (such as cost and environmental concerns), nuclear power provides real benefits for the captains and crews of aircraft carriers, and this means that any change had better offer significantly greater benefits. In order to resolve this question, NNS has been conducting a power plant study for CVX at their Carrier Innovation Center. There they are looking at gas turbines, turbine-electric motors, marine diesels, fossil-fueled boilers, and nuclear power as candidate CVX power plants. While the study is still in the early stages, don't be surprised if nuclear power winds up the winner. Steam turbines are a highly compact and efficient means of powering large warships, and nuclear reactors are more compact and efficient than boilers for producing that steam.
- **Weapons**—CVX will probably have a mix of Mk. 41 and 49 launchers very like CVN-77's. However, laser weaponry is advancing so fast that the first CVX or some of its sister ships may well be equipped with a first-generation laser CIWS. The Air Force will deploy a similar system aboard a modified Boeing 747-400 in a few years, and a shipboard system would probably be a highly effective counter to the new generation of supersonic antiship weapons now being deployed around the world.
- **Catapults**—Though for over a half century steam catapults have been successfully shooting aircraft off carriers, they nevertheless have significant drawbacks. For one thing, the high-pressure steam lines that power the catapults are complex and take up a lot of internal volume. For another, the saturated steam they carry is vicious stuff if a line cracks or breaks or is damaged. Finally, if a leak develops or the pressure is incorrectly set, steam catapults will occasionally ''cold shoot'' aircraft into the water. All of these problems have led to a major CVX initiative to replace the old steam units with a catapult using another technology. For instance, the electromagnetic technology that was to be used on the rail guns being designed for the Strategic Defense Initiative back in the 1980s might well

work on carriers. However, an internal-combustion technology looks like a better prospect. Here jet fuel would power a contained fuel-air detonation in a piston to fire the aircraft on its way. Internal-combustion catapults are simple and reliable in concept, and could use the existing jet fuel system on the flight deck.

- **Automated Weapons Handling**—Since weapons stowage, movement, buildup, loading, and arming eat up an enormous portion of a carrier's personnel, a high priority in the CVX design is to automate the weapons handling and loading on future aircraft carriers. One idea already under consideration involves using an unpowered, but human-controlled, bomb cart and loader that makes clever use of counterweights and levers to upload even the largest pieces of Navy ordnance. Other ideas include robotic inventory/handling control of weapons in the magazines.
- **Advanced Flight/Hangar Deck Management**—One of the Navy's biggest challenges is to improve the efficiency of operations on the flight and hangar decks. Specifically, they want to reduce the number of personnel involved in operations on the flight/hangar decks, to improve the quality of the work environment, and to increase the rate of sortie generation for the embarked air wing. Along with the "pit stop" systems planned for use on CVN-77, robotic servicing equipment will probably be used for fueling, arming, aircraft handling/positioning, and for monitoring systems.

If the CVX-78 program manages to stay on track, the first ship of the class will be commissioned sometime in 2013, and a second unit will probably be added to the fleet about four or five years later. Beyond that, it's anybody's guess. We're talking about aircraft carriers that will be operating in a world fifty years from now. What will the world and the military balance of 2050 look like? I wish I knew. But if the people at NNS and NAVSEA have done their homework, the carriers being built and planned today will provide useful platforms to base the combat aircraft of tomorrow well past the halfway mark of the 21st century.

Tools of the Trade: Birds and Bombs

One day when I was a young man just beginning to design airplanes, the great person who founded the company that bore his name, Donald Douglas, took me by the shoulder and taught me a lesson that was simple, though vital to success. At the time, we were trying to generate business from the U.S. Navy. ''Navy planes take a beating,'' he said. ''They slam down on the carriers when they land and get roughed up by the unforgiving elements of the high seas. If we want the Navy to buy our airplanes, we must build them rugged. They have to take punishment and still work.''

Aircraft Design (Ed Heinemann, 1985)

It is a matter of historical record that some things on carrier aircraft are terribly simple, and can't be easily replaced. The Curtis biplane that Eugene Ely first landed on the *Pennsylvania* in 1911 was equipped with many of the same items used by modern carrier aircraft. In particular, it had a small tailhook and a beefed-up tail structure so that the sudden shock of deceleration from the primitive arresting system would not tear the aircraft apart. However, good as these ''shade tree'' solutions to getting on and off carriers were, they were just a start. Future naval aircraft would have even more systems to adapt them to the unique problems and challenges of the ocean environment. Hard as it is on sailors and ships, the ocean is a terror for pilots and aircraft, and the challenges it offers to airplane designers are unlike anything found on land.

First and most obvious are the problems of moisture and corrosion, which can literally eat a plane or helicopter from the inside out. Then there are the limitations of the ship's confined spaces for operating and storing aircraft, and the need to reduce the aircraft's ''footprint'' while on the flight deck. These aircraft must also be able to operate in what has to be an ''expeditionary'' environment, where crews may lack the maintenance and repair facilities of a land base. Then there is the matter of assisting the aircraft into and out of the air without destroying them. And like all military aircraft, these flying machines must be capable of carrying useful payloads an adequate distance with acceptable performance and a good survival rate.

With this in mind, it's not hard to understand why only a handful of companies worldwide have successfully built aircraft for naval service. Carrier aircraft are odd hybrids, combining the qualities of conventional planes that fly off concrete runways with the unique ability to operate off the confined spaces of warships. While naval aircraft perform virtually all the missions that land-based aircraft do, they are also tasked with a number of missions unique to the sea services. For example, the U.S. Air Force (USAF) takes a well-deserved pride in dropping laser-guided bombs (LGBs) down the center of buildings, but the U.S. Navy has aircraft that can do that too. In addition, these same Navy craft can hunt submarines, defend ships against missile attacks, and transfer supplies between vessels. These are just some of the many jobs unique to naval aviation, and Navy aircraft have to be equipped to handle the fullest possible range of roles and missions. This has generally made naval aircraft among the most capable and flexible designs of their design generations. Perhaps the best example of this was the classic F-4 Phantom II, which served not only with the Navy and Marine Corps, but also the USAF and over a dozen foreign countries. Such diversity and capability is not easy, and it comes at a high price.

In general, naval aircraft are both heavier and more complex than equivalent land-based craft. In an era where the cost of new aircraft is directly tied to their weight, USN aircraft generally are more expensive—which usually means smaller production runs and higher financial and technical risks for the manufacturers. Very few companies have been able to meet all of these challenges and turn a profit. For decades, just a few manufacturers have dominated the American naval aviation scene. Airframes made by Grumman, McDonnell Douglas, and Sikorsky were for many years all that you could find on the decks of U.S. carriers. In fact, the rare bird from a company like Lockheed or General Dynamics (traditional USAF contractors) was considered an aberration, a sign that the favored incumbent had made an error during the design competition. As a result, naval aircraft design grew inbred and lacked some of the innovation seen in land-based designs. Back in the 1970's, the Navy was fully briefed on the results of the USAF's Have Blue program. This was the flying prototype of the 1970's that led to the development of the Lockheed F-117A Nighthawk stealth fighter. But the USN chose to ignore the new technology in favor of more conventional aircraft—only one example of such lost opportunities. Another lost chance came when Texas Instruments began to develop its third-generation Paveway III LGB and the Navy stuck with the older-generation Paveway II-series bombs. With just these two decisions, the USN denied itself the two most effective weapons of the Gulf War.

By making a string of similar decisions, naval aviation leadership fostered a two-decade-long Dark Age that denied them some of the best that modern aerospace technology had to offer. The result was the near-mortal wounding of naval aviation as a community in the early 1990's, just at the time that they were being forced to find new roles, new missions, and even new enemies in the post-Cold War world. In an era when military power was becoming more "precision" oriented, naval aviation still valued how well a pilot could deliver a "stick" of unguided iron bombs. As of this writing, it has been over fifteen years since the Navy has taken delivery of a completely new tactical aircraft for fleet use. During that same period, over a half-dozen other major aircraft programs have been canceled or terminated. Desert Storm found the fleet ill-equipped for the first major post-Cold War conflict, and the part it did play was poorly publicized to a world hungry for the high-tech images of LGBs hitting their targets with eye-splitting precision.[46]

Even worse, following the Persian Gulf war, it began to appear that the top leadership of U.S. naval aviation could not even buy the aircraft and weapons they would need to fit into the new "littoral warfare" strategy planned for the 21st century. There was even an attempt by the top leaders of the USAF to replace carrier aviation with a concept called "Virtual Presence." This was the notion that long-range bombers based in the continental U.S. and armed with precision weapons could threaten potential enemies enough that forward-based forces like carrier battle groups would not be necessary.[47] "Virtual presence" was a nice idea, especially if you wanted to justify the purchase of additional B-2A Spirit stealth bombers. Unfortunately, it was completely unrealistic in a world where "presence" really is the sight of a gray-painted USN ship near where a crisis is breaking. Clearly, naval aviation had to "get well" so that it could fulfill its essential task in the national security of the U.S.

All Fall Down: Naval Aviation in the 1980s

Earlier (see the third chapter), we saw how the culture of naval aviators has been forced to deal with changes in the society of the nation they serve. Unfortunately, there was more than just a morale problem to be dealt with. Material problems were also at the heart of the questioning of the credibility of naval aviation by the national leadership. Not that these were new problems—they first started over two decades ago. Naval aviation's downward slide really began back in the 1970s, when the administration of President

46 While Naval aviators did have some precision weapons such as Paveway II LGBs and the new AGM-84E Standoff Land Attack Missile (SLAM), their stockpiles were small, and lacked the capabilities of the newest systems like the Paveway III LGB and GBU-15 electro-optical guided bombs. So rapidly were these stocks used up that the Navy had to borrow a supply of Paveway II LGB kits from the USAF so that they could continue to strike precision targets.

47 The "Virtual Presence" campaign was designed to support additional procurement of the B-2 Spirit stealth bomber, hopefully with the funds that could be diverted by canceling production of additional aircraft carriers and their aircraft. Saner views took hold, and the production of the B-2 was capped at twenty-one.

Jimmy Carter cut off the funds for services to upgrade their equipment, an action that was coupled with an almost complete moratorium on the buying of replacement weapons and spare parts for aircraft. Carriers frequently went on cruises short of airplanes with only partially filled magazines, requiring the "cross-decking" of planes, munitions, and equipment from ships headed home. Naval aviation was being forced to eat its "seed corn" to fulfill the missions it had been assigned. Though the Carter Administration did eventually reverse policy and spend some badly needed funds on procurement for the sea services, by then it was too late. The damage had been done.

The next Administration—that of President Ronald Reagan and his Secretary of the Navy, John Lehman—attempted to rebuild naval aviation in the 1980's. Lehman was a smart, energetic man, with a strong sense of purpose. But he could not *instantly* do *everything* that needed to be done, so priorities had to be set. His vision of a "600 Ship Navy," for example, meant that since naval vessels had the longest procurement time, the largest portion of early funds from the huge Reagan-era defense expenditures would have to go into shipbuilding. He did find funds to replenish the weapons and spare parts inventories, however, and within a few years, the existing aircraft fleet was flying and healthy. But the question of how to build the right mix of aircraft in adequate numbers was a problem that would defy even Secretary Lehman's formidable powers of organization, persuasion, and influence. Under his "600 Ship" plan, the numbers of carriers and air wings (CVWs) were to be expanded and updated. An active force of fifteen carriers would be built up, with fourteen active and two reserve CVWs to fill their decks. To provide some "depth" to the force, the reserve CVWs would be given new aircraft, so they would have the same makeup and equipment as the active units.

Unfortunately this plan contained the seeds of a disaster. The basic problem was airframes—or more specifically, the shortage of them. Because of financial constraints, the Navy had not bought enough aircraft in the 1970's to flesh out sixteen CVWs. Furthermore, the sea services were already heavily committed to the replacement of their force of F-4 Phantom fighters and A-7 Corsair II attack jets with the new F/A-18 Hornet. Normally, the Navy tries to stagger such buys, so that only one or two aircraft types are being modernized at any given time. Now, however, Secretary Lehman was faced with buying or updating *every* aircraft type in the fleet virtually simultaneously. Either way, the cost would be astronomical.

During this same time, the Soviet Union, under the new leadership of Mikhail Gorbachev, was not quite the "evil empire" it had been under Khrushchev, Brezhnev, and Andropov. Meanwhile, the growing federal budget deficits began to take their toll on the defense budget. At a time when the Navy's budget needed to be increasing, the decline of the Soviet Empire and growing domestic problems at home made a continued arms buildup seem unnecessary, and so the Navy was not able to obtain the funding it needed.

When John Lehman left the Administration in 1986 for a career in the private sector, the budget for procuring new aircraft was already being slashed. Far from building sixteen fully stocked CVWs, the Navy's focus now became building just one new type of aircraft for the 1990s. That one airplane, the A-12 Avenger II, came close to destroying naval aviation. Few people outside the military are aware of the A-12 program. Though not actually a "black" program, the shadow of secrecy that shrouded it was at least charcoal gray.[48] The A-12 was designed to replace the aging fleet of A-6 Intruder all-weather attack bombers, but the exact roots of the aircraft are still something of a mystery, though some details have come to light.

Back in the 1980s, the first major arms reduction accord signed between the Reagan and Gorbachev governments was a controversial agreement known as the Intermediate Nuclear Forces (INF) treaty. The INF treaty completely eliminated several whole classes of land-based nuclear weapons, and severely restricted others. Under this agreement, both sides would remove land-based nuclear missiles based in Europe, and aircraft capable of nuclear weapons delivery would be limited and monitored. This was a significant reduction in theater nuclear stockpiles, and at least gave the appearance of a reduced threat of regional conflict. The appearance was not quite the reality, however, because both sides wanted to maintain as large a regional nuclear stockpile as possible. As might be imagined, both sides began looking for loopholes.

U.S. defense planners immediately noticed that sea-based nuclear-capable aircraft and cruise missiles were not counted or monitored under the INF accord—which meant that the existing fleet of A-6's and F/A-18's could immediately provide an interim replacement for the lost nuclear missile fleet. As good as that was, it wasn't good enough. What the nuclear planners really wanted was a carrier aircraft that would hold even the "hardest" targets in the Soviet Union and Warsaw Pact countries "at risk," and that would do it with impunity.

The Navy was thus directed by the Department of Defense (DoD) to develop such an aircraft. The DoD wanted an aircraft that could replace a variety of attack bombers, including the A-6 Intruder, F-111 Aardvark, and even newer aircraft like the F-117A Nighthawk and F-15E Strike Eagle. The program would be developed in total secrecy, and would take advantage of the new technology of passive electromagnetic stealth, much like the F-117 Nighthawk and the B-2A Spirit. It would carry a two-man crew, have the same levels of stealth as the B-2A, and carry a new generation of precision munitions (some possibly with nuclear warheads) guided by the new NAVISTAR Global Positioning System (GPS). Plans had the first units being assigned to the Navy and Marine Corps, with the Air Force getting their A-12's later in the production run.

The Navy had problems with the A-12 from the very start. First, thanks

48 "Black" procurement programs are designed to be so secret that they are not officially acknowledged in the federal budget. Only a select group of legislators and administrators are allowed to know of these projects, and the clearances required to work on them are above Top Secret.

to its lack of interest in the Have Blue program, the Navy knew very little about stealth—a problem that was magnified by the strange rules of "Black" programs, which required them to almost reinvent the technology from scratch. USAF contractors were not allowed to transfer their experience with the F-117 and B-2 programs to the Navy and to potential contractors for the A-12. Even companies like Lockheed and Northrop, who already had stealth experience, were restricted from transferring their corporate knowledge to their own teams developing A-12 proposals. Furthermore, the Navy program management lacked experience in taking a small "Black" research project and turning it into a large, multi-billion-dollar production program. From the beginning, progress was slow and costs were high.

The winning entry in the A-12 competition came from the General Dynamics/McDonnell Douglas team, utilizing a strange-looking design that had been under development by General Dynamics since 1975. Because of its triangular flying-wing shape, it was quickly nicknamed "the flying Dorito." Designated the A-12 Avenger II (after the famous World War II torpedo bomber), it was designed to carry up to 10,000 lb/4,535 kg of ordnance in internal weapons bays. It also would have had enough unrefueled range to hit targets in Eastern Europe if launched from a carrier in the Mediterranean Sea. Unfortunately, the A-12 would never make it off the shop floor, much less onto a carrier deck.

From the start of the A-12 engineering and development effort, there were disagreements between the Navy program managers and the contractor team over a number of issues. The plane was too heavy, for one thing, and there were difficulties creating the composite layups that made up the A-12's structure. Costs escalated rapidly. While the Navy has never officially acknowledged this, it appears that every other major Naval aircraft program was either canceled or restructured in order to siphon money to the troubled A-12. What is known is that during the time when the A-12 was suffering its most serious developmental problems, the upgraded versions of the F-14 Tomcat fighter and A-6 attack bomber were canceled outright, and several other programs took severe budget hits. The situation reached the critical point in 1990, when the A-12 and a number of other major aircraft programs were publicly reviewed in light of the recent fall of Communism in Eastern Europe. By this time the Avenger program was a year late and perhaps a billion dollars over budget. Even so, in his major aircraft program review presentation to Congress, then-Secretary of Defense Dick Cheney declared the A-12 to be a "model" program.

Nine months later, he radically changed his tune. Though what the DoD and Navy were thinking at this time remains something of a mystery, the pending commitment of an additional half-billion dollars to the A-12 program certainly had much to do with the decision. Whatever the reason, Secretary Cheney ordered the program canceled in January of 1991, just as the Desert Storm air campaign was getting under way. So sudden was this action that several thousand General Dynamics and McDonnell Douglas employees were

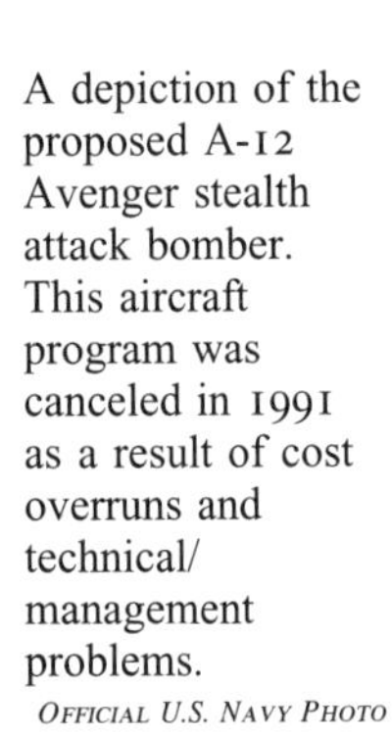

A depiction of the proposed A-12 Avenger stealth attack bomber. This aircraft program was canceled in 1991 as a result of cost overruns and technical/management problems.
Official U.S. Navy Photo

simply told to put down their work and go home. All told, the Navy had spent something like $3.8 billion, and did not have a single plane to show for it.[49] Even worse was the total wrecking of the Navy's aircraft acquisition plan, which had seen so many other new aircraft programs canceled to support the A-12.[50]

It did not take long for the fleet to begin suffering the consequences of the A-12 debacle. The Navy tried to make a fresh start with a program called A/FX (Attack/Fighter, Experimental), which was designed to replace the A-6 and the F-14 fleets, both of which were aging rapidly. A/FX would have made use of the systems developed for the A-12, but would not attempt to achieve the level of stealth planned for the Avenger. Unfortunately, in the tight budget climate of the early 1990's, there was little support or money for the A/FX program, and it died before a prime contractor team was selected. Another blow to the naval aviation community came at the beginning of the Clinton Administration, when Secretary of Defense Les Aspin, as a cost-cutting measure, decided to prematurely retire the entire fleet of A-6E/KA-6D Intruder attack/refueling aircraft.[51] Within months, the entire

49 As if all this was not absurd enough, there was the problem that DoD and the Navy improperly canceled the A-12 program, claiming that the GD and MDC had somehow "defaulted" on the contract. Normally, such cancellations are of government "convenience," allowing the contractors to recover their losses and costs for shutting down the program. However, DoD and the Navy contended that the contractors had failed to do their jobs properly, and thus actually *owed* the government around $1.3 billion in money already paid. As might be imagined, this rapidly became a matter for high-priced lawyers, and resulted in an expensive show trial that the government decisively lost. While the government and contractors continue arguing over the details, it looks like the Navy will have to cough up something like $3.8 billion over the roughly $4.8 billion already spent to pay for its improper cancellation of the program.

50 The shame of it is that the design of the A-12 appears to have been sound, and while it would have been expensive at between $150 and $175 million a copy, that seems quite reasonable when compared to the Lockheed Martin F-22A Raptor stealth air-superiority fighter, which is priced almost identically!

51 To put an ironic perspective on this decision, General Charles "Chuck" Horner, who commanded U.S. Air Forces in Desert Storm, has told me that the only Navy strike aircraft he valued during the Desert medium-attack community was wiped out, leaving the F/A-18 as the Navy's only strike aircraft, and only a single high-performance Naval aircraft was in development: an evolved/growth version of the Hornet. With nothing else on the horizon, Naval aviation was going to have to bet the farm on a machine called the F/A-18E/F Super Hornet.

New Paradigms: The Road Back

By late 1995, naval aviation had hit rock bottom. Military analysts were beginning to believe that the Navy had forgotten how to develop and buy new weapons and aircraft. In fact, many were questioning if the Navy should let the USAF buy their aircraft, since they seemed so much better at it. The real doomsayers were projecting the end of naval aviation as we know it sometime in the early 21st century, when the existing aircraft would wear out and have to be retired. But these people did not know the true character of naval aviation leadership. Though the Navy's aviation problems were deadly serious, in 1996 naval aviation took the first steps toward putting itself back on a healthy course.

Even before he became Chief of Naval Operations, Admiral Jay Johnson was already working toward this goal. He started by appointing two of his most trusted officers, Rear Admirals Dennis McGinn and ''Carlos'' Johnson (no relation to the CNO), to key leadership positions as the heads of NAVAIR and the Naval Aviation Office in the Pentagon known as N88. Soon they started to shake things up. They began to promote a new vision for naval aviation, in direct support of the Navy's ''Forward from the Sea'' doctrine, and to develop a *realistic* long-range plan for upgrading Naval aviation and developing new capabilities. The two men also saw the need to put a few good naval aviators in key positions within the Pentagon so that the procurement program problems of the past would not be repeated. They knew that people with real talent would need to be in some of the key staff jobs to help get new ideas into naval aviation.

As a consequence of this kind of thinking, the Navy Strike Warfare Directorate (N880—the group that defines future specifications and capabilities for new naval aircraft and weapons systems) came under the inspired leadership of a talented F/A-18 Hornet driver, Captain Chuck Nash. While he probably could have gone on to command his own CVW, he chose the good of the service over his own ambitions, and took charge of N880 in the Pentagon.

It was Chuck Nash who really started to shake things up for naval aviation in 1996. Under his leadership, support from the fleet was focused on the new Super Hornet, in an effort to ensure that there would be at least *one* new airframe to anchor the carrier air wings of the early 21st century.

Storm air campaign was the A-6E Intruder. It could operate at night, deliver LGBs and other PGMs, and had enough fuel capacity to minimize the impact upon the limited tanker resources of the Allied coalition.

At the same time, Nash increased Navy support for other developmental aircraft programs like the V-22 Osprey and Joint Strike Fighter (JSF), as well as a new Common Support Aircraft (CSA) to replace the S-3 Viking, E-2 Hawkeye, and C-2 Greyhound airframes.

To shore up the existing force of carrier aircraft, he helped start a program to equip the fleet of F-14 Tomcat interceptors with the same AAQ-14 LANTIRN targeting pod used on the USAF F-15E Strike Eagle. LANTIRN pods allow Tomcats to carry out precision strikes with LGBs and other weapons ashore, a completely new mission for them. In order to arm the Tomcats, the Navy was directed to procure a stock of highly accurate Paveway III-series LGBs, as well as the deadly BLU-109/I-2000 penetrating warheads. Nash's office also began to contract for modifications to existing precision weapons like the AGM-84E SLAM, so that their range, lethality, and service lives might be further extended.

Finally, N880 took a leadership position with the other services on a new generation of precision-strike weapons. These would be guided to their targets by GPS navigation systems, and then given final guidance by a new family of self-locking, all-weather seeker systems.

By the time he retired in early 1998, Chuck Nash had done more for Naval aviation as a captain than most admirals. As a result of the programs inspired by the likes of Jay Johnson, Dennis McGinn, "Carlos" Johnson, Chuck Nash, and *many* others, there is now real hope and drive in naval aviation. A new air wing structure has been defined, and plans for aircraft procurement are now clear for the next quarter century.

Today the climate in the fleet and naval aviation program offices is very different. Much like their counterparts at NAVSEA, the leaders at the Naval Air Systems Command (NAVAIR) are now looking toward the future rather than back toward the past. Their goal is to produce the aircraft and weapons that will fly off the new generation of carriers that are due in the middle of the second decade of the next century. For the first time in a generation, Naval aviation leaders are not content to run programs and buy updated versions of old aircraft and weapons. Naval aviation's vision is now on the cutting edge of weapons technology.

To this end, a new aircraft, the F/A-18E/F Super Hornet, is being tested and headed into the fleet, while existing aircraft like the F-14 Tomcat, EA-6B Prowler, and S-3B Viking have been modified to take on new roles and missions. These will help maintain the credibility of naval aviation until the new aircraft types arrive in a few years. New weapons, with greater precision and utility than those used in Desert Storm, are on their way as well. The sea services, along with the other branches of the U.S. military, are in the early stages of developing the replacement for today's aircraft through the new JSF program. There are even visionary studies for the first generation of Unmanned Aerial Combat Vehicles (UCAVs), which will likely be seen in ten to twenty years. What a difference just a few years make!

An F-14 Tomcat delivering a GBU-24 Paveway III laser-guided bomb during tests. The addition of new air-to-ground strike systems have turned the Tomcat into a potent fighter bomber.

Raytheon Strike Systems

The Plan: Naval Aviation in the 21st Century

The plan for naval aviation as it heads into the 21st century is designed to take carrier aviation from the current post-Cold War CVW structure to one that reflects the perceived needs of the Navy in 2015. To do this, NAVAIR has put together a three-stage program of procurement and reorganization that relies heavily on the success of the past—and that learns from the mistakes that were made. Back in the early 1970's, the so-called "CV Air Wing" organization was created to reduce the number of carriers and air groups in the fleet. This type of CVW was an all-purpose unit, with capabilities in antiair warfare (AAW), antisubmarine warfare (ASW), antisurface warfare (ASUW), and land attack. Its structure is laid out below:

Aircraft Type	Number	Mission(s)
F-14 Tomcat	24	Fleet Air Defense/Air Superiority/Reconnaissance
F/A-18 Hornet	24	Air Superiority/Light Strike/Suppression of Enemy Air Defenses
A-6E/KA-6D	10/4	Medium Attack/Aerial Refueling
E-2C Hawkeye	4	Airborne Early Warning
EA-6B Prowler	4	Electronic Warfare
S-3B Viking	8 to 10	Antisubmarine Warfare/Surface Surveillance

Aircraft parked on the busy flight deck of the USS *George Washington* (CVN-73). Efficient deck handling of aircraft can make or break the daily air tasking order of a battle group.
JOHN D. GRESHAM

Aircraft Type	Number	Mission(s)
EKA-3B "Electric Whale"	2	Electronic/Communications Intelligence/Aerial Refueling
SH-60F/HH-60G Seahawk	6/2	Antisubmarine Warfare/Combat Search and Rescue
Total	88 to 90	

As the table shows, the "CV" air wing had a primary emphasis on defense against air and submarine attack. It could also dish out a great deal of punishment against enemy naval forces, though its ability to strike land targets was more limited. It was this air wing structure that John Lehman tried to flesh out with his aircraft procurement plan in the 1980's. But because of the fallout from the A-12 fiasco, the aircraft necessary to fill out sixteen such units were never purchased, and the fleet made frequent draws on Marine F/A-18 Hornet and EA-6B Prowler squadrons in order to sustain the heavy deployment schedule of the late Cold War years.

After the end of the Cold War, the following air wing organization was created, and is in use today around the fleet:

Aircraft Type	Number	Mission(s)
F-14 Tomcat	14	Air Superiority/Precision Strike/Reconnaissance
F/A-18C Hornet	36	Air Superiority/Precision Strike/Suppression of Enemy Air Defenses
E-2C Hawkeye	4	Airborne Early Warning/Surface Surveillance
EA-6B Prowler	4	Electronic Warfare/Suppression of Enemy Air Defenses
S-3B Viking	6	Sea Control/Antisubmarine Warfare/Aerial Refueling

Aircraft Type	Number	Mission(s)
ES-3B Shadow	2	Electronic/Communications Intelligence/Aerial Refueling
SH-60F/HH-60G Seahawk	4/2	Antisubmarine Warfare/Combat Search and Rescue
Total	72	

This CVW structure reflects a number of realities, most importantly the fact that there will only be eleven CVWs (ten active-duty and one reserve) for twelve carriers, greatly reducing the number of new aircraft required to sustain carrier aviation into the 21st century. Also, this 1990's CVW has a new orientation: to project precision-striking power onto targets ashore. Both the F-14's and F/A-18's are equipped with precision-targeting and reconnaissance systems, as well as a wide variety of Desert Storm-era PGMs. All of these systems give the new CVWs much more punch than before, and while the number of fighter/attack aircraft has been greatly reduced, this new air wing actually can strike twice the number of precision targets that a Cold War CVW could hit. It will acquire even greater power when the new generation of GPS-guided PGMs arrives over the next few years.

The next big move will occur in the early years of the 21st century. Starting somewhere around 2001, the Navy will commission its first combat squadron of F/A-18E/F Super Hornets, replacing the F-14 Tomcat squadron in CVWs. The Navy will then be able to rapidly retire the elderly F-14As, some of which will be over three decades old when they head to the boneyard. During this same period, the SH-60B/F and HH-60G fleet will be remanufactured into a common variant known as the SH-60R. The surviving H-60 airframes will then be consolidated into a single version that can be used either on carriers or escorts. The Navy will also buy a number of CH-60 airframes, which will take over from the old UH-46 Sea Knight in the Vertical Replenishment (VERTREP) mission aboard supply ships, as well as the special operations/combat search and rescue (SO/CSAR) mission of the HH-60G.

Despite all these changes, the dominant airframe of this air wing will continue to be late-model F/A-18C Hornets, which will soldier on well into the 21st century. With these changes, the typical CVW of 2001 to 2015 will probably look like this:

Aircraft Type	Number	Mission(s)
F-14 Tomcat or F/A-18E/F Super Hornet	14	Air Superiority/Precision Strike/Reconnaissance
F/A-18C Hornet	36	Air Superiority/Precision Strike/Suppression of Enemy Air Defenses

Aircraft Type	Number	Mission(s)
E-2C Hawkeye	4	Airborne Early Warning/Surface Surveillance
EA-6B Prowler	4	Electronic Warfare/Suppression of Enemy Air Defenses
S-3B Viking	6	Sea Control/Antisubmarine Warfare/Refueling
SH-60R/CH-60 Seahawk	6/2	Antisubmarine Warfare/Combat Search and Rescue
Total	72	

Again, the key attribute of this CVW will be striking power against land-based precision targets. However, with a new generation of self-designating, GPS/INS-guided PGMs, it will be able to dish out truly devastating damage to targets afloat or ashore, and in almost any kind of weather.

The final step in the CVW modernization plan is shown below, and will begin to appear around 2011:

Aircraft Type	Number	Mission(s)
JSF	10	Air Superiority/Precision Strike/Reconnaissance
F/A-18E/F Super Hornet	36	Air Superiority/Precision Strike/Suppression of Enemy Air Defenses
CSA AEW Variant	4	Airborne Early Warning/Surface Surveillance
EF-18F Electric Hornet	4	Electronic Warfare/Suppression of Enemy Air Defenses
CSA Sea Control Variant	6	Sea Control/Antisubmarine Warfare/Refueling
CSA ESM Variant	2	Electronic/Communications Intelligence/Refueling
SH-60R/CH-60 Seahawk	6/2	Antisubmarine Warfare/Combat Search and Rescue
Total	72	

This is an air wing that is almost entirely composed of aircraft that now exist only on paper. Even so, it has several clear advantages over earlier CVW structures, including the fact that this projected CVW has just four basic airframes: the JSF, F/A-18E/F, the CSA, and H-60. This means lower operating and maintenance costs as well as a simpler logistics chain. It will also have the Navy's first true stealth strike fighter (the JSF), a new EW/SEAD aircraft (the proposed EF-18F Electric Hornet), as well as new sea control, ESM, and AEW aircraft based upon the new CSA airframe. This likely will be what will go aboard the new CVX when it is commissioned around 2015. Once all eleven CVWs have their first squadron of JSFs, the Super Hornets will begin to be retired, and eventually there will be four JSF squadrons aboard each carrier with ten aircraft each.

None of this will come cheaply or overnight. Just maintaining the existing fleet of aircraft is expensive, and buying something like two thousand new F/A-18E/F Super Hornets, JSFs, CSA derivatives, and any other major airframe that comes along will cost between $20 and $30 billion. And that's without even beginning to address the spare parts, engines, weapons, and other necessities that these aircraft will consume in their operational lifetimes. Meanwhile, naval aviators will continue to fly the aircraft they've flown for most of their careers. The designs of not a few of these aircraft, in fact, date from before many of the men and women who fly them were born.

Northrop Grumman F-14 Tomcat: King of the Air Wing

You always know when you see an F-14 Tomcat that it is a *fighter*. It is a big, noisy, powerful brute of an airplane that lacks *any* pretense of stealth or subtlety. For over two decades, the F-14 Tomcat has been the king of American carrier flight decks, yet only recently has it realized its full combat potential. It is also one of the most difficult and dangerous of Naval aircraft. As the plane that Tom Cruise "piloted" in the movie *Top Gun*, it has become the symbol of naval aviation in American popular culture. More tellingly, to date the Tomcat has a *perfect* air-to-air combat record. Now in the twilight of its career, the F-14 is being asked to buy time for the rest of naval aviation to get its collective act together.

The origins of the F-14 lay back in the 1950's when American intelligence agencies identified a growing family of Soviet air-launched cruise missiles as a potential threat to NATO fleet units. Carried to their launch points by heavy bombers, aircraft like the Tu-16 Badger or Tu-95 Bear, they could be launched well outside the range of enemy SAMs and antiaircraft (AAA) guns. Designated by NATO intelligence analysts as AS-1 "Kennel," AS-2 "Kipper," AS-3 "Kangaroo," AS-4 "Kitchen," AS-5 "Kelt," and AS-6 "Kingfish," these long-ranged, radar-guided pilotless jet- or rocket-powered weapons packed enormous ship-killing power. Armed with 1,000-kg/2,200-lb warheads (or high-yield nuclear warheads), they were capable of destroying a destroyer or frigate with a single hit. By way of comparison, the single AM-39 Exocet air-to-surface missile (ASM) that sank the British guided-missile destroyer HMS *Sheffield* (D 80) in 1982 had a warhead just one tenth that size. Since a single large bomber might carry two or three such monster ASMs, finding a way to defend the fleet against them became a high-level priority.

Experience in World War II against Japanese Kamikaze planes (which were essentially manned ASMs) showed that the best way to protect a fleet was to shoot down the missile-carrying enemy bombers before they could launch their missiles. Thus the response to the ASM threat was the acceler-

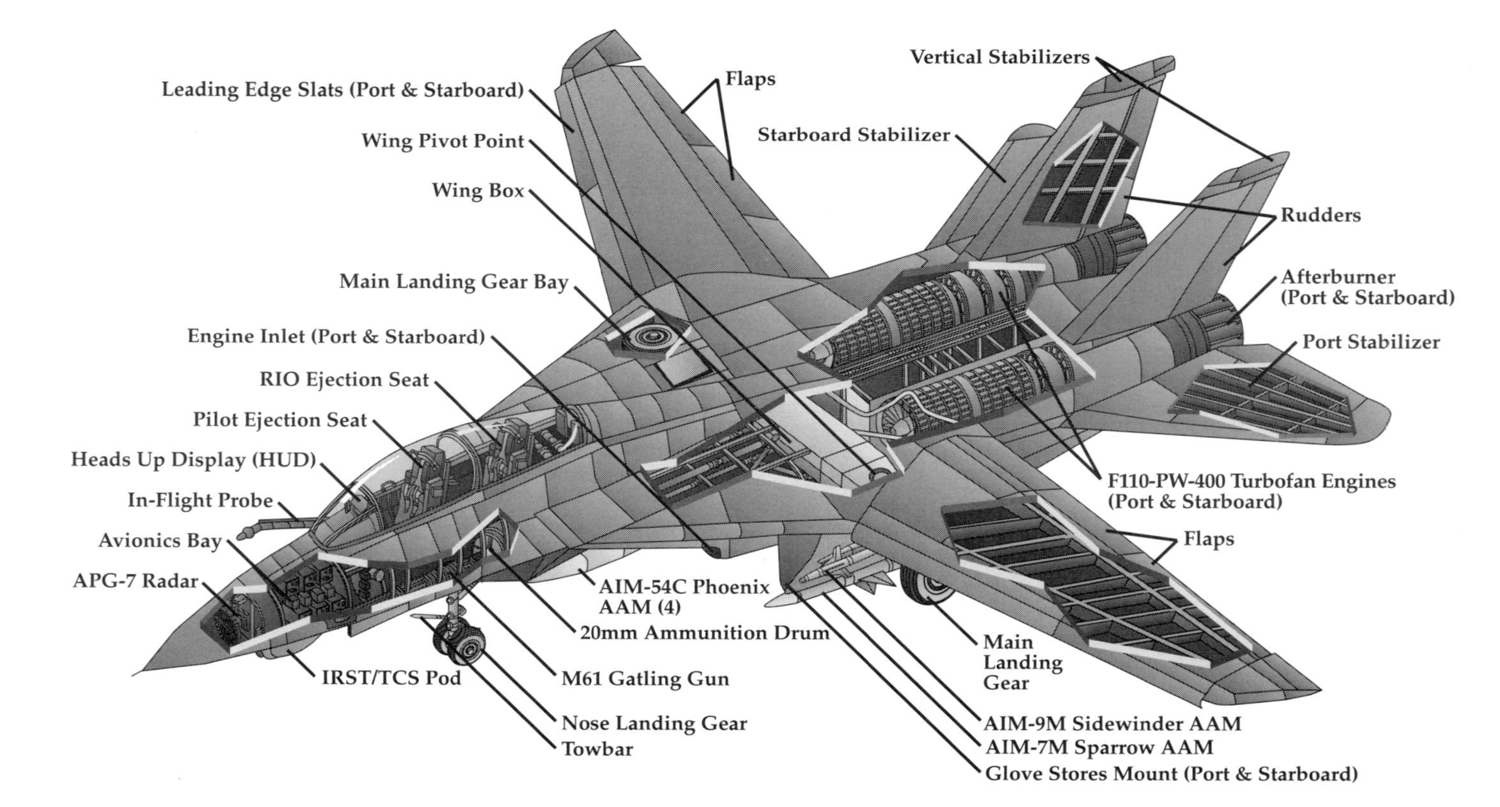

A cutaway view of a Northrop Grumman F-14D Tomcat fighter bomber.

Jack Ryan Enterprises, Ltd., by Laura DeNinno

ated development of extremely long-range air-to-air missiles (AAMs), which could maintain an outer ring in a layered defense system. Any missiles that "leaked" through the outer ring would then face an inner barrier of patrolling fighters, ship-launched SAMs, and point-defense missiles launched from surface ships. This was supposed to be the U.S. strategy until the end of the Cold War—a scheme that envisioned an extremely high-performance, long-ranged AAM that could be carried by a relatively slow but long-endurance carrier aircraft, the Douglas F6D Missileer. The Missileer would have carried eight long-range Bendix Eagle AAMs, along with powerful airborne radar. The F6Ds would have acted as airborne SAM sites, and would have been placed hundreds of miles ahead of a carrier group to intercept incoming bombers. However, fiscal realities now began to effect the Navy's plans.

The F6D program was canceled in December 1960, mostly due to the fact that it was a single-mission aircraft only for fleet air defense. Even so, the Eagle missile was eventually resurrected as the Hughes AIM-54 Phoenix, which today is carried by the F-14. Already strapped for funds, the Navy decided that its next fighter should do the job of the F6D, as well as provide air superiority and other missions. Then high-level politics stepped in. In the early 1960's, then-Secretary of Defense Robert MacNamara, frustrated by seemingly endless inter-service rivalries and hoping to save money, tried to force the Air Force and Navy to procure common types of aircraft. Out of this dream came the TFX (Tactical Fighter, Experimental) program—which became the Air Force's F-111 swing-wing bomber. To meet its fighter missions, the Navy was directed to develop a variant of the F-111 that would be suitable for carrier operations. It was expected that it would accomplish its fleet air defense and air-superiority missions with the planned F-111B, which would replace the classic F-4 Phantom II.

The problem was that the "navalized" F-111B (which was built by Grumman in partnership with General Dynamics, the USAF "prime" contractor) was just too heavy, fragile, and complex for carrier operations, and its landing speed was too high for a safe landing on a carrier deck. Furthermore, the F-111B, with little maneuverability and thrust from its overworked engines, was not much of a fighter. For all of these reasons, the Navy rejected the F-111B, and the program was scrapped, though not without a fight. In those days, one did not go against a man as powerful as Secretary MacNamara without paying a price. The Navy paid in blood. In a scene reminiscent of the 1940's "Revolt of the Admirals" a generation earlier, a senior naval aviator, Rear Admiral Tom "Tomcat" Connelly, sacrificed his own career by standing up to MacNamara in Congressional testimony. He stated flatly in an open session, "Senator, there is not enough thrust in all of Christendom to make a fighter out of the F-111!" With this legendary remark, the F-111B died, and the F-14 Tomcat was born.

Politics aside, the Navy still had the problem of those Soviet ASM armed bombers to deal with. As if to amplify the problem further, the Russians had deployed a new supersonic swing-wing bomber in the late 1960s that caused a near panic in U.S./NATO defense planners: the Tu-22M Backfire. The eventual answer to the Navy's problem came after a series of fighter studies funded by the Navy and run by Grumman. The plan was to wrap a completely new, state-of-the-art airframe around the basic avionics, weapons, and propulsion package that had been intended for the F-111B (including the Phoenix missile system), and then run a series of product improvements upon the new bird. One of the aircraft's most notable features would be a variable geometry "swing-wing" design that would allow it to "redesign" itself in flight. For good slow-speed performance during landing and cruise the wings would be set forward, and be swept back for supersonic dashes.

It was an ambitious design for the late 1960s. The new fighter would not only carry up to six of the massive AIM-54 Phoenix missiles and the AWG-9 radar to guide them, but it would *also* be a superb dogfighter. In Vietnam the F-4 Phantom II had severe shortcomings during close-in air-to-air engagements. The Phantoms weren't very maneuverable, were easy to see (both big and smoky), and didn't have much range. The new fighter would be very different.

The Request for Proposals went out in 1968, and a number of airframe manufacturers submitted responses to build the new bird. However, with their fighter study and F-111B experience, Grumman had a clear edge, and early in 1969 they won the contract to build what would become known as the F-14. Quickly, Grumman got to work and began to cut metal, and the new bird rapidly came together. The first flight of the F-14A prototype occurred almost a month ahead of schedule, on December 21st, 1970, at Grumman's Calverton plant on Long Island. Though three of the preproduction aircraft were lost in testing (including the prototype on its second flight), the program progressed well. The new fighter moved along on schedule, with the first two fleet squadrons, VF-1 (the "Wolfpack") and VF-2 (the "Bounty Hunters"), standing up in 1974. In honor of Admiral Connelly's role in its creation, the Navy named the new bird the "Tomcat."[52]

The Tomcat is a two-seat, twin-engined fighter that measures 62 feet, 8 inches/19.1 meters in length. Its height to the tip of the vertical stabilizer is 16 feet/4.88 meters. The maximum wingspan is 64 feet, 1.5 inches/19.54 meters at a minimum sweep angle of 20°. Minimum wingspan in flight is 38 feet, 2.5 inches/11.65 meters at a maximum flight sweep angle of 68°. For storage in the cramped confines of the flight hangar decks, the wings can "oversweep" (only on deck for stowage) to an angle of 75°, overlapping the horizontal tail surfaces and reducing the span to only 33 feet, 3.5 inches/ 10.15 meters. The Tomcat's empty weight is 40,150 lb/18,212 kg, with a maximum takeoff weight of 74,500 lb/33,793 kg. The F-14 is by far the heaviest aircraft flying on and off a carrier these days. You can actually *feel* an aircraft carrier shudder whenever one is catapulted off.

The famous Grumman "Iron Works" has a well-earned reputation for

52 This also conformed to the usual Navy practice of giving feline names to Grumman fighters.

producing the most durable and robust aircraft in the world. Much of the plane's structure, including the critical ''wing box'' (containing the swing-wing mechanism), is made of titanium, a metal lighter than aluminum, stronger than steel, and notoriously difficult to weld. The Tomcat's horizontal tail surfaces were built from boron-epoxy composite—a very costly and advanced material that was used for the first time on any aircraft.

The F-14 is the Navy's only ''variable geometry'' aircraft, a trait it inherited from its predecessor, the F-111B. While complex, the swing wing was a valid engineering solution to a difficult design problem for the Navy. The F-14 had to be both a long-range interceptor that could ''loiter'' (fly slow and wait) and a high-performance fighter for air-superiority missions. If one aircraft was to do both jobs and still be capable of operating off aircraft carriers, it had to be able to literally ''redesign'' itself in flight. This was the job of the swing wing. The Tomcat's wings sweep forward for increased lift in low-speed flight, particularly the critical takeoff and landing phases of a carrier-based mission, but when the wings sweep back for reduced drag at high speed, the F-14 can move like a scalded cat.

Unlike other variable-geometry aircraft like the F-111 Aardvark and MiG-23/27 Flogger, the F-14's wing sweep is controlled automatically by a computer known as the ''Mach Sweep Programmer.'' This means that the pilot does not have to worry about it—the plane dynamically reconfigures itself from moment to moment for the optimum solution to the complex equations governing lift and drag. The wings then pivot on immensely strong bearings, moved by jackscrews driven by powerful hydraulic motors, giving the flight crew the best possible ''design'' for any situation they are in. The result is an aircraft that is always being optimized, whether it is making a low-level, high-speed reconnaissance dash, or digging into a cornering turn pulling ''lead'' on an enemy fighter. Along with the swing wings, the F-14's engineers managed to provide the flight crew with a full array of control surfaces, including full-span flaps along the trailing edge, leading edge slats, and spoilers on the upper surface of the wings. The speed brake is positioned far aft, between the twin vertical stabilizers. In fact, it was the seemingly random movement of these surfaces that caused Landing Signals Officers (LSOs) to dub the F-14 ''the Turkey'' during tests.

Visually, the F-14 is an imposing aircraft. The topside of the Tomcat's forward fuselage and two huge engine pods blend into a flat structure called the ''pancake,'' which supports the tail surfaces and the tailhook. The pancake itself is a form of ''lifting body,'' and provides a significant amount of the aircraft's total lift. The large canopy offers superb all-around visibility—a great improvement over previous Navy fighters like the F-4 Phantom, which had a deadly blind spot to the rear. This was one of the design criteria that helped make the Tomcat a much better dogfighter than the F-4, or the MiGs that it was designed to kill. The two-person flight crew (a pilot and Radar Intercept Officer or ''RIO'') enters the cockpit using a retractable boarding ladder and cleverly designed ''kick-in'' steps. Both positions have Martin-

Baker ''zero-zero'' ejection seats, meaning that they can actually save an air crew if the aircraft is sitting still (zero speed) on the ground (zero altitude). Three rearview mirrors are positioned around the canopy frame to help the pilot with rear visibility.

The design of the pilot's station was quite advanced for the early 1970's, with the most important data being displayed on an integrated ''Air Combat Maneuvering panel.'' The Tomcat was also equipped with the Navy's first heads-up display (HUD) projected into the pilot's forward field of view, and the first use of the ''Hands-on-Throttle-and-Stick'' (HOTAS) in the cockpit. The control stick and throttles are studded with buttons that govern weapon selection, radar modes, and other functions. HOTAS allows pilots to keep their eyes *outside* the cockpit during a dogfight. The rest of the cockpit is not so advanced. Since the F-14 was designed a decade ahead of ''glass cockpit'' aircraft (like the F/A-18 Hornet), most of the control panels are traditional dial-type ''steam gauge'' indicators. Unlike USAF fighters, though, the RIO's backseat position does not provide flight controls (unless you count the ejection seat). A large circular display screen—the Tactical Information Display—dominates the RIO's position, with a smaller Detail Data Display panel above it. These provide readouts for the AWG-9 radar/ fire control system, as well as weapons control. Again, circular ''steam gauges'' dominate the RIO's cockpit.

When they arrived upon the aviation scene, the sensor and weapons systems of the Tomcat were a revolution.[53] The heart of the F-14 weapons system (in the -A and -B models) is the Raytheon-Hughes Airborne Weapons Group Model Nine (AWG-9) fire-control system. Composed of powerful radar, weapons-computer, signal-processor, and other components, the AWG-9 made the F-14 the most powerful fighter in the world. Unfortunately, it never really got a chance to show its awesome capability in combat. Designed for the extremely long-range, multiple-target engagements that were projected for the Cold War at sea, the F-14 spent a generation waiting for a battle that never came. The AWG-9 requirement was to simultaneously track up to two dozen airborne targets (in an environment that might have hundreds), while actually engaging (that's Navy for ''shooting'') six of them at once. The actual tracking ranges against various-sized targets are highly classified, but the AWG-9 has regularly tracked fighter-sized targets out beyond 100 nm/ 185 km.

Since F-14 operations have always been constrained by strict rules of engagement (ROE) that require visually identifying the target, long-range shots with radar-guided AAMs have been rare. The five enemy air-to-air ''kills'' that the Tomcat has scored to date were all achieved at fairly short

53 When Soviet intelligence obtained the specifications for the F-14 in the early 1970's, the numbers actually terrified the Russian fighter pilots. So desperate were the Soviets to counter the F-14, and the other third-generation Western fighter designs, that they began to spend exhorbitant amounts of money on new fighter designs, and on intelligence efforts to obtain technical information that they could copy into their new aircraft.

A VF-102 F-14B Tomcat aboard the USS *George Washignton* (CVN-73) in 1997. Fully loaded, it carried fuel tanks and "iron" bombs, as well as AIM-9 Sidewinder and AIM-54 Phoenix air-to-air missiles.

Official U.S. Navy Photo

ranges, the killing missile shots all occurring with visual range of the targets. In recognition of these ROE realities, the F-14 carries a pod under the radome holding a television camera system (TCS). The TCS is equipped with a zoom lens that can be used to identify targets visually at fairly long ranges. As an added bonus, it feeds an onboard videotape recorder, which provides the flight crew an excellent visual record of their engagements.

From the very start of its career, the F-14 was intended as an air-to-air killer, with little effort or money expended to give it an air-to-ground capability. The Tomcat's claws were designed to give it the ability to kill at every range, from close in to over 100 nm/185 km, which is still something of a record.

The weapon with the longest range is the mighty Raytheon-Hughes AIM-54 Phoenix AAM. An outgrowth of the original Eagle AAM that was to have armed the F6D, the AIM-54 first flew in the 1960's. With a range in excess of 100 nm/185 km, the AIM-54 was the first deployed AAM equipped with its own active onboard radar-guidance system. This gave it the capability of being launched in a "fire-and-forget" mode, allowing the launching aircraft to turn away to evade or begin another engagement after firing. It also means that up to six AIM-54's can be launched at up to six different targets at once. Once launched, the missile climbs in a high-altitude parabolic trajectory, reaching speeds approaching Mach 5. When a Phoenix gets near a target, a huge 133.5-lb/60.7-kg high-explosive warhead ensures that it dies quickly. It was this capability that Navy planners wanted to utilize had the Soviet bomber/ASM missile threat ever been encountered in wartime. The Phoenix has had several versions, each one designed to keep pace with Soviet improvements in their own weaponry; the AIM-54C is the latest.

Along with the AIM-54, the Tomcat is equipped with three other weapons for killing aerial targets. The first of these, the Raytheon AIM-7M Sparrow, is an updated version of the semiactive radar-guided AAM that has been in service since the 1950's. Weighing some 503 lb/228 kg, this medium-range (out to twenty-plus nm/thirty-seven-plus km) AAM requires continuous "illumination" from the AWG-9 radar to hit its target. Once there, the eighty-eight-pound/forty-kilogram blast-fragmentation warhead can kill any aerial target that it hits. However, the AIM-7 has always been a difficult weapon to employ, because of its need for constant radar illumination of the target. There were plans to replace the Sparrow on the F-14 with the new AIM-120 Advanced Medium Range Air-to-Air Missile (AMRAAM). Unfortunately, budget cuts at the end of the Cold War, combined with the fact that the Tomcat already had a long-range fire-and-forget AAM in the Phoenix, caused this to be canceled.

Shorter-range missile engagements are handled by the classic AIM-9M Sidewinder AAM, which utilizes infrared (heat-seeking) guidance to find its targets. The current AIM-9M version is badly dated, and almost obsolete compared with the Russian R-73/AA-11 Archer, Matra R.550 Magic, or Rafael Python-4. These missiles are not only controlled via helmet-mounted sighting systems, but also can be fired up to 90° ''off-boresight'' (i.e., the centerline of the firing aircraft). This shortcoming will be rectified in the early 21st century with the introduction of the new AIM-9X.

The last of the Tomcat's air-to-air weapons was the one that designers of the F-4 Phantom thought unnecessary in the age of AAMs: a 20mm cannon. During the Vietnam War, Navy pilots complained bitterly about the MiG kills that they missed because of the Phantom's lack of a close-in weapon (it was armed only with AIM-7/9 AAMs). When the specification for the F-14 was being written, ''Tomcat'' Connelly made sure that it had a gun to deal with threats inside the minimum range of AAMs. The gun in the F-14 is the same one in most U.S. fighters, the classic six-barreled 20mm M61 Vulcan. Able to fire up to six thousand 20mm shells per minute, it can literally ''chop'' an enemy aircraft in half.

With the exception of the internal six-barrel 20mm M61 Gatling gun, all the Tomcat's weapons are carried externally. For mechanical simplicity, there are no weapon pylons on the movable portions of the wings, since these would have to swivel to stay pointed directly into the airflow. Because of this, drop tanks and other external stores must be accommodated under the fuselage and engines, or on the structure of the wing ''glove'' inside the pivot. Four deep grooves known as ''wells,'' shaped to the contours of AIM-7 Sparrow AAMs, are sculpted into the flat underbelly of the fuselage in the tunnel between the engine pods. When the huge (984-lb/447.5-kg) AIM-54 Phoenix missiles are carried, they are mounted on removable pallets that cover the Sparrow wells. Up to four of the AIM-54's can be carried here, along with another pair on the ''glove'' pylons. However, these pylons are more normally configured with rails for an AIM-9 Sidewinder and AIM-7 Sparrow AAM.

The reason for this is an arcane number called ''bringback weight,'' which represents the maximum landing weight of an aircraft on a carrier deck. The bringback weight is a combination of the aircraft's ''dry'' weight with the minimum safe fuel load (for several attempts at landing) and whatever ordnance and stores are being carried. An F-14 loaded with six of the big Phoenix AAMs and a minimum fuel load is above the allowable bringback weight, which means that the largest external stores load allowed are four AIM-54's, two AIM-7's, a pair of AIM-9's, two external fuel tanks, and the internal M-61 20mm Gatling gun. A normal ''peacetime'' weapons load is composed of two of each kind of missile, the gun, and two fuel tanks. Other kinds of weapons mixes are designed around particular kinds of missions, including air superiority and strike escort.

A fighter lives or dies by its engines, and the F-14 fleet suffered for many years from an inadequate power plant, the Pratt & Whitney TF-30-P-412. This was the first turbofan engine designed specifically for a fighter, and was inherited from the F-111B program. Originally intended for the subsonic F-6D Missileer and used in the Vought A-7 Corsair II attack bomber, it was augmented with an afterburner (as the TF30-P-100) for the supersonic F-111, and adapted as a "temporary" expedient for the F-14A. Turbofan engines are more fuel-efficient and powerful than turbojets, but are "finicky" about the airflow into their first stage of compressor blades. Turbulent "dirty" air, such as the wake of another aircraft, can cause compressor stalls, flameouts, and, too often, loss of an aircraft. The TF-30's sensitivity to dirty air was well understood by the Grumman designers, who provided the engines with huge inlets and a system of air valves or "ramps." These are a complex system of hydraulically controlled mechanical plates deployed at high speed, creating internal shock waves that slow the incoming air to subsonic velocity.

Though these fixes tamed the TF-30 for the Tomcat's introduction, the Navy had plans for something better. This was to have been the Pratt & Whitney F-401, in what would have been known as the F-14B. Once again, however, developmental problems and escalating costs prevented it from entering service. This left the entire force of F-14A's equipped with the TF-30 engine, which has killed more aircraft and crews than enemy fire ever did.

For over two decades Tomcat crews have tried to get the most out of their finicky TF-30's (even as they lived in dread of them). To feed these huge power plants, the Tomcat carries plenty of fuel, allowing long-range missions or long loiter time on patrol. Internal fuel capacity is 2,385 U.S. gallons/9,029 liters, and two external drop tanks can be mounted under the engine inlets, each with a capacity of 267 U.S. gallons/1,011 liters. To extend its range even further, a NATO-standard retractable refueling probe is fitted on the starboard side of the forward fuselage. Even so, in these days of littoral warfare, the F-14's rarely have to "hit" a tanker to conduct their missions. This is increasingly important, for the retirement of the fleet of KA-6D Intruder tankers means the only remaining refueling aircraft in the carrier air group are the overtaxed S-3 Vikings.

Along with its air-to-air duties, the Tomcat was designed to take on another—and perhaps its most vital—task. This is the dangerous job of photo-reconnaissance for the battle group and local theater commanders. About fifty Tomcats of all models have been specially modified to carry the Tactical Air Reconnaissance Pod System (TARPS) pod under the fuselage. This large external store (17 feet/5.2 meters long and about two feet/.6 meters in diameter) contains three different sensors. These include a conventional frame camera that looks forward and down, a "panoramic" camera that captures the ground picture from horizon to horizon on either side of the

A D/TARPS reconnaissance pod mounted under the fuselage of a VF-102 F-14B Tomcat.

John D. Gresham

aircraft, and an infrared line-scanner that sweeps the terrain directly below the aircraft. Normally, four F-14's in each CVW are fitted to carry the TARPS pod (in addition to their normal avionics fit), and at least six crews get special training to fly them.

TARPS is the best low-to-mid-altitude photo-recon system in the world, and is a significant national strategic asset, able to capture imagery at a level of detail much greater than the high-flying U-2 or reconnaissance satellites. During the 1991 Gulf War, TARPS was especially valuable for post-strike battle-damage assessment (BDA), and was much favored over the USAF RF-4C (which has since been retired). Currently, TARPS is being upgraded to provide battle group commanders with a whole new capability: near real-time photo-reconnaissance. By replacing one of the existing film cameras with a digital unit, and tying it into the existing UHF radio system, an airborne F-14 equipped with the new pod can send a picture with good resolution back to the carrier while still in the air. With a delay of only about five minutes from the time the picture is taken to its viewing by intelligence staff, the new system (called Digital TARPS or D/TARPS) can give a battle group commander the necessary information to rapidly hit a mobile target. This is a capability long sought by military leaders of all services, and is being improved all the time.

Even though it has fought in few actual battles, the F-14 has had an active service life. The first operational deployment came in September 1974, with Pacific-based squadrons VF-1 and VF-2 on board *Enterprise* (CVN-65). The Tomcat's first known combat action came on the morning of August 19th, 1981, when two Libyan Su-22 "Fitter" interceptors made the mistake of engaging a pair of patrolling Tomcats from VF-41 (the "Black Aces") flying from the *Nimitz* (CVN-68). Using their superb maneuverability, the two Tomcats evaded a Libyan AAM and downed the Fitters with a pair of short-range AIM-9L Sidewinder shots. A few years later, in October 1985, four Tomcats from VF-74 (the "Bedevilers") and VF-103 (the "Sluggers"), embarked on USS *Saratoga* (CV-60), intercepted an Egyptian 737 airliner carrying the terrorists who had hijacked the Italian passenger ship *Achille Lauro*. By March of 1986, Tomcats were back on the front lines when Libya fired S-200/SA-5 Gammon SAMs at F-14's from *America* (CV-66) and *Sar-*

atoga (CV-60) patrolling over the Gulf of Sidra. In response, the carrier groups attacked the SAM sites and sank a number of threatening Libyan patrol boats. Later that year, F-14's provided cover for Operation Eldorado Canyon, the bombing raids on Tripoli and Benghazi. January 1989 saw another confrontation with the Libyans when a pair of VF-32 Tomcats engaged and destroyed a pair of MiG-23 Flogger-Bs. When the MiG-23's came out and acted in a threatening manner, they were quickly dispatched in a barrage of Sparrow and Sidewinder AAMs.

During the 1990/91 Persian Gulf crisis, most of the duties of the Tomcats embarked on the deployed carriers involved regular Combat Air Patrol (CAP) and reconnaissance missions, with none of the glamor accorded to the land-based USAF F-15's. Day after day, the Tomcats flew cover for the carrier and amphibious groups in the Red Sea and Persian Gulf, and supported the embargo of Iraq. Part of the reason they had few opportunities to show their capabilities was the reluctance of the Iraqi Air Force to come out over water and be slaughtered. But the big reason was the Navy's failure to develop the necessary systems and procedures to integrate carrier air groups as part of a joint air component command. Key among these was the ability to conduct Non-Cooperative Target Recognition (NCTR), which utilizes various classified radar techniques to identify enemy aircraft by type. This allows fighters with Beyond Visual Range (BVR) AAMs like the AIM-7 and AIM-54 to fire their missiles at long ranges. Because USAF F-15's had these systems and the Tomcats did not, it was the Eagle fleet that was used against the Iraqi Air Force over their homeland.

The only Tomcat air-to-air kill of the war was scored with a Sidewinder by an F-14A from VF-1 over an unfortunate Iraqi Mi-8 Hip helicopter. The bad news was that an F-14B, from VF-103 on the *Saratoga*, was downed by an Iraqi V-75/SA-2 Guideline missile on a TARPS reconnaissance run over Wadi Amif. The one bright point throughout Desert Storm for the F-14 community was the timely and accurate battle-damage assessment provided by TARPS-equipped F-14's.

The fall of the Soviet Union and Warsaw Pact meant that a large part of the threat that the F-14 had been created to defend against was gone. The big Russian bombers and their massive ASMs were rapidly scrapped, and the Tomcat community was left scrambling for a role in the New World Order. Tomcats were not able to perform many of the missions that would make them useful to regional commanders in chief in the new age of ''joint'' warfare. In particular, the AWG-9's lack of NCTR capabilities made the Tomcats also-rans compared with F-15's.[54] But the biggest drawback for Tomcats was the huge cost of buying and maintaining them. Because it was the most expensive aircraft on a carrier deck to procure,

54 Compared to the F-14's single helicopter air-to-air ''kill'' during the war, F-15's scored thirty-five victories. Much of this was due to the advanced NCTR systems of the USAF aircraft, which made them better able to employ their long-range weapons with the certainty required to avoid possible ''blue-on-blue'' incidents.

operate, and maintain, the Navy saw cutting the Tomcat population as a way to save money. Ironically, this occurred just as the F-14 was finally getting the engine and systems upgrades it had always needed to make it the fighter it could have been.

Back in the 1980's, John Lehman's original aircraft acquisition plan had included upgrades to the Tomcat fleet. The first phase of this effort was to re-engine a large part of the existing fleet of F-14A's, and upgrade its avionics. This was to be accomplished by modifying the -A model Tomcats to carry a pair of the new General Electric F110-GE-400 advanced turbofan engines. The F110 (also used in the Air Force F-15E and F-16C/D fighters) had greater thrust and none of the vices of the TF-30. It came to the F-14 in 1986. The new F110-equipped Tomcat, designated F-14B (originally the F-14A+), entered service in April of 1988. Some of the -B models were re-engined F-14As, while the rest were newly built. The contrast with the old TF-30-powered Tomcat was spectacular. The F-110-engined Tomcats are the fastest of their breed, with better acceleration and performance in dogfights than most other fighter types.

There is a story about several of the prototype F-14Bs visiting NAS Oceana near Norfolk, Virginia. On the other side of the Chesapeake Bay were the F-15's of the USAF's 1st Fighter Wing at Langley AFB, their premier air-to-air fighter unit. Normally, the F-15's easily defeated the F-14As with their anemic TF-30's; but this time the high-spirited Naval aviators decided to play a trick on their blue brethren and challenge the USAF pilots to an air-to-air "hassle" over an offshore training range. The Naval aviators showed up in the souped-up Tomcats, and left the two Eagle drivers running away screaming, "Who were those guys!" Clearly, the F-110 made the new-generation Tomcats a very different cat. The new bird still had one significant shortcoming, though. It was still equipped with the original 1960's-vintage AWG-9 radar and avionics systems.

The Tomcat community had always dreamed of making a final break with the old F-111B systems and producing an F-14 with a new generation of digital avionics. At one point, an F-14C model with more advanced electronics was proposed, but it was never developed. Finally, in the fall of 1990, the dream was realized in the form of the F-14D. Like the earlier F-14B program, some of the -D-model Tomcats were rebuilds of earlier -A-model aircraft, while the rest were new production airframes. The -D model has the same F110 engines as the -B, but adds a new radar (the Hughes APG-71) and a host of avionic, computer, and software improvements.

The APG-71 is a vast improvement over the earlier AWG-9, and is based upon the APG-63/70-series radars used on versions of the F-15 Eagle. This is a state-of-the-art, multi-mode radar with a variety of capabilities. In addition to the basic air-to-air functions of the AWG-9, the APG-71 is capable of both Low Probability of Intercept (LPI—making it difficult to detect with passive sensors) and Non-Cooperative Target Recognition (NCTR) modes. In addition, the APG-71 has the ability to perform advanced ground mapping in heavy weather, a feature that would come in handy when the Tomcat community got interested in air-to-ground operations in the 1990's.

Though the F-14D is the ultimate Tomcat, equipped with everything that a crew could want in a fighter today, budget cuts meant that less than fifty -Ds were built, just enough for two or three squadrons. When new production and conversions of -B- and -D-model F-14's were terminated, plans were made to phase out the aircraft. It began to look like the Tomcat might go the way of the A-6/KA-6 Intruders—straight to the boneyard—just as the aircraft had finally gotten the engines and avionics that the crews had always dreamed of. The hunger to cut costs within the Department of Defense in the early 1990's meant that a number of valuable aircraft types were retired, regardless of the consequences, and the F-14 almost suffered the same fate.

Fortunately for the Tomcat community, even allowing for the downsizing of post-Cold War CVWs, there was a shortage of tactical carrier aircraft. Meanwhile, new missions were found for the F-14. Now that there were no longer regiments of missile-armed Soviet bombers to defend against, the Navy planned to provide the Tomcat community with a rudimentary capability to drop "iron" (unguided) bombs (called "Bombcat" conversions) and perhaps fire AGM-88 High-Speed Anti-Radiation Missiles (HARMs) against enemy radars. At the same time, members of the F-14 community were teaching their old Tomcats a few new tricks. While the majority of the Navy's aviation-procurement dollars were headed toward F/A-18 Hornets, the Tomcat operators found ways to squeeze a few of the scarce greenbacks to preserve their mounts. To better understand what they did, you need to know a bit about how many Tomcats of various models were built. Here is a look at the total production run of the F-14 program:

F-14 Tomcat Production

Model	New-Build	Conversions	Total	Comments
F-14A (USN)	557	N/A	557	Does not include one IIAF aircraft that was taken over by USN.
F-14A (IIAF)	80	N/A	80	One aircraft embargoed and eventually delivered to USN.
F-14B (USN)	38	47	85	Converted aircraft were selected F-14A-model aircraft.
F-14D (USN)	37	18	55	Converted aircraft were selected F-14A-model aircraft. Does not include four full-scale-development aircraft that were later retired.
Total	712	65		

A total of 712 Tomcats were delivered to the Navy, the first in October 1972 and the last in July 1992.[55] While no USN F-14 has been lost in air-to-air combat, more than 125 have been lost in accidents—mostly engine-related (Iranian losses are unknown, at least in open sources). At the end of 1997 some 250 F-14's remained in U.S. Navy service. Most of the USN F-14As are now between ten and twenty years old, and have only had rudimentary upgrades to their structures and avionics. The two F-14As that shot down the Libyan MiG-23's in 1989 still had the same APR-25 radar-warning receivers (RWRs) that had been installed when they were built in the 1970's. These RWRs were so elderly they could not detect the signals from the MiGs' radars, which also dated back to the early 1970's. Because of their age, NAVAIR has decided to sacrifice the -A-model Tomcats to the boneyard, and preserve the fleet of remaining -B- and -D-model F-14's. It is unlikely that any F-14As will be in service past 2001, when the first F/A-18E/F Super Hornet squadron stands up. That leaves approximately 130 F-14Bs and -Ds to flesh out the ten remaining squadrons that will serve into the first decade of the 21st century.

All of these aircraft have F-110 engines, and are being given avionics upgrades such as the installation of new GPS receivers and radios. Tomcat crews have also been provided with Night Vision Goggles (NVGs) to give them improved low-level situational awareness in darkness. But the jewels of the upgrade program are the D/TARPS program (mentioned earlier) and an air-to-ground weapons-delivery system: the AAQ-14 LANTIRN targeting pod. This is a self-contained system equipped with a Forward Looking Infrared (FLIR) thermal-imaging system, a laser range finder, laser spot tracker, and laser illuminator. The AAQ-14 pod, one of two used on the USAF's F-15E Strike Eagle, has proven to be the best of its kind in the world today. It can detect targets on the ground from their thermal signatures, and then deliver LGBs and other weapons. The Navy version of the LANTIRN pod has an additional feature: a beer-can-shaped Litton GPS/Inertial Navigation System (INS), which provides the F-14 with the necessary navigational/positional accuracy to deliver the new generation of PGMs that are coming into service. Carried on the starboard wing "glove" pylon, the LANTIRN is controlled by the RIO, and can deliver LGBs day or night with greater accuracy than any other aircraft in the fleet.

These improvements, however, did not come easily. They cost a great deal of money, which the senior leaders at NAVAIR controlled. Focused on acquiring the F/A-18, the NAVAIR "Hornet Mafia" was sworn to eliminate anything from the budget that might detract from that effort. On the

55 Along with the U.S. Navy, there was a single foreign customer for the Tomcat: the Imperial Iranian Air Force (IIAF). The IIAF Tomcat sale was approved by the Shah of Iran, based upon the capability of the AIM-54 Phoenix to hit the fast, high-flying MiG-25 Foxbat-R reconnaissance aircraft that had been intruding across the border shared with the then-Soviet Union. Of the eighty IIAF Tomcats ordered, all but one was delivered, with the last F-14 being embargoed and eventually delivered to the USN. Very little has been published in open sources about the air battles of the Iraq-Iran War (1980–88), though some of the F-14's are reportedly still flying today.

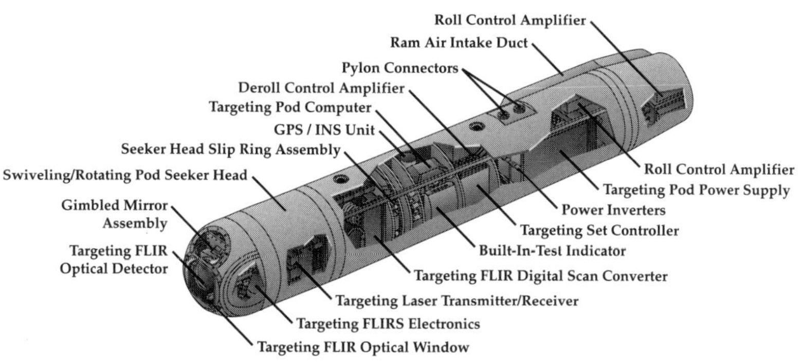

A cutaway view of a Lockheed Martin AAQ-14 LANTIRN Targeting Pod.
Jack Ryan Enterprises, Ltd., by Laura DeNinno

other hand, there was also a ''Tomcat Mafia'' down at NAS Oceana (where all the F-14 squadrons had been consolidated), which was able to find small parcels of money, as well as support from out in the fleet. Also, contractors like Lockheed Martin, the manufacturer of the AAQ-14 LANTIRN pod, spent their own money to develop systems for use on the Tomcat. They worked better than anyone had imagined. Suddenly, regional Cincs wanted all the Tomcats they could get. The incomparable navigational accuracy of the GPS-equipped LANTIRN made them excellent ''quick-look'' reconnaissance birds, especially against mobile targets like SCUD missile launchers. Now, the twenty to thirty F-14's that are deployed at any given time are precious national assets and are doing far more than merely carrying their load until the first squadrons of Super Hornets arrive early next century. They remain the most versatile and powerful aircraft in the fleet. ''Tomcat'' Connelly would have been proud that his dream has proved so adaptable.

F/A-18 Hornet: The Now and Future Backbone

Originally conceived as a low-cost replacement for two aging naval aircraft (the F-4 Phantom and A-7E Corsair), the F/A-18 Hornet fighter-bomber was designed to fulfill a number of widely different roles. It functions as both the Navy's primary light-strike bomber and as a fighter for the Navy and Marines. Though some think the Hornet does neither job very well, others consider it the finest multi-role aircraft in the world. Some will tell you that the F/A-18 is a short-legged burden on naval aviation, while others will make a case that it is the backbone for *all* of naval aviation. I would tell you it is all of these things, and many more. The drawback with any multi-role combat aircraft is that it tries to do too much for too many different people. On the other hand, when such a complex beast works, it works out quite well indeed. Read on and I'll explain.

The origins of the Hornet program date back to the mid-1970's, when the Navy was beginning to suffer ''sticker shock'' from the costs of buying

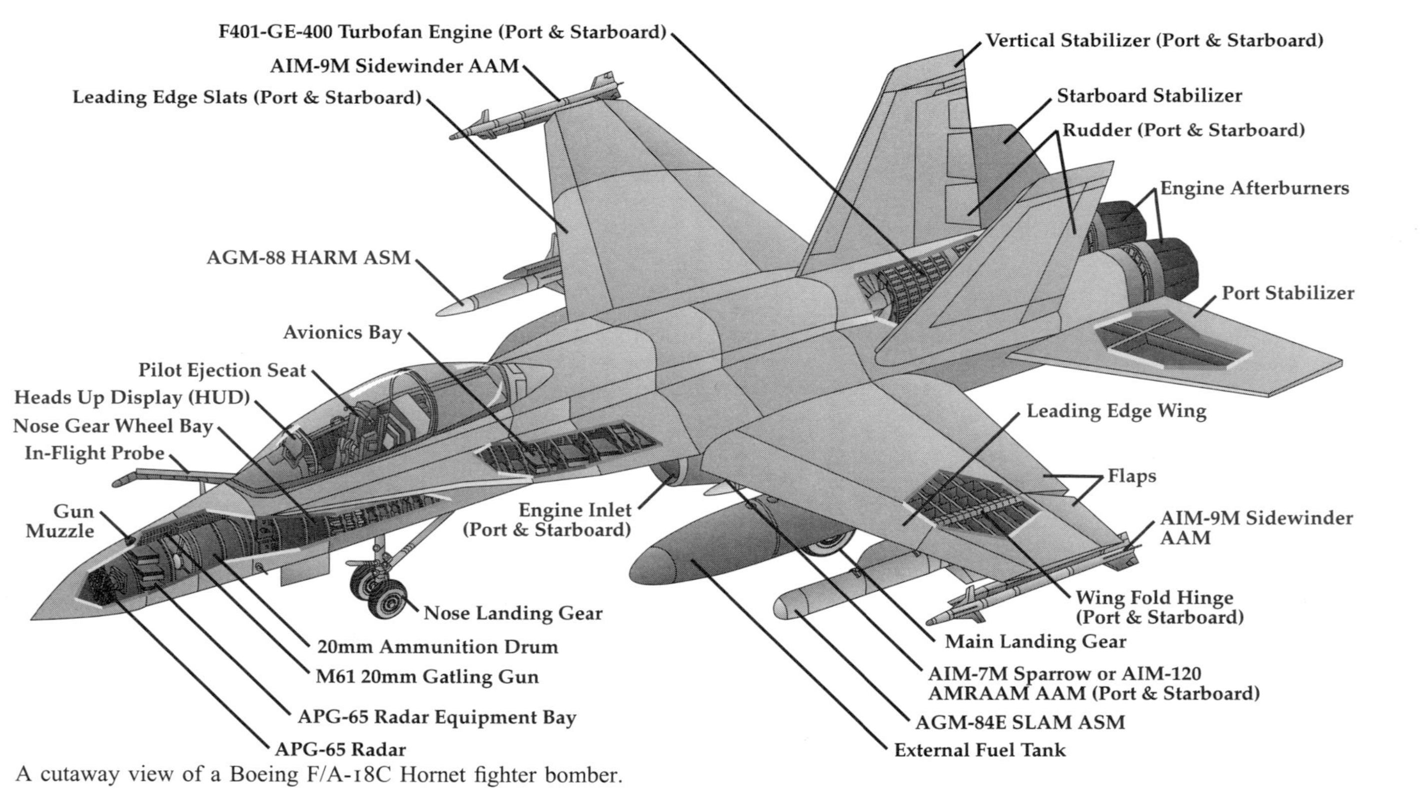

A cutaway view of a Boeing F/A-18C Hornet fighter bomber.

Jack Ryan Enterprises, Ltd., by Laura DeNinno

new aircraft for its carrier force. The double-digit inflation of the early 1970's was driving the price of new combat aircraft up at a dangerous rate, bringing about a reassessment of the kinds and numbers of aircraft the U.S. military could afford. After a start was made on the modernization of the F-14, A-6, and S-3, the Navy looked to the problem of replacing the existing force of A-7 light-attack bombers. Since every CVW had two squadrons each of the A-7's (with a dozen aircraft per squadron), this represented a huge aircraft buy. At the same time, the Navy and Marines had to replace about a dozen squadrons of elderly F-4 Phantom fighters, which operated from carriers and bases. From these twin needs came what was known as the VFAX (Navy Fighter/Attack, Experimental) requirement. The hope was that a *single* aircraft might be designed to fulfill *both* the fighter and light-strike roles, and thus save money by reducing the number of airframes. About the same time, the USAF was evaluating a pair of "lightweight" fighter designs, and was preparing to procure one of them. Since the USAF was going to use this aircraft as a multi-role fighter-bomber, the Congress and Department of Defense directed that the Navy and Marines should use a version of the same aircraft. That is where the troubles began.

The two competing lightweight fighter designs, the General Dynamics (GD, now part of Lockheed Martin) YF-16 and the Northrop (also now part of Lockheed Martin) YF-17, had a "fly-off" competition at Edwards AFB in California. When it was over, the YF-16 was declared the winner, and has proved to be an outstanding combat aircraft. The USAF and our allies have bought thousands of the little fighters, and continue to do so to this day. Unfortunately, many of the qualities that made the USAF love the F-16 were unacceptable in a carrier-based aircraft. For example, the Navy prefers twin-engined aircraft for their redundancy and ability to accept battle damage. The F-16 has only a single engine, and is too lightly built to carry some of the equipment needed for carrier operations. Since the Navy had been directed to base the VFAX aircraft on the contenders from the USAF lightweight-fighter competition, it chose to run a "paper" competition that would allow it to evaluate and choose the airplane it would buy.

Meanwhile, both GD and Northrop decided that since neither had recent experience building carrier aircraft, they would look for a partnership with an aircraft company that did. Thus GD in Fort Worth teamed up with its crosstown neighbor Vought, while Northrop adopted McDonnell Douglas (MDC) in St. Louis as its partner. At the end of the evaluation process, the Navy chose a derivative of the twin-engined, twin-tailed YF-17, which it judged was better suited to the rigors of duty aboard aircraft carriers. This award to MDC/Northrop provoked a loud protest from the losing Vought/GD team, which had thought the *original* DoD/Congressional direction was an ironclad guarantee that *they* would win. Though it took an inspired campaign of political pressure and technical documentation by the Navy to preserve the decision, the MDC/Northrop team held on to their win. But there is more to the story.

Winning a contract is one thing. Building the aircraft specified is another thing entirely; especially when it is the most advanced of its type ever built. The Navy and Marine Corps were asking a great deal more from the new aircraft than the USAF was of the F-16, and that complicated matters greatly. For instance, the new bird, now designated the F/A-18 Hornet (the F/A stood for Fighter/Attack), would have to carry a great deal more equipment than the USAF bird. This included a multi-mode radar capable of providing guidance for the large AIM-7 Sparrow AAMs and FLIR targeting pods it was to be equipped with. The Hornet would also have to lug around a lot of extra weight in the form of beefed-up structure (representing about 4,000 lb/1,818.2 kg, approximately 20% of the Hornet's total weight), to allow it to operate on and off carriers. These requirements proved to be far beyond the modest abilities of the YF-17. The Navy was in fact asking not simply for a Navy version of the original Northrop design, but for a brand-*new* aircraft. Simply scaling up the YF-17 was not going to do.

To further compound the difficulties presented by this design, there was no *true* prototype of the F/A-18. The first Hornets to fly were preproduction aircraft, which went directly into operational testing at NAS Patuxent River, Maryland. This meant that any normal problems that might have shown up (and been eliminated) in a prototype were now found in the preproduction birds. This proved to be a costly mistake. In fact, some problems (such as structural cracks) did not show up until the Hornet was actually into squadron service with the fleet. There were also troubles with the aerodynamics around the ''cobra hood'' and leading-edge extensions, which had to be modified fairly late in the development process. Luckily, the ability of the F/A-18's new digital fly-by-wire (FBW—the first ever on a carrier-capable aircraft) flight-control system to be reprogrammed made the fix relatively easy. The worst problem, though, was the scarcity of internal fuel tankage.

One of the most important measures of a combat aircraft's range is expressed by a number called the fuel fraction; that is, the weight of internal fuel expressed as a percentage of an aircraft's takeoff weight. Normally, combat aircraft designers like to build aircraft with a fuel fraction of between .30 and .35. This gives enough gas to fly a decent distance, drop bombs or dogfight, and then return to the base or boat with a minimum of refueling from airborne tankers. In the design of the Hornet, the fuel fraction was woefully low. The origins of this problem dated from the original YF-17 design. *That* aircraft had been a technology demonstrator that did not require the kind of fuel load a *combat* aircraft would normally carry. Thus, the Northrop designers had not installed large internal fuselage tanks. In the process of ''scaling up'' the YF-17 into the Hornet, the MDC designers had failed to take this into account. For some reason that still defies explanation, the F/A-18 was given the same fuel fraction as the original YF-17—around .23. As a result, the Hornet would *never* be able to fly all of the missions that had been specified in the original VFAX requirement. For example, when operating in a bombing mode, the F/A-18 cannot possibly fly the same weapons loads as far as the A-7E Corsair, which it replaced.

The Hornet's "short legs" came to light just as the Navy was about to make the production decision for the aircraft. It took more than a little hand-wringing and more than a few briefings to Navy, Marine, and Congressional leaders to make the case to put the F/A-18 into production. The NAVAIR rationalization was that since the aircraft had shown such good performance in so many other areas of flight test, the really-long-range-strike-mission requirement could be compromised. For example, the new APG-65 multi-mode radar was quickly hailed as one of the best in the world, and the weapons system integration made the Hornet an ordnance-delivery dream. Besides, the test and fleet pilots loved flying the new bird. They could see its potential, and were willing to accept a few shortcomings to get the Hornet into the fleet. So the decision to buy the first production batch of Hornets was made, and the first deliveries to VFA-125 at NAS Lemore, California, began in 1980. With this part of the story told, let's take a closer look at the F/A-18.

At first glance, the Hornet looks very much like the F-14 (twin engines and tails), but the similarities are only superficial. The F/A-18 is more than a decade ahead of the Tomcat in technology. A sizable percentage of the Hornet's structure, for example, is composed of plastics and composite structures. The twin General Electric F404-GE-400 engines utilize the same engine technology as the F110, and give the Hornet exceptional agility. Aerodynamically, the fixed wing of the F/A-18 is optimized for dogfighting, with six stations on the wings for ordnance (as well as AIM-9 Sidewinder AAMs on the wingtips). At the midpoint of each wing is a folding hinge, which allows the deck crews to reduce the "footprint" of the F/A-18 on the limited space of the flight and hangar decks. On the fuselage are two recessed wells for AIM-7 Sparrow and AIM-120 AMRAAM AAMs, as well as various types of sensor and data-link pods. There also is a centerline station suitable for a small external fuel tank. The nose of the Hornet is a very busy place, with the APG-65 multi-mode radar mounted just ahead of a bay, which houses the M61 20mm Gatling gun. Normally, placing a vibration sensitive instrument like a radar close to a fire-spitting device like a cannon would be suicidal in an aircraft. Unfortunately, the F/A-18's limited internal space gave MDC designers no choice. That this unlikely pairing of systems in the nose actually works speaks volumes about the care that designers gave every component of the Hornet.

The Navy has a real aversion to doing new things, and frequently prefers to let other services pioneer technology and ideas. However, for the F/A-18 to fulfill its missions, the Navy had to try some things that nobody had done before. One of these was to make the Hornet an effective dual-role (fighter and attack) aircraft, with only a single crewman. The only way to make this possible was to use an advanced cockpit design, a generation ahead of any used by any other combat aircraft. Like other fighters of its generation, the F/A-18 has a bubble canopy, with the pilot sitting with his/her shoulders above the cockpit rails in an ACES-series ejection seat, which provides the necessary "zero-zero" capability needed for safety in flight and deck operations. After that, the novelty begins.

To design the Hornet cockpit, MDC brought a unique talent to bear. Engineer Eugene Adam, acknowledged to be the finest cockpit designer in the world, led the MDC cockpit design team that produced the "front office" for the F/A-18. For years, Adam had advocated a "glass" cockpit, composed only of computer screens, which could be configured in any way desired by the pilot. With computer screens, a wide variety of data could be displayed at any time, depending upon what the pilot was doing at a given moment. Such a system was installed in the cockpit of the Hornet, which is made up of a series of square computerized Multi-Function Displays (MFDs) with buttons around the bezels that allow the pilot to select the data they want. To complement the MFDs, there were a second-generation HUD and HOTAS controls on the throttles and control stick. This made it possible for the pilot to switch from "Fighter" to "Attack" mode with just a flick of a switch. So advanced was the Hornet at the time of its introduction that it even included the first onboard GPS receiver seen in the fleet. These systems are backed up by one of the best avionics suites ever installed in a tactical aircraft.

The result was a cockpit still considered to be among the world's finest. Perhaps best of all, it was a cockpit with room for improvements and upgrades. Soon, there will be a new helmet-mounted sighting system, which will allow the pilot to cue the radar and weapons-targeting systems by just looking at a target. The new AIM-9X version of the classic Sidewinder AAM will be the first to use this new feature.

Naval aviators love to tell me how much "fun" the Hornet is to fly, and this has had a positive effect on its image in the fleet. Pilots especially love the responsiveness of the FBW control system and the integrated "glass" cockpit. The F/A-18 can even land itself, using a system called "Mode-1" to automatically fly the bird to a perfect "OK-Three" landing. Maintenance personnel love it too, since its digital electronics are so reliable that aircraft are rarely down for equipment failures. There is a "down" side, though. Because of the F/A-18's small internal fuel fraction, it almost always carries a pair of large fuel tanks under the wings, and frequently another one under the centerline of the fuselage. This leaves just four wing stations for actual weapons carriage. Since the two outer wing stations are load-limited (they are outboard of the wing fold line), these are usually reserved for additional AAMs, leaving just the two middle wing stations for carrying air-to-ground ordnance.[56]

If the Hornet is tasked for a bombing mission, the two fuselage stations will normally be filled with a single AIM-120 AMRAAM, and an AAS-38

56 During Desert Storm, this was usually four Mk. 83 1,000-lb/454.5-kg general-purpose bombs or a pair of Mk. 20 Rockeye cluster bombs. Today, the Hornet tends to carry PGMs like Paveway LGBs.

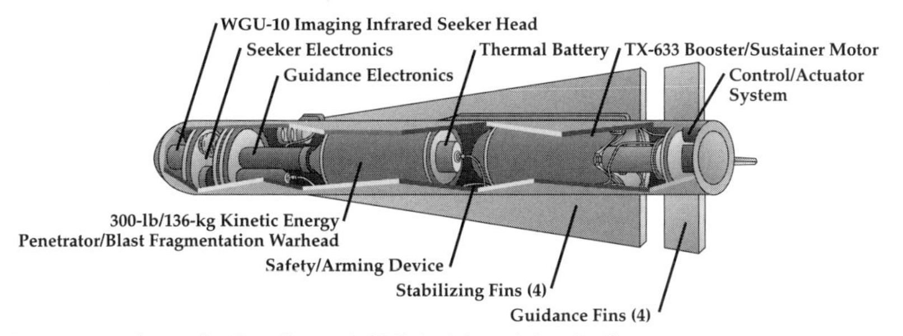

A cutaway view of a Raytheon AGM-65 Maverick missile.
Jack Ryan Enterprises, Ltd., by Laura DeNinno

Nighthawk FLIR/laser targeting pod. This configuration allows the F/A-18 to pick up targets in darkness or low visibility, and then deliver PGMs (like Paveway-series LGBs) or "iron" ordnance onto them with accuracy. Unlike the LANTIRN system used on the F-14, F-15, and F-16, Nighthawk (built by the Loral Division of Lockheed Martin) is designed to be operated by just a single crewman.[57] This means that a Hornet driver can pick up a target using the Nighthawk FLIR, "lock" it up, and then trust the pod to automatically track the target and handle the release and delivery of the weapon. While early versions of the Nighthawk lacked the laser designator and had some reliability problems, the current version is doing a fine job in the fleet. More than any other piece of equipment, the Nighthawk pod has transformed the image of the F/A-18 around the world. Where once it was seen only as an "iron" bomber, now it carries a reputation for deadly precision.

The Hornet can also employ other PGMs like the AGM-88 HARM antiradar missile, the AGM-65 Maverick tactical ASM, the AGM-84D Harpoon antishipping missile, and the new AGM-84E Standoff Land Attack Missile (SLAM).[58] SLAM is a relative newcomer to the fleet, having first been introduced and employed during Desert Storm in 1991. Since that time, SLAM has seen action in Bosnia in 1995, and has become one of the finest standoff strike weapons in the world. What makes it such a winner is the use of the basic (and very dependable) AGM-84 Harpoon engine, airframe, and warhead package, which is now married to a new guidance and seeker package. This new system combines a GPS/INS unit, an imaging infrared (IIR) seeker head from an AGM-65 Maverick ASM, and a man-in-the-loop datalink unit from the old Walleye guided bomb.

The result is a weapon that achieved perhaps the most spectacular hit of Desert Storm. On its first combat "shot," run against a heavily defended

57 While the LANTIRN system is used on single-seat F-16C fighter-bombers, it tends to be limited to striking fixed, preplanned targets only. For missions requiring a search for a target, two-person aircraft like the F-14 Tomcat or F-15E Strike Eagle are necessary.

58 The A/U/RGM-84 Harpoon has been in service since the late 1970's, originally having been designed to deal with patrol boats and other surface combatants at ranges of up to 60 nm/100 km. Thousands have been bought by dozens of nations for use on aircraft, ships, and submarines. It remains the most common and popular antishipping missile in the world today.

An AGM-84E SLAM missile being launched from an F/A-18C Hornet. SLAM was used during Operations Desert Storm and Deliberate Force, where it performed with amazing precision and lethality.
BOEING MILITARY SYSTEMS

Iraqi weapons plant near Baghdad, two SLAMs were launched several minutes apart. The first missile, taking its basic guidance from the GPS/INS unit, flew to the target and locked up the desired aimpoint without difficulty. It then flew directly into the building wall, detonated, and made a very large hole. Several minutes later, the second SLAM flew through the hole created by the first missile and destroyed the equipment inside. Further success for the SLAM came during Operation Deliberate Force in Bosnia. The outstanding performance of SLAM has made it one of the most feared PGMs in the U.S. arsenal.

In fleet service in its early years, the Hornet showed the shortcomings that had been seen in testing. While the F/A-18's range limitations became obvious at once, for example, this could be improved simply by altering the altitude and speed (called the flight profile) that it flew during missions. Thus, the aircraft's range could be stretched simply by flying it at higher altitudes, where the F404 engines were more efficient. Still, some of the Hornet's original specifications would never be met, especially those of acceleration and range. Still, in the crucible of combat it passed the ultimate test. This first came in 1986, when a number of Hornet squadrons took part in operations against Libya. In ACM engagements against the MiGs and Mirages of the Libyan Air Force, the Hornets had no trouble staying on the tails of the opposing warplanes. They also helped suppress the Libyan air defenses with HARM missiles, another role they took over from the A-7E. The Hornet provided the Navy with one other pleasant surprise: its incredible reliability compared with other Navy aircraft like the F-14 and A-6. This meant that the Hornet was cheaper to operate, and could be flown more often than other comparable aircraft—so often, in fact, that the early F/A-18As wore out faster than expected, and had to be replaced earlier than planned. This led to an improved variant, the F/A-18C/D, which arrived in the fleet during 1986.

The -C/D model gained some weight over the -A/B Hornets, but unfortunately did not carry any more gas. The radar, avionics, engines, and other systems were significantly improved, however, including provisions to carry the AIM-120 AAM and IIR version of the AGM-65 Maverick ASM. The new Hornet also had a new-generation monitoring system that allowed maintenance crews to diagnose problems automatically and even predict when individual components and "black boxes" might fail. There were also provisions for the new Hornet to be operated at night with night-vision goggles (NVGs), and a new radar: the APG-73 (planned for the new F/A-18E/F Super Hornet).

The F/A-18C/D was the Hornet that the Navy and Marines had wanted all along; and the Marine Corps bought six squadrons of modified -D models as night-attack aircraft to replace their force of retired A-6's. The Hornet was also becoming something of a success in the export market. The first foreign customer was Canada, which bought 138 CF-18's to conduct continental air defense as part of the North American Air Defense (NORAD) Command. Australia (seventy-five), Kuwait (forty), Spain (seventy-two), Switzerland (thirty-four), Finland (sixty-four), Thailand (eight), and Malaysia (eight) also bought various models of the F/A-18 to upgrade their air forces. All told, around 1,500 Hornets have been built to date.

By the time of the Iraqi invasion of Kuwait in 1990, the Hornet had been in service for almost a decade and was ready for its biggest combat challenge. Almost as soon as the U.S. began to react to the invasion, F/A-18 units were in the front lines of Desert Shield. Eventually, five carrier groups and an entire Marine Air Wing with Hornets as their backbone deployed into the theater. The Canadians also contributed a squadron of their CF-18's to the effort. In Desert Storm the F/A-18 proved to be a deadly air-to-air killer. On January 17th, a pair of VF-81 Hornets from the USS *Saratoga* (CV-60) downed a pair of Iraqi F-7's (Chinese MiG-21 clones) with a salvo of "in-your-face" AAM shots. The two F/A-18's were loaded for a bombing mission at the time, but quickly switched to the air-to-air mode, shot down the enemy fighters, then went on to complete their bombing mission. The rest of the war was mainly spent delivering "iron" bombs onto battlefield targets in Kuwait and Iraq. In this mission, the success of the Hornets was something less than total.

Part of the problem was the weather, which was the worst on record in the region. Because many of the bombing sorties required visual identification of the targets, some of these had to be aborted due to the cloud cover. There was also a requirement that bombs be delivered from medium altitude (above 10,000 feet/3,048 meters), making the accuracy of the results uncertain. Had the Hornet been armed with LGBs and other precision weapons, this problem would not have arisen. Unfortunately, the new version of the Nighthawk pod (with the laser designator and tracker) had not yet come into service.

There also was the fuel problem. Since most of the Hornets were based on carriers in the Red Sea, they required several in-flight refuelings in order to reach their targets in Iraq and Kuwait. This placed a severe burden on the limited airborne tanker resources of General Horner's Central Command Air Force (CENTAF). This meant that the F/A-18's were sometimes left off the daily Air Tasking Order (ATO) in favor of other aircraft, like USAF F-16's, which were based closer to their targets. Eventually, the Navy moved a total

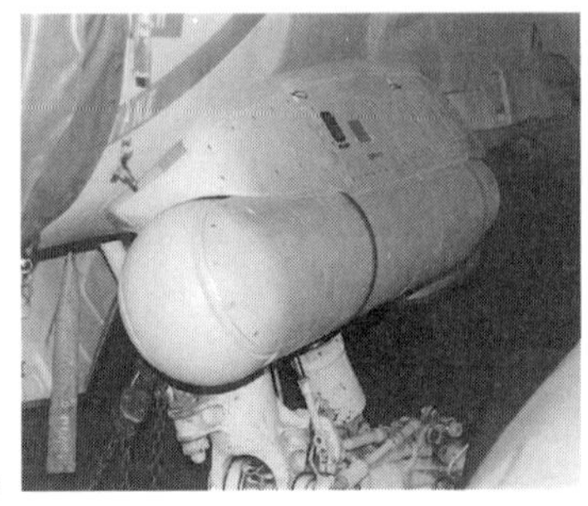

An AAQ-38 Nighthawk laser targeting pod, mounted on the starboard fuselage station of an F/A-18C Hornet. This pod allows Hornet crews to deliver laser-guided bombs and other precision munitions.

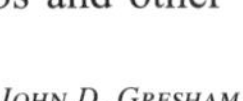

John D. Gresham

of four carrier groups into the Persian Gulf itself, to bring the Hornets close enough to their targets to do some real good.

By the time Hornets next went into combat (in Bosnia in 1995), a number of improvements had been made. The -C/D-model Hornets had been rearmed with new AIM-120 AMRAAM AAMs, SLAM ASMs, and Paveway LGBs guided by their new Nighthawk targeting pods. And this time, their carrier, the USS *Theodore Roosevelt* (CVN-71), operated closer to shore than was the practice in Desert Storm and they were given adequate tanker support from NATO/USAF resources. Now that they were properly supported and armed, the PGM-armed Hornets (including a squadron of Marine F/A-18D Night Attack variants) were the heart of Operation Deliberate Force in 1995, and did all that was asked of them. In fact, Navy and Marine Corps Hornets dropped and launched the bulk of the PGMs that were used during the Bosnia strikes.

Today the Hornet is the backbone of U.S. carrier aviation, and will remain that way for at least the next decade. Every CVW is being equipped with three F/A-18 squadrons (each with twelve aircraft), which means that fully half of the aircraft on U.S. carrier decks today are Hornets. There will soon be significant Hornet upgrades, with the introduction of new PGMs, as well as a new version of the classic AIM-9 Sidewinder. Even so, there can be little doubt that the F/A-18's short legs, limited weapons load, and design compromises will continue to be a lightning rod for critics. Still, the folks who fly the Hornet love their mounts. Though it's a flying compromise, it's easy to fly, forgiving for new pilots, and capable of many different missions.

EA-6B Prowler: The Electric Beast

Looking like a flying metal tadpole, the EA-6B Prowler will probably be the last survivor of a long line of Grumman carrier aircraft that date back to before the Second World War. Its mission is electronic warfare (EW), which explains why the aircraft looks like a flying antenna farm. As many as thirty (or more) antennas are smoothly faired into the fuselage or packed into the "football" (actually, it looks more like a Brazil nut), a fiberglass radome at the top of the vertical stabilizer. These devices allow the Prowler to throw

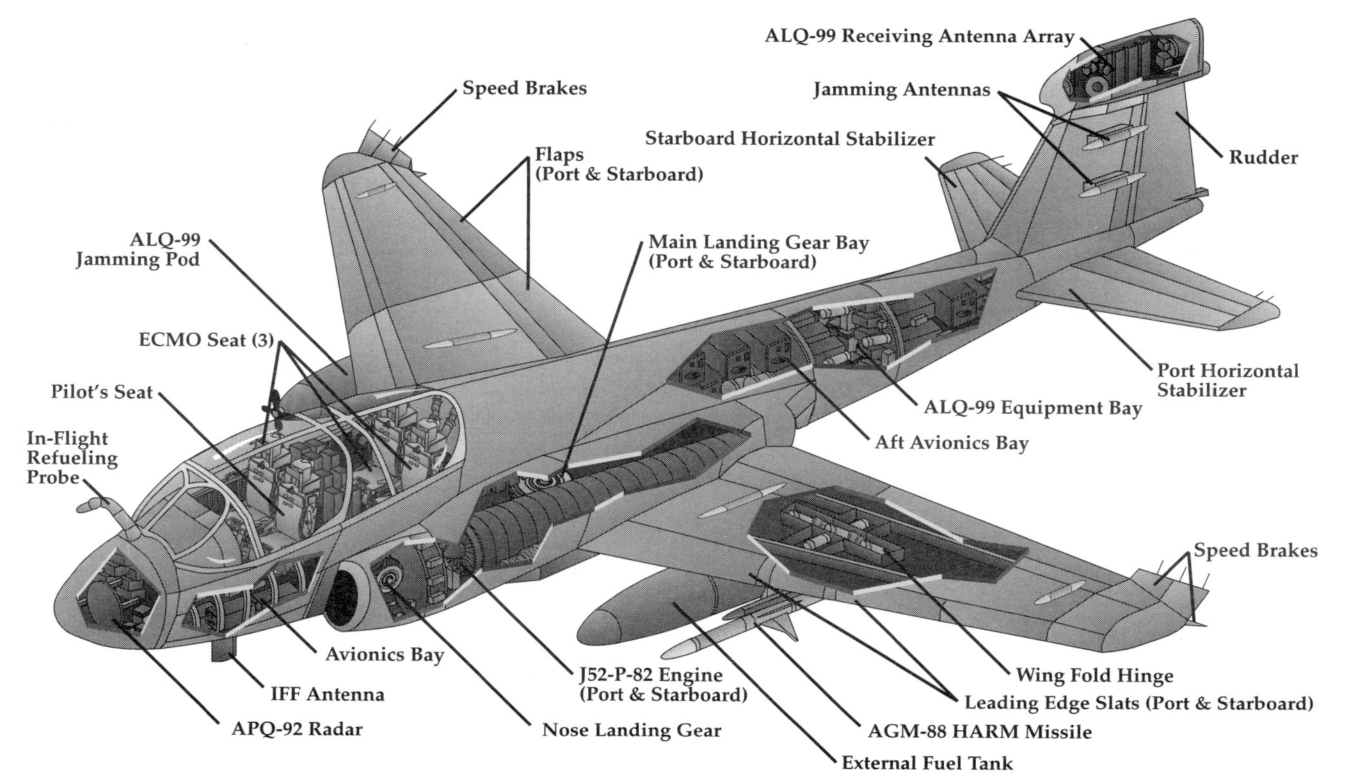

A cutaway view of a Northrop Grumman EA-6B Prowler electronic warfare aircraft.

Jack Ryan Enterprises, Ltd., by Laura DeNinno

an invisible veil of protection over the aircraft and ships of the carrier battle group. They detect, classify, and locate enemy radar, electronic data links, and communications, then jam them with precisely crafted and targeted interference. And as an added bonus, since 1986 Prowlers have also been capable of making "hard kills" using the AGM-88 High-Speed Anti-Radiation Missile (HARM), which homes in on radar transmitters and shreds them with a blast-fragmentation warhead.[59]

Today, the Prowler is the finest tactical EW aircraft in the world. It's so good that the USAF is quietly retiring its own fleet of EF-111 Raven EW aircraft and employing EA-6Bs in joint (USN/USAF) squadrons. All this is even more impressive when you consider that the thirty-year-old Prowler design has been around for almost half of the six decades that radar has been used in military operations; and with regular updates, it has at least another ten or fifteen years to go.

Electronic warfare (intercepting and jamming enemy signals) began with the first military use of radio in the Russo-Japanese War (1905), and reached a high degree of sophistication during the Second World War, as Axis and Allied scientists and technicians fought for control of the electromagnetic spectrum. EW aircraft have been in use since World War II, with modified USN TBF/TBM Avengers being among the first such aircraft. The start of the Vietnam War saw two carrier-capable EW birds in service with the Navy, though both were already getting old. The EA-1E was a modification of the classic Douglas AD-1 Skyraider, while the EKA-3B "Electric Whale" (which also served as a tanker aircraft) was a development of the Cold War-era A-3 Skywarrior attack bomber. As American aircraft began to fall to radar-controlled AAA guns, SAMs, and MiGs over Vietnam, the need for a third-generation EW aircraft became almost desperate. Out of this need came the development of what would become the EA-6 Prowler.

The original airframe of the Prowler was derived from the A-6 Intruder, which was the Navy's first *true* all-weather, day or night, low-level medium-strike aircraft. The Intruder saw extensive combat in Vietnam, the Cold War, and Desert Storm, and was immortalized in Stephen Coonts's 1986 novel, *Flight of the Intruder*. The Prowler's immediate ancestor, the EA-6A, was a modified two-seat "Electric Intruder" developed to fill a Marine Corps requirement for a jammer aircraft that could escort strike missions into the high-intensity threat of North Vietnam's integrated air defense system. Hard-won experience showed that what was really needed for such missions were more EW operators and jammers aboard the aircraft. From this came the all-new EA-6B Prowler, which is an all-weather, twin-engine aircraft manufactured by Northrop Grumman Aerospace Corporation as a modification of the basic A-6 Intruder airframe. The first flight of the EA-6B was on

59 The Raytheon (formerly Texas Instruments) AGM-88A HARM missile is 13 ft, 7 in/4.2 meters long, 10 in/25.4 cm in diameter, and weighs 798 lb/363 kg. Range depends on the speed and altitude of the launching aircraft, but a standoff of 50 nm/92.6 km is typical for Prowler missions.

May 25th, 1968, and it entered operational service in July of 1971. Just a few months later, the Prowler entered combat over Vietnam with VAQ-132, based on aircraft carriers in the Gulf of Tonkin.

The Prowler is big for a "tactical" aircraft. The overall length is 59 feet, 10 inches/17.7 meters. It has a wingspan (with the wings unfolded) of 53 feet/15.9 meters, and sits 16 feet, 3 inches/4.9 meters high on the deck. It is also quite heavy, with a maximum gross takeoff weight of 61,000 lb/27,450 kg, much of which is fuel. The Prowler has a cruising speed of just over 500 knots/575 mph/920 kph, an unrefueled range of over 1,000 nm/1,150 mi/1,840 km, and a service ceiling of 37,600 feet/11,460 meters.

The EA-6B can hardly be called a "high performance" tactical aircraft. Although it is quite stable in flight and relatively easy to fly, the Prowler is somewhat underpowered. The two non-afterburning Pratt & Whitney J52-P408 turbojet engines lack the kind of thrust available to F-14 or F-18 crews (11,200 lb/5,080 kg of thrust each), which presents the pilot with a number of challenges during every mission (especially on takeoff and landing). Due to the complexity of its systems, the EA-6B is also a relatively high-maintenance aircraft—about one mission in three returns with a "squawk" or malfunction requiring unscheduled maintenance. On the plus side, the side-by-side twin cockpit arrangement gives maximum efficiency, visibility, and comfort for the four-person crew. This is important during long missions, which can last up to six hours with in-flight refueling. The canopies are coated with a microscopically thin (and very expensive) transparent layer of gold leaf, which reflects microwave energy and protects the crew from getting cooked by their own high-energy jammers.

The Prowler's crew includes a pilot and up to three Electronic Countermeasures Officers (ECMOs). The senior officer on board—either the pilot or one of the ECMOs—is normally the mission commander. In fact, a Prowler squadron commander is often an ECMO rather than a pilot. ECMO-1, who mans the position to the pilot's right, handles navigation and communications, while ECMO-2 and -3 (they sit in the rear cockpit) manage the offensive and defensive EW systems. Within the squadron, there are normally more crews than aircraft, due to the workload of flying, administration, and mission planning. In a low-threat environment, a crew of three is considered sufficient—with one ECMO remaining behind on the boat to plan the next mission, catch up on paperwork, or perform any of the countless additional duties that Naval aviators must juggle when they are deployed.

The Prowler's EW capabilities depend largely on the ALQ-99 electronic countermeasures system. This is not a single piece of equipment, but a complex and ever-changing mix of computers, jammers, controls and displays, receivers, and transmitters. Some of these components are built into the airframe, while others are packaged in pods. All are externally identical, but each is optimized for specific frequency bands. Up to five such pods can be carried—two under each wing and one under the fuselage. A more typical mission configuration is two or three pods, with the other stations occupied by fuel tanks or AGM-88 HARM missiles. Each pod generates its own electrical power, using a ''ram air turbine'' or RAT (a compact generator spun by a small propeller). To generate full power for jamming, the aircraft must fly above a minimum speed (225 knots). Using the RATs brings a slight drag penalty; the Prowler loses about 1% of its maximum combat radius for each pod carried. Still, the pods and missiles are the reason why the Prowler exists. Without the electronic smoke screen provided by the EA-6B's jamming pods, losses to enemy defensive systems would be many times greater than they have been.

Normally, the EA-6B is used to provide a combination of services for strike packages inbound to a target area. If active SAM sites are nearby, the ECMOs will use the ALQ-99 to provide targeting for the HARMs, which are deadly accurate when fired from a Prowler. Once the HARMs are gone, the EA-6B orbits away from the target area and uses the ALQ-99 jammer pods to ''knock back'' enemy radars and other sensors that might engage the strike group. Other missions include electronic surveillance, as the ALQ-99 is a formidable collection system for electronic intelligence (ELINT). Because they are considered ''high value units'' by enemy defenders, one or two fighters usually provide them with an escort, just in case the locals get nosey. In fact, no Prowler has ever been lost in combat, though about forty have been destroyed in accidents. The worst of these was a horrific crash while landing aboard the *Nimitz* (CVN-74) back in 1979, which killed the entire crew as well as a number of deck personnel in the ensuing fire.

EW is an unusual facet in the spectrum of warfare. For every measure there is a countermeasure, and the useful life span of a system in actual combat is often only a few months. Because a new ''generation'' of electronic warfare technology emerges every few years, if you fall a generation behind you are ''out of the game.'' This helps to explain the bewildering variety of upgrades and variants that mark the Prowler's long career. Production of new-built Prowlers ended several years ago, but about 125 remain in active service today. This is just enough for twelve Navy, four Marine Corps, and four ''joint'' squadrons of EA-6Bs. Normally, each deploys with four aircraft. Navy and joint USAF/USN Prowler squadrons are home-based at NAS Whidbey Island, Washington, while the Marine units live at MCAS Cherry Point, North Carolina. The joint EA-6B squadrons are a new phenomenon in the post-Cold War world, an expression of budget realities that no longer allow the services to duplicate aircraft types with the same mission. Although the Navy and USAF developed very different EW concepts and doctrine over the years, the Air Force has agreed to retire its only tactical jammer aircraft, the EF-111 Raven. Now the two services will ''share'' five joint ''expeditionary'' Prowler squadrons, which will operate with mixed Navy and Air Force ground and flight crews. Despite the predictable concerns about USAF officers commanding Navy squadrons (or vice versa), this program is well under way and looks to be a real winner.

Like their brethren in the Tomcat community, EA-6B crews have

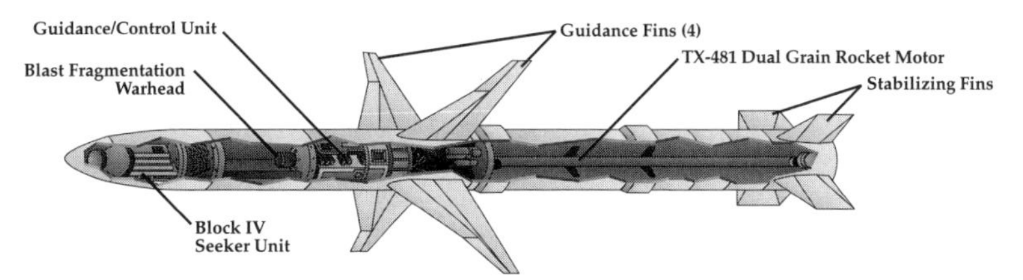

A cutaway view of a Raytheon AGM-88 HARM anti-radiation missile.
Jack Ryan Enterprises, Ltd., by Laura DeNinno

learned some new tricks in recent years, like shooting AGM-88 HARM missiles at enemy radars. Prowlers have even been used as command and control aircraft, functioning as strike leaders for other planes on bombing missions. Other improvements include plans to start another upgrade program known as ICAP (Improved Capability) III. This will take the basic EA-6B package as it currently exists (known as Block 89) and add improved computers, signal processors, and jammers, as well as a GPS receiver, new radios and data links, and other new avionic systems. ICAP III-equipped Prowlers should begin to appear in a few years. As for future EW aircraft on carriers, long-range plans have been developed for a two-seat EW version of the new F/A-18E/F Super Hornet. A highly automated follow-on version of ALQ-99 would be fitted to this bird, as well as more advanced HARMs and other systems. However, since there is no money for this bird in the current budget, the old Prowlers will have to soldier on for at least another decade or two.

E-2C Hawkeye: Eyes of the Fleet

Put a sensor of sufficient resolution high enough, and you will see enemy forces before they can harm you. This is the guiding principal behind most early warning systems, from reconnaissance satellites to Unmanned Aerial Vehicles (UAVs). For naval leaders, there is no more important ''high ground'' than that occupied by Airborne Early Warning (AEW) aircraft. The first U.S. Navy AEW birds date back to World War II, when converted TBF/TBM Avengers were modified to carry a small airborne radar and operator for the purpose of detecting incoming Japanese Kamikaze aircraft far enough out for fighters to be vectored to intercept them. After the war, special purpose-built AEW aircraft were developed. These were designed to deal with the new generation of jets and ASMs faced by Cold War-era Naval forces. The first of these was the Grumman E-1 Tracer, a development of the S-2F Tracker ASW aircraft. For almost a decade the E-1 worked as the primary carrier-based AEW aircraft for the USN; but the operational conditions of the Vietnam conflict showed the numerous shortcomings of the Tracer, including poor over-land radar performance and limited endurance and service altitude. Though they served aboard modified *Essex*-class (SCB-27C/CV-9) carriers until 1976,

A cutaway view of a Northrop Grumman E-2C Hawkeye Airborne Early Warning aircraft.

JACK RYAN ENTERPRISES, LTD., BY LAURA DENINNO

The business end of a VRC-40 C-2A Greyhound COD aircraft on the deck of the USS *George Washington* (CVN-73). These aircraft are used to ferry personnel, cargo, and supplies to and from carrier battle groups.
JOHN D. GRESHAM

there was a clear need for a more advanced AEW aircraft for the fleet. That aircraft was the E-2 Hawkeye.

One of the last propeller-driven aircraft in the CVW, the E-2C Hawkeye is the Navy's all-weather, carrier-based tactical AEW aircraft. The E-2C uses computerized sensors for early warning, threat analysis, and control against air and surface targets. It provides the carrier battle group with all-weather AEW services, as well as command, control, and communications (C^3) functions for the carrier battle group. Additional missions include surface surveillance, strike and interceptor control, Combat Search and Rescue (CSAR) guidance, Over-the-Horizon (OTH) targeting, and communications relay. Designed to a 1955 specification, and upgraded through at least six generations of electronic technology, the Hawkeye remains in production today. The E-2C has also been adopted by the French Navy, and at least five other countries that do not even have aircraft carriers. This is a tribute to the cost-effective mix of robust airframe, compact sensor and avionics suite, and turboprop power plants. Unit cost: $51 million—cheap for the protection it provides. Before you gag on that number, consider that a new F/A-18E/F Super Hornet will cost you even more per copy, and I don't know any battle group commander who would not like a few more of the precious E-2Cs.

One thing all that money does *not* buy is beauty. As you walk up to a Hawkeye, pieces of it seem to be going everywhere. Wings are folded back on the fuselage, with the big radar rotodome perched up top like a tethered flying saucer. Though it is not gorgeous to look at, the E-2C has a functional elegance, doing the same kind of mission as its larger USAF cousin, the Boeing E-3 Sentry. That it does this on an airframe a fifth the size, and off a carrier deck, is a measure of its sophistication and value. When the Grumman engineers designed the E-2, they started with a perfect cylinder. Into this they packed all the electronics, fuel, two pilots, and three radar controllers. The finishing touch came when they mounted the rotating radar dome (called a ''rotodome'') on top, and attached a pair of long wings mounting a pair of Allison T-56-A427 turboprop engines with five thousand shaft horsepower each.

Dimensionally, the Hawkeye is 57 feet, 6 inches/17.5 meters long, with a wingspan of 80 feet, 7 inches/28 meters, and a height of 18 feet, 3 inches/ 5.6 meters to the top of the radar dome. Though it is the largest aircraft flying on and off carriers today, it is not the heaviest. At a maximum gross takeoff weight of 53,000 lb/23,850 kg (40,200 lb/18,090 kg "dry"), the E-2C is actually lighter than the F-14 Tomcat. The wings have the longest wingspan of any carrier aircraft in the world; and when folded, they use the classic Grumman "Stow-Wing" concept, which has them folding against the fuselage. The tail is composed of a horizontal stabilizer with four vertical stabilizers to give the Hawkeye the necessary "bite" to move the heavy bird around the sky. Though it has only ten thousand horsepower behind the twin props, the Hawkeye is capable of speeds over 300 knots/345 mph/552 kph, and can operate at altitudes of 30,000 feet/9,144 meters. Because Hawkeyes are unarmed, no battle group commander would be considered sane if there were less than two fighters protecting his E-2C. Hawkeyes are true "high value units" and are *always* a target for enemy fighters.

On board, the crew of five is busy, for they're doing a job that on the larger E-3 Sentry takes several dozen personnel. The pilot and copilot fly precisely positioned and timed racetrack-shaped patterns, designed to optimize the performance of the E-2C's sensors. In back, the three radar-systems operators are tasked with tracking and sorting the contacts detected by the Hawkeye's APS-145 radar. This Westinghouse-built system is optimized for operations over water and can detect both aircraft and surface contacts out to a range of up to 300 nm/345 mi/552 km. To off-load as much of the workload as possible, a great deal of the raw data is sent back to the task force's ships via a digital data link. With this off-board support, the three console operators are able to control a number of duties, including intercepts, strike and tanker operations, air traffic control, search and rescue missions, and even surface surveillance and OTH targeting.

Along with the 141 E-2Cs produced for the USN, the Hawkeye has had considerable export success. No less than six foreign governments have bought them: Israel (four), Egypt (six), France (two for their new carrier *Charles de Gaulle*), Japan (thirteen), Singapore (four), and Taiwan (four). There are more Hawkeyes in use throughout the world than any other AEW aircraft ever built.

There also has been one major variant of the Hawkeye, a transport version known as the C-2A Greyhound. Basically an E-2 airframe with a broader fuselage and the radar rotodome deleted, it can deliver cargo and passengers hundreds of miles/kilometers out to sea. Known as a COD (for Carrier Onboard Delivery) aircraft, it replaced the elderly C-1 Trader, which is itself a variant of the earlier E-1 Tracker. With its broad rear loading ramp and fuselage, the C-2 can carry up to twenty-eight passengers, twenty stretcher cases, or cargo up to the size of an F-110 engine for the F-14.

The Hawkeye has had a long run in USN service. The original -A model was first flown in October 1960, to provide early warning services for the

E-2C Hawkeye AEW aircraft on the deck of the USS *George Washington* (CVN-73). They generally parked alongside the island structure, on a spot called "the Hummer Hole."
JOHN D. GRESHAM

new generation of supercarriers then coming into service. In January 1964, the first of fifty-nine E-2As were delivered to their squadrons, and were shortly headed into combat in Southeast Asia. These were later updated to the E-2B standard, which remained in use until replaced by the E-2C in the 1970's. The first E-2Cs entered USN service with Airborne Early Warning Squadron (VAW) 123 at NAS Norfolk, Virginia, in November of 1973. The -C-model Hawkeye was produced in order to provide the F-14 Tomcat with an AEW platform matched to the new fighter's capabilities. Though visually identical to the earlier models, the E-2C was equipped with new-technology digital computers that provided a greatly increased capability for the new Hawkeye. These gave the operators the ability to track and intercept the dozens of Soviet bombers and hundreds of ASMs and SSMs that were expected to be fired at CVBGs if the Cold War ever turned "hot."

In any event, the E-2Cs never directed the massive air battles they had been designed for. Instead, the Hawkeye crews spent the declining years of the Cold War flying their racetrack patterns over the fleets, maintaining their lonely vigil for a threat that never came. Carrier-based Hawkeyes were not strangers to combat, however. E-2Cs guided F-14 Tomcat fighters flying combat air patrols during the 1981 and 1989 air-to-air encounters with the Libyan Air Force, as well as the joint USN/USAF strike against terrorist-related Libyan targets in 1986. Israeli E-2Cs provided AEW support during their strikes into Lebanon in 1982, and again during the larger invasion the following year. More recently, E-2Cs provided the command and control for successful operations during the Persian Gulf War, directing both land strike and CAP missions over Iraq and providing control for the shoot-down of the two Iraqi F-7/MiG-21 fighters by carrier-based F/A-18's. E-2 aircraft have also worked extremely effectively with U.S. law enforcement agencies in drug interdictions.

Today the entire Hawkeye fleet is being upgraded under what is called the Group II program. Along with thirty-six new-production aircraft, the entire USN E-2C fleet is being given the improved APS-145 radar, new computers, avionics, data links, and a GPS/INS system to improve flight path and

A VS-32 S-3B Viking ASW aircraft on the deck of the USS *George Washington* (CVN-73) with wings folded. The S-3B has rapidly taken over many critical roles in carrier operations, espcially in-flight refueling of other aircraft.

JOHN D. GRESHAM

targeting accuracy. This means that a single Hawkeye can now track up to two thousand targets at once in a volume of six million cubic miles of airspace and 150,000 square miles of territory. Current plans have the Hawkeye/Greyhound fleet serving until at least the year 2020, when a new airframe known as the Common Support Aircraft (CSA) will be built in an AEW version. By that time, the basic E-2 airframe will have served for almost six decades!

Lockheed Martin S-3B Viking: The Vital ''Hoover''

Antisubmarine warfare (ASW) is probably the most complex, frustrating, operationally challenging, and technically secretive mission that any aircraft can be called upon to perform. To locate, track, classify, and destroy a target as elusive as a nuclear submarine in the open ocean often seems virtually impossible. And against a quiet modern diesel boat in noisy coastal waters, the odds are even worse. In fact, the ASW mission doesn't have to be *that* successful. It has succeeded as long as enemy subs are forced to go deep, run quiet, and keep their distance from a Naval task force or convoy. It is a matter of record that the most effective weapon against submarines during the Second World War was the ASW patrol aircraft. Such aircraft have continued to do this job ever since.

Today, the USN operates two fixed-wing ASW aircraft. One is the venerable four engined P-3C Orion, which operates from land bases. The other is its ''little brother'' from the Lockheed Martin stable, the S-3B Viking, which is carrier-capable. Airborne ASW has long been a Lockheed specialty. Their land-based Hudson and Ventura patrol bombers played a key role in World War II against German U-boats. More recently, their P-2V Neptune and P-3 Orions have kept vigil over the world's oceans, watching for everything from submarines to drug-running speedboats. The so-called ''sea control'' mission is thankless work, with nearly day-long missions, most of which are flown over inhospitable and empty seas. The boredom arising from these missions in no way reduces their importance. A maritime nation that cannot monitor and control the sea-lanes it uses is destined to sail at the whims of other powers.

Early on, carrier aviators knew that they too needed the services of such aircraft, and began to build specially configured ASW/patrol aircraft shortly after the end of World War II. The first modern carrier-based ASW aircraft was Grumman's twin-engine S-2 Tracker, which entered service in 1954 and remained in the fleet for over twenty-five years with more than six hundred built.[60] In 1967, the growing sophistication of the Soviet submarine threat led the Navy to launch a competition for a radically new generation of carrier ASW aircraft. Known as the VSX program, it was designed both to replace the Tracker and to provide a utility airframe for other applications. In 1969, the design submitted by Lockheed and Vought was declared the winner and designated S-3. The prototype S-3A first flew on January 21st, 1971, and the type entered service in 1974 with VS-41 at NAS North Island, California. By the time S-3A production ended in 1978, 179 had been delivered.

The S-3 Viking is a compact aircraft, with prominent engine pods for its twin TF-34-GE-2 engines. This is the same basic non-afterburning turbofan used on the Air Force's A-10 "Warthog," and its relatively quiet "vacuum-cleaner" sound gives the Viking its nickname: the "Hoover." The crew of four sits on individual ejection seats, with the pilot and copilot in front, and the tactical coordinator (TACCO) and sensor operator (SENSO) in back. A retractable aerial refueling probe is fitted in the top of the fuselage, and all S-3B aircraft are capable of carrying an in-flight refueling "buddy" store. This allows the transfer of fuel from the Viking aircraft to other Naval aircraft. Because ASW is a time-consuming business that requires a lot of patience and equipment, the Viking is relatively slow, with a long range and loiter time. This means the S-3 is pretty much a "truck" for the array of sensors, computers, weapons, and other gear necessary to find and hunt submarines. But don't think that the Viking is a sitting duck for anyone with a gun or AAM. The S-3 is surprisingly nimble, and it's able to survive even in areas where AAW threats exist.

There are three primary ways to find a submarine that does not want to be found. You can listen for sounds, you can find it magnetically (something like the way compass needles find north), or you can locate a surfaced sub with radar. Since sound waves can travel a long way underwater, a sub's most important "signature" is acoustic. But how can an aircraft noisily zooming through the sky listen for a submarine gliding beneath the waves? The answer, developed during World War II, is the sonobuoy. This is an expendable float with a battery-powered radio and a super-sensitive microphone. "Passive" sonobuoys simply listen. "Active" sonobuoys add a noise-

60 In 1953, some of the Navy's older aircraft carriers were redesignated as Antisubmarine Aircraft Carriers, with air groups specifically tailored for ASW. These were mostly made up of ASW helicopters and S-2F Trackers.

maker that sends out sound waves in hope of creating an echo. By dropping a pattern of sonobuoys and monitoring them, an ASW aircraft can spread a wide net to catch the faint sounds of the sub's machinery, or even the terrifying ''transient'' of a torpedo or missile launch.

Another detectable submarine signature is magnetism. Since most submarines are made of steel, they create a tiny distortion of the earth's magnetic field as they move.[61] The distortion is *very* small, but it is detectable. A ''magnetic anomaly detector'' (MAD) can sense this signature, but it is so weak that the aircraft must practically fly directly over the sub at low altitude to do so.[62] In order to isolate the MAD from the plane's own electromagnetic field, it is mounted on the end of a long, retractable ''stinger'' at the tail of the aircraft.

Eventually, every submarine must come to periscope depth to communicate, snorkel, or just take a quick look around. Although periscope, snorkel, and communications masts are usually treated with radar-absorbing material, at close range sufficiently powerful and sensitive radar may obtain a fleeting detection. Finally, there are more conventional means of detection. For example, an airborne receiver and direction finder may pick up a sub's radio signals, if it is foolish or unlucky enough to transmit when an enemy is listening. And sometimes the telltale ''feather'' from a mast can be seen visually or through an FLIR system.

The integrated ASW package of the initial version of the Viking, the S-3A, was designed to exploit all of these possible detection signatures. Sixty launch tubes for sonobuoys are located in the underside of the rear fuselage. In addition, the designers provided the ASQ-81 MAD system, an APS-116 surface search radar, a FLIR system, a passive ALR-47 ESM system to detect enemy radars, and the computer systems that tie all of these together. Once a submarine has been found, it is essential that all efforts be made to kill it. To this end, the S-3 was not designed to be just be a hunter; it was also a killer. An internal weapons bay can accommodate up to four Mk. 46 torpedoes or a variety of bombs, depth charges, and mines. Two wing pylons can also be fitted to carry additional weapons, rocket pods, flare launchers, auxiliary fuel tanks, or a refueling ''buddy store.''

All this made the S-3A one of the best sub-hunting aircraft in the world, which was good enough in its first decade of service. By 1981, though, the -A model Viking clearly needed improvement in light of the growth in numbers and capabilities of the Soviet submarine fleet. In particular, the improved quieting of the Russian boats made hunting even more of a challenge. In order to improve the S-3's avionics, sonobuoy, ESM and radar data processing, and weapons, a conversion program was started. The result was the S-3B, which

---

61 The Soviet ''Alfa''- and ''Sierra''-class SSNs, along with a few experimental boats, had hulls welded from titanium, a very non-magnetic metal. The Russians can no longer afford the exotic construction methods required to build such boats.

62 This can be as low as five hundred feet, according to some open-source publications. For obvious reasons, MAD performance specs are highly classified.

The prototype ES-3A Shadow on a test flight. The sixteen Shadows provided the fleet with electronic reconnaissance and surveillance services until recently.

John D. Gresham

upgraded basic -A model airframes to the new standard. The first S-3Bs began to arrive in the fleet in 1987, and they quickly showed both their new sea control abilities and capability to fire AGM-84 Harpoon antiship missiles. This is the version that serves today.

One of the original hopes for the S-3 was to provide a basic airframe for a number of other aircraft types. Unfortunately, the small production run of the Viking has limited its opportunities for other roles. A small number of early S-3As were modified by removing all the ASW equipment and fittings for armament, allowing them to carry urgent cargo and mail and providing seats for a crew of three and up to six passengers (with minimal comfort). Designated US-3A and possessing a much longer range than the normal C-2A Greyhound COD aircraft, a total of five served in the Pacific fleet until they were recently retired. A dedicated tanker version, the KS-3A, was tested in 1980, but never went into production.

The single most important variant was the ES-3A "Shadow," an electronic surveillance (ESM) and signals intelligence (SIGINT) platform, which replaced the venerable EKA-3B "Electric Whale." Externally, the Shadow is quite distinctive, with a prominent dorsal hump and a retractable radome. About 3,000 lb/1,360 kg of ASW gear was removed and 6,000 lb/2,721 kg of electronics were packed into the weapons bay. While the Shadow is unarmed, it can also carry external fuel tanks and "buddy" refueling stores. Sixteen of these aircraft are split between two squadrons: VQ-5 (the "Sea Shadows") in the Pacific Fleet and VQ-6 (the "Ravens") in the Atlantic. Detachments of two or three aircraft normally deploy with every carrier air group, providing ESM, SIGINT, and OTH support for the CVBG. Unfortunately, recent budget cuts have targeted the shadow community which appears to be headed for disestablishment. Plan on seeing the ES-3 head for the boneyard in 1999.

The S-3 community has changed a great deal since the end of the Cold War. As long as the Soviet Union maintained the world's largest submarine fleet, the ASW squadron was an integral part of the carrier air group. But today, that ''blue-water'' submarine threat has receded. This hardly means that the S-3's can be retired and their crews given pink slips. On the contrary, the VS squadrons have taken on a whole new set of roles and missions, making them more valuable than ever. After the premature retirement of the KA-6D fleet in 1993, they took on still another role, becoming the primary aerial refueling tanker for the CVW. This has not proved to be the best solution to the aerial refueling problem, since an S-3B can only off-load about 8,000 lb/3,628 kg of fuel, as compared to over 24,000 lb/10,886 kg for the KA-6D. With the thirsty F/A-18's needing at least 4,000 lb/1,814 kg every time they go on a long CAP or strike mission, even the ES-3 Shadows are being used as tankers! To reflect all this, the previous ASW designation of their squadrons has been changed to ''Sea Control,'' which uses the ''VS'' nomenclature.

The S-3B community currently includes ten operational squadrons, administratively divided between two Sea Control Wings: one for the Atlantic Fleet and one for the Pacific. A single Fleet Replacement Squadron, VS-41, based at North Island NAS, California, serves as the advanced training unit. During Operation Desert Shield and Desert Storm, S-3 squadrons flew maritime patrols to help enforce sanctions against Iraq. In fact, the only complaint I've ever heard about this wonderful aircraft is that the Navy bought too few of them. Another two hundred would have been invaluable today, but the poor choices on the part of naval aviation leaders scuttled that idea. At the end of 1997, about 120 S-3's remained in service. Eventually, all of their tasks will be taken over by the future Common Support Aircraft that is scheduled to enter service around 2015.

Sikorsky H-60 Seahawk: A Family of Winners

Fixed-wing aircraft that hunt submarines on the prowl have one major vice: They move too fast. One solution is to use an aircraft that can stand still, dip a sonar into the water, and just listen for a while, the way a surface ship or submarine can. Then, if needed, it can rapidly dash to another spot, and do it all again. In other words, you need ASW helicopters. The Germans were the first to use helicopters for this purpose. During World War II they used them to hunt Russian submarines in the Baltic Sea. Following the war, it was only a matter of time and technological development until a true ASW helicopter was developed. After several false starts in the 1950's, Sikorsky developed the SH-3 Sea King. One of the finest helicopters ever built, it was equipped with a dipping sonar and homing torpedoes, and had plenty of range and power. However, by the mid-1970's it was clear that the old SH-3 was heading into its last legs as the USN's premier sub-hunting helicopter.

Meanwhile, the USN had operated another fleet of ASW choppers, so-called ''light'' helicopters, which can operate off small platforms on escort ships. Starting in the late-1960's, this mission was filled by the Kaman SH-2 Seasprite LAMPS I (Light Airborne Multi-Purpose System). For three de-

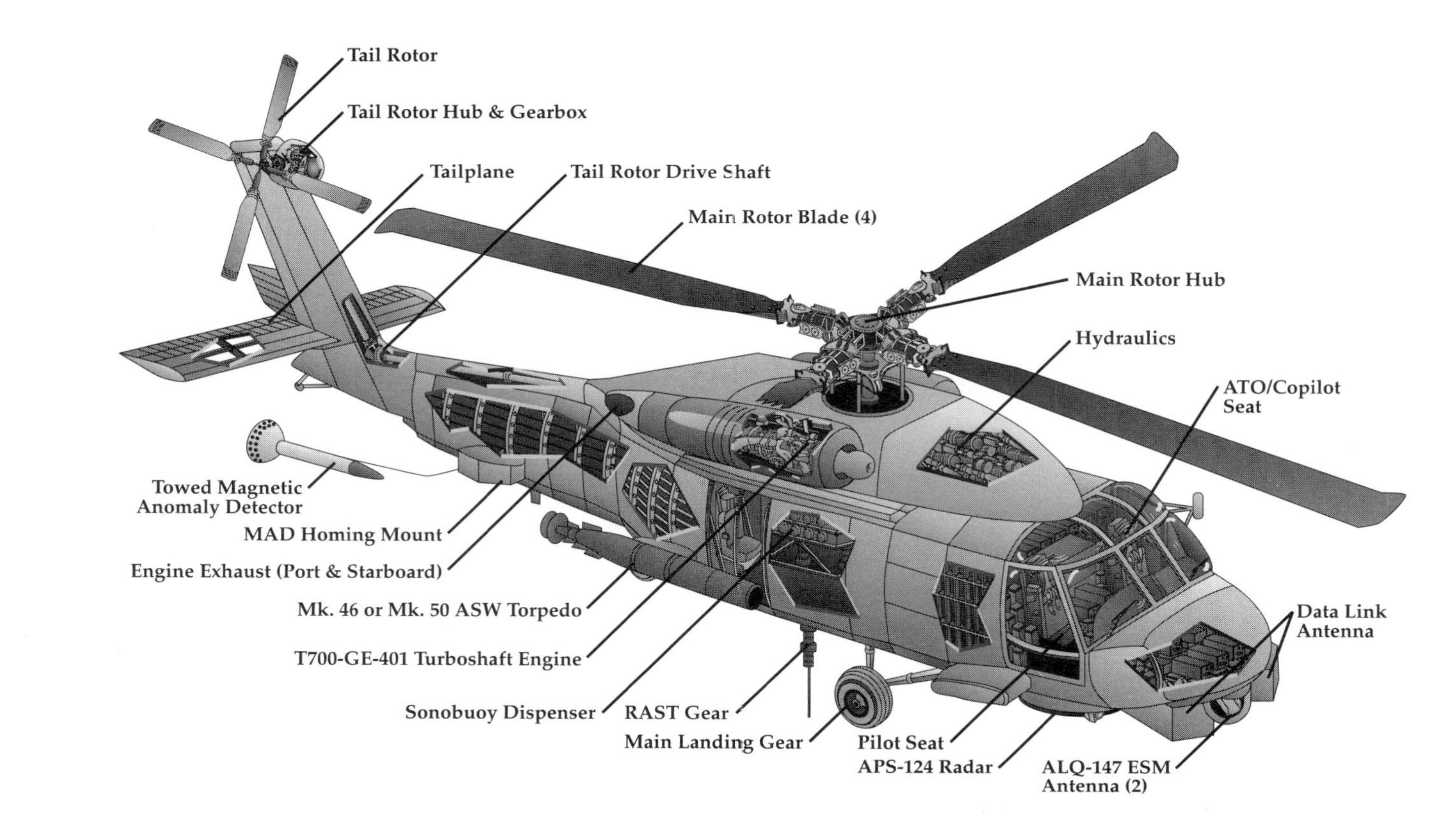

A cutaway view of a Sikorsky SH-60B Seahawk LAMPS Mk. III Helicopter.

Jack Ryan Enterprises, Ltd., by Laura DeNinno

cades, SH-2's have operated off the Navy's smallest ships (such as the now-retired *Knox*-class (FF-1052) frigates), and are still being produced for foreign navies. While the SH-2 was a good start, it lacked the range and payload to hunt front-line Soviet submarines. The Navy wanted a LAMPS helicopter that could hunt the new generation of Soviet submarines coming into service, and began development in the early 1970s.

In 1977, the Navy awarded a contract to IBM Federal Systems and Sikorsky to build a new light ASW helicopter system called Light Airborne Multi-Purpose System—Mark III (LAMPS III).[63] The helicopter itself was called the SH-60B Seahawk. The SH-60B was developed from Sikorsky's UH-60 Blackhawk transport helicopter, which had recently won the Army's competition to replace the venerable UH-1 ''Huey.''[64] This saved a *lot* of development money for the Navy and gave them an airframe with excellent growth potential.

Equipped with sonobuoys, MAD, radar, and other detection gear, the SH-60B would be the helicopter equivalent of the S-3B for escort ships. The LAMPS III birds would be based aboard the new generation of *Ticonderoga*-class (CG-47) Aegis cruisers, *Spruance* (DD-963) and *Kidd*-class (DDG-993) destroyers, and *Oliver Hazard Perry*-class (FFG-7) frigates. These ships were being designed with enlarged helicopter hangars and landing platforms, and a combat center with two-way data links to process information from the SH-60's onboard sensors. When they first deployed in 1984, the LAMPS III-capable ships were the most powerful ASW escorts in the world. In a task force or convoy, they would form an ''outer zone'' barrier against any submarines trying to attack.

Meanwhile, it was time to replace the SH-3, the protectors of the ''inner zone'' of ASW defenses for the CVBG. Once the SH-60Bs had been well launched, it was a logical jump to build a Sea King replacement from the existing Seahawk airframe. In 1985 the USN contracted with Sikorsky for development and production of seventy-four ''CV-Helo'' versions of the H-60. They would be equipped with a new lightweight dipping sonar and some avionics improvements over the earlier-B-model Seahawks. These improvements came at a price, however: the loss of most of the LAMPS equipment, including the sonobuoy launchers and data links. The new SH-60F came into service in 1989, and began to replace the elderly SH-3's aboard the carriers. At this same time, in response to an ongoing initiative to expand the special warfare capabilities of the USN, another H-60 variant went into development. The HH-60H version of the Seahawk provided a whole new range of capabilities for battle groups commanders, including Combat Search and Rescue (CSAR) and the covert insertion and retrieval of Special Forces like the famous Sea-Air-Land (SEAL) teams.

---

63 Because of the scope of the original LAMPS Mk. III system with ship-mounted data links and processors, the Navy felt that the aircraft was only a secondary component. Therefore, IBM was selected to integrate the entire aircraft/ship system.

64 For more on the UH-60 Blackhawk, see my book *Armored Cav* (Berkley Books, 1994).

An HH-60G special operations/SAR helicopter landing on the deck of the USS *George Washington* (CVN-73).
JOHN D. GRESHAM

Having three aircraft all based upon the same H-60 airframe has saved lots of scarce naval aviation dollars. All share the same 1,690-horsepower General Electric T700 turboshaft engines, as well as a common rotor system (with a diameter of 53 feet, 8 inches/16.4 meters) and transmission. In fact, the primary differences between the -B, -F, and -H versions are in the various mission-equipment packages. With an overall length of 64 feet, 10 inches/ 19.75 meters, height of 17 feet/5.2 meters, and maximum gross weight of 21, 884 lb/9,908 kg, the Seahawk is a compact and nimble aircraft. It handles well on wet, rolling decks, even those of small escort ships. To assist ships' crews in handling, Seahawks have a cable system called RAST (Recovery, Assist, Secure, and Traversing), allowing ships' crews to haul it down safely in heavy seas. Developed from the Canadian ''Beartrap'' system, RAST has a tracked receiver on the helicopter platform, which ''captures'' a small cable hanging from the bottom of the helicopter. Once the receiver has snagged the cable, the helicopter is hauled down, and then towed into the ship's hangar.

The armament of the Seahawks, while limited, is well tailored for their assigned missions. The normal weapons load for the ASW versions is a pair of Mk. 46 or Mk. 50 lightweight torpedoes. Extra fuel tanks can also be carried to extend the Seahawk's range. The -B model is also equipped to fire the Norwegian-built AGM-119 Penguin Mk. 2 Mod. 7 ASM. With a range of up to 18 nm/33 km and a passive infrared seeker, it can take out a patrol boat or small escort ship, even in close proximity to a shoreline or neutral shipping traffic. All the variants of the Seahawk can be fitted with light machine guns, and have rescue hoists for hauling in downed air crews or other personnel.

The various models of Seahawk have helped maintain the sometimes-dicey peace in the post-Cold War world. In the Persian Gulf, for instance, LAMPS III birds have been monitoring maritime traffic and the maritime embargo of military materials into Iraq. At the same time, the -F models have kept a wary eye on the three Project 877/Kilo-class diesel boats of the Iranian Navy, and -H model Seahawks have been transporting inspection teams to

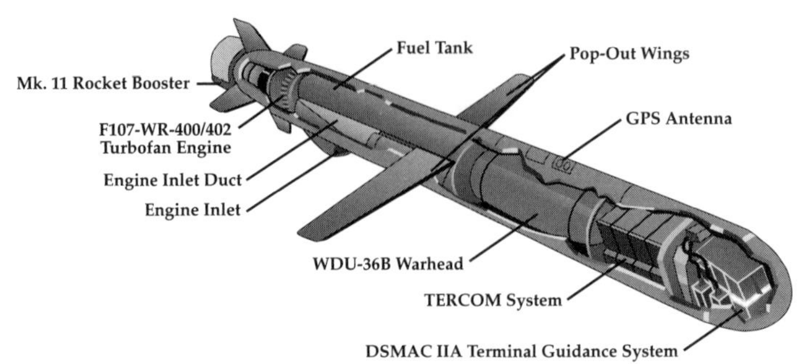

A cutaway view of a Raytheon BGM-109 Tomahawk Land Attack Missile.
Jack Ryan Enterprises, Ltd., by Laura DeNinno

ships and conducting CSAR missions. Seahawks have been active in supporting our operations in Bosnia as well. In fact, you probably could not even operate a modern USN task force without Seahawks. This is emphasized by the continuing popularity of the H-60 to export customers around the world. So far, Spain, Japan, Australia, and Taiwan have all bought their own versions of the Seahawk to operate off various classes of escort.

The future of the Seahawk community is looking decidedly upbeat these days, mostly due to the modernization plan that has recently been announced. Shortly, the two hundred or so surviving -B, -F, and -H-model Seahawks will be sent back to the Sikorsky factory in Stratford, Connecticut, to be remanufactured to a common SH-60R standard. All Seahawks will now carry the LAMPS III and -F sensor packages (both sonobuoys and dipping sonar), as well as improved engines and avionics. This upgrade should make it possible for the -R Seahawks to last into the 21st century until the next generation of sea control helicopter is designed.

Unfortunately, the use of the HH-60H airframes to produce -R-model birds will create a shortfall for the CSAR/special operations force. At the same time, the elderly fleet of UH-46 Sea Knight Vertical Replenishment (VERTREP) helicopters is about to fall out of the sky from wear and tear. Recognizing this, the Navy has ordered the development and production of an entirely new model of Seahawk, the CH-60, which will take over the CSAR/special operations duties previously assigned to the -H model, as well as the VERTREP mission of the Sea Knight. The first prototype is currently flying, and low-rate production has been approved for up to two hundred of the CH-60 variants. First deliveries to the fleet will begin in FY-1999.

Raytheon BGM-109 Tomahawk: The "Other" Strike Aircraft

Not all the aircraft that fly from the CVBG are manned. Another strike weapon available to battle group commanders for hitting targets ashore is the BGM-109 Tomahawk cruise missile. The Tomahawk is an all-weather submarine- or ship-launched land-attack cruise missile, with a variety of warheads. Stowed in vertical launch tubes or containers, it can be launched from long range, and can strike with pinpoint precision (less than three meters/ten feet from the aimpoint). In the U.S. Fleet, everyone calls it the TLAM (pronounced ''tea-lamb''), which is an acronym for Tomahawk Land Attack Missile, to distinguish it from the discontinued TASM, or Tomahawk Anti-Ship Missile. Conceived in the 1970's for a nuclear ''Doomsday'' scenario, TLAM has been reborn in the '90's as *the* big stick of U.S. policy.

TLAM looks rather like a cigar with stubby pop-out wings and tail fins. A solid-fuel booster rocket (which is attached to the rear of the missile and looks like an oversized coffee can) hurls the missile out of its launch canister/container. TLAM is 18 feet, 3 inches/5.6 meters long (20 feet, 6 inches/6.25 meters with the booster), 20.4 inches/51.8 cm in diameter (it fits inside a standard 21-in/533mm torpedo tube), has a deployed wingspan of 8 feet, 9 inches/2.7 meters, and weighs 2,650 lb/1,192.5 kg (3,200 lb/1,440 kg with the booster). It flies at a speed of approximately Mach .75/550 kn/880 kph, and has a range of 870 nm/1,000 mi/1,610 km for the basic land-attack version. The standard payload for a TLAM is a 1,000-lb/454-kg-class ''unitary'' warhead that has blast, fragmentary, and penetration effects. There are also versions with other types of warheads, including small submunitions for use on area targets like SAM sites and airfields. TLAMs are not as stealthy as F-117's or B-2's, but they are still almost undetectable by an enemy, thanks to the missile's small radar cross-section and low-altitude flight path.[65] And because the turbofan engine emits very little heat energy, infrared detection is no easier.

The current TLAM inventory has a complex family tree of variants and modifications, extending through three distinct generations or ''Blocks.'' These are distinguished mainly by the different guidance and warhead systems shown in the table below:

BGM-109 Tomahawk Variant Chart

Block	Designation	Guidance	Warhead	Notes
I	BGM-109A/TLAM-N	Inertial / TERCOM	W-80 Thermonuclear Warhead	No Longer in Service
I	RGM-109B/TASM	Radar Seeker / Inertial	1,000 lb/454 kg HE Unitary	No Longer in Service
IIA	BGM-109C/TLAM-C	TERCOM/ Inertial/DSMAC	1,000 lb/454 kg HE Unitary	In service
IIB	BGM-109D/TLAM-D	TERCOM/ Inertial/DSMAC	166 BLU-97/B CEMs	In service

65 On the nose of all Block II and III TLAMs is a small ''lip'' that helps reduce the radar signature of the missile by deflecting, rather than reflecting, incoming radar waves. Also, RAM is used at various places around the TLAM airframe to further reduce the missile's radar return.

The launch of a BGM-109 Tomahawk Land Attack Missile (TLAM) from the Aegis cruiser USS *Shiloh* (CG-67) during Operation Desert Strike in 1996. Cruise missiles like the Tomahawk are frequently integrated with manned airstrikes to help suppress enemy air defenses.

OFFICIAL U.S. NAVY PHOTO

Block	Designation	Guidance	Warhead	Notes
III	BGM-109C/TLAM-C	GPS/TERCOM / Inertial/DSMAC	1,000 lb/454 kg Equivalent WDU-36B HE Unitary with Reactive Titanium Case	In service—Produced from Existing Block I Missile Stocks. 100 with One-Way Missile Status Satellite Data Link.
III	BGM-109D/TLAM-D	GPS/TERCOM / Inertial/DSMAC	166 BLU-97/B CEMs	In service—Produced from Existing Block I Missile Stocks.

The nuclear-armed TLAM-N was taken out of service by a Presidential executive order shortly after the end of the Cold War in 1991. Similarly, the collapse of the Soviet Fleet at the end of that conflict meant that the long-range (greater than 300 nm/555 km) antishipping capabilities of the TASM were no longer required. Following their withdrawal from service, the TLAM-N and TASM airframes were remanufactured into new Block III missiles (the Navy often does this with so-called ''legacy'' systems). The Block III missiles have been recently given the new BDU-36B penetrating warhead, with a case composed of highly reactive titanium for penetrating a good thickness of reinforced concrete, as well as exceptional incendiary effects. In about a hundred of the Block IIIs, there is also a one-way satellite data link that at various times during the flight sends updates on the missile's status and position back to the firing units and command centers. The Block III's precision navigational systems use a combination of guidance modes to give them the same kind of accuracy (less than three meters/ten feet from the aim point) as an LGB.

When a Tomahawk is launched, the Mk. 111 rocket booster fires, thrusting it vertically into the air (after burnout, the booster is discarded). The wings and guidance fins are then deployed and a cover plate is blown off the inlet duct of the tiny Williams International F107-WR-402 turbofan engine. The F107 burns a special high-energy, high-density liquid fuel called JP-8, which gives it more range per gallon than normal JP-5. As soon as the missile has stabilized, it begins to fly a preprogrammed route to its first navigational waypoint just prior to landfall. Once over land, the missile flies along its programmed flight path to the target. Most of the time, the flight path is monitored by an inertial guidance system, which senses the drift from winds and small flight errors. In order to compensate for any ''drift'' in the inertial system itself, the TLAM utilizes a system called Terrain Contour Mapping (TERCOM) to match the terrain below with data from pre-surveyed strips of land stored in the missile's computer. Should the flight path deviate from the planned course, it will be corrected, and the missile will continue to the next TERCOM strip.[66]

When the missile reaches the target area, the precision Digital Scene Matching Area Correlation (DSMAC) system takes control. This utilizes a downward-looking infrared camera with an infrared illumination system (for consistent lighting at night) that matches up features on the ground and makes any necessary corrections to the missile's flight path. Though the DSMAC system does not actually ''home'' onto the target, it does provide enough accuracy to fly a TLAM through the goalpost uprights on a football field. In order to improve the existing Inertial/TERCOM/DSMAC guidance package, a GPS receiver has been installed in the new Block III missiles. In the event of a rapidly planned strike, GPS eliminates the need for TERCOM maps; and with GPS, the atomic clocks aboard the satellites provide a precision Time-of-Arrival (TOA) control capability. Using this, the missile's arrival at the target can be timed to the second. Once the TLAM is over the target area, the missile's job is to put the payload onto the desired target. It can fly or dive into the impact point (a bunker or building), explode over a ''soft'' target (such as an aircraft or radar), or spread a load of submunitions over a desired area.

While the existing stockpile of Block II and III TLAMs are capable of doing a fine job, there are plans to make them even better. Admiral Johnson would like to drive the cost of TLAM strike missiles down, and the way to manufacture them more cheaply is to re-engineer the design to take advantage of new structures, materials, and computer/software advances. This proposed TLAM variant is the so-called ''Tactical Tomahawk,'' which would probably cost around $575,000 a copy. Tactical Tomahawk would be equipped with a two-way satellite data link, which would allow it to be re-targeted in flight. The new TLAM will also be equipped with a camera system, allowing the

---

66 The development and production of the thousands of TERCOM terrain maps that were necessary to hit targets around the world is a mind-numbing job. So much so that it took the Defense Mapping Agency (now part of the National Imagery and Mapping Agency) six months to build the TERCOM maps needed to give TLAM planners just three routes (one each from the Persian Gulf, Red Sea, and Mediterranean Sea) for the missiles to fly into the Baghdad area.

One of the prototype/preproduction F/A-18 Super Hornets during a test flight. The Super Hornet will replace early-model F-14 Tomcats in the early 21st century.
BOEING MILITARY SYSTEMS

missiles to conduct their own damage assessments. Expect to see this new variant in the a few years.

Once upon a time, the TLAM filled naval aviators with anxieties. They feared that the Tomahawk had "This machine wants your job!" written on the side. But their fears have faded, and today most of them view the TLAM the way a hunter sees his favorite hunting dog—good and faithful beasts that are willing to go places where human beings should not go, and do things that human beings really should not do. Still, naval aviators like to joke that in the next war no more Navy Crosses will be handed out; the cruise missiles will have hit the really difficult targets! Every bomb carries a political message. Today, TLAM is probably America's most effective bomb-carrying political messenger. The "Gunboat Diplomacy" of the 19th century has become "Tomahawk Diplomacy" in the 20th and 21st.

The Future: Boeing F/A-18E/F Super Hornet

The shortcomings of the existing F/A-18 Hornet are well understood, and have long caused Naval aviators to wish for their resolution. Meanwhile, the 1993 retirement of the A-6E/KA-6D fleet and the failure to produce a replacement for it have meant that NAVAIR has been hard pressed to get *any* kind of new aircraft onto U.S. carrier decks. At one point the feeling seemed to be that since the Navy was unable to produce *new* aircraft, perhaps it might be able to field a highly modified one. Back in 1991, the Navy leadership decided to build an upgraded version of the Hornet, which would replace the F-14 and early versions of the F/A-18. This redesigned F/A-18 would (hopefully!) resolve the Hornet's fuel-fraction problem as well as other shortcomings and provide an interim aircraft until a more advanced and suitable long-term solution to the Navy's aircraft procurement need could be developed. Thus was born the F/A-18E/F Super Hornet, the key to the Navy's current naval aviation upgrade plan.

As planned, the F/A-18E (single seat) and -F (two-seat trainer) are more than just -C/D models with minor improvements. They are in fact brand-new airframes, with less than 30% commonality with the older Hornets. The airframe itself has been enlarged to accommodate the internal fuel load that was lacking in the earlier F/A-18's. With a fuel fraction of around .3 (as opposed to the .23 of the earlier Hornets), much of the range/endurance problems of the earlier birds should be resolved. The twin engines are new General Electric F414-GE-400's, which will each now deliver 22,000 lb/9,979 kg of thrust in afterburner. There is also a new wing, with enough room for an extra weapons pylon inboard of the wing fold line on each side, which should help resolve some of the complaints about the Hornet's weapons load. To ensure that the Super Hornet can land safely with a heavier fuel/weapons load than earlier F/A-18's, the airframe structure and landing gear have also been strengthened. Since most of the-E/F's weapons load is planned to be expensive PGMs, which must be brought back if not expended, this is essential.

The Super Hornet will also be the first USN aircraft to make use of radar and infrared signature-reduction technologies. Most of the work in this area can be seen in the modified engine inlets, which have been squared off to reduce their signature and coated with radar-absorbing material. This should greatly increase the survivability and penetration capabilities of the new bird.

Finally, the Super Hornet will be the first naval aircraft to carry a new generation of electronic-countermeasures gear including the ALE-50, a towed decoy system that is proving highly effective in tests against the newest threats in the arsenals of our potential enemies.

To back up the new airframe and engines, the avionics of the new Hornet will be among the best in the world. The radar will be the same APG-73 fitted to the late-production models of the F/A-18C/D. An even newer radar, based on the same fixed-phased-array technology as the APG-77 on the USAF's F-22A Raptor, is under development as well. To replace the sometimes troublesome Nighthawk pod, Hughes has recently been selected to develop a third-generation FLIR/targeting system for the Super Hornet, which will give it the best targeting resolution of any strike aircraft in the world.

The cockpit, designed again by the incomparable Eugene Adam and his team, will have a mix of "glass" MFDs (in full color!), and an improved user interface for the pilot. One part of this will be a helmet-mounted sighting system for use with the new AIM-9X version of the Sidewinder AAM. Other weapons will include the current array of iron ordnance and PGMs, as well as the new GBU-29/30/31/32 JDAMS, AGM-154 JSOW, and AGM-84E SLAM-ER cruise missile.

There will also be provisions for the Super Hornet to carry larger external drop tanks as well as the same "buddy" refueling store used by the S-3/ES-3 to tank other aircraft.

All this capability comes at a cost, though. At a maximum gross weight of some 66,000 lb/29,937 kg, the Super Hornet will weigh more than any other aircraft on a flight deck, including the F-14 Tomcat.

When McDonnell Douglas (now part of Boeing Military Aircraft) was given the contract to develop the Super Hornet, they set out to have a high level of commonality with the existing F/A-18 fleet. Early on in the design process, though, it became apparent that only a small percentage of the parts and systems could be carried over to the new bird. Despite this lack of *true* commonality, the Super Hornet was the only new tactical aircraft in the Navy pipeline, and so the Navy went forward with its development.

Today, the aircraft is well into its test program, with low-rate production approved by Congress.[67] At around $58 million a copy (when full production is reached), the Super Hornet will hardly be a bargain (-C/-D-model Hornets cost about half that). On the other hand, when stacked next to the estimated $158-million-dollar-per-unit cost of the USAF's new F-22A Raptor stealth fighter, the Super Hornet looks like quite a deal! Considering the current budget problems within the Department of Defense, there is a real possibility that one program or the other might be canceled. Since the Super Hornet is already in production (the F-22A has just begun flight tests), it may have an edge in the funding battles ahead.

If the Super Hornet survives the budget wars, current plans have the Navy buying at least five hundred of them in the next decade. This means they will begin to replace early model F-14As when the first fleet squadron stands up and goes to sea in 2001. Meanwhile, there is advanced work on several Super Hornet derivatives, including a two-seat all-weather strike version (that would restore the lost capabilities of the A-6 Intruder) and an electronic combat version of the F/A-18F (the so-called "Electric Hornet") that would replace the EA-6B Prowler.

The Future: Joint Strike Fighter (JSF)

Airmen and other warfighters often get testy when they hear somebody trying to sell them a "joint" project. All too often, "joint" has meant, "Let's pretend to cooperate, so the damned bean-counters and politicians won't slash our pet projects again." One of the longest-running of these joint dreams has looked to find a common airframe that *all* the services could use to satisfy their tactical fighter and strike requirements. The newest incarnation of this dream is called the Joint Strike Fighter (JSF). The lure of potential multi-billion-dollar savings from such a program is the basis for the JSF program, which is an attempt to reverse the historic trend of escalating unit cost for combat aircraft. Taxpayer "sticker shock" at the price of aircraft like the F-22 Raptor and F/A-18E/F Super Hornet is threatening to unleash a political backlash against the entire military aerospace complex. Thus the JSF program

67 Current Navy plans have some thirty F/A-18E/F Super Hornets being built in Fiscal Year 1999, which will provide enough of the new aircraft to constitute the first advanced training and fleet squadrons.

An artist's concept of the Lockheed Martin Navy variant of the proposed Joint Strike Fighter (JSF).

Lockheed Martin, Fort Worth

is aiming for a flyaway cost in the $30-to-$40-million range, for the first time emphasizing affordability rather than maximum performance.

The contracting battle for JSF will pit Lockheed Martin against Boeing (newly merged with McDonnell Douglas), with the winner possibly becoming the builder of the last manned tactical aircraft of all time. With a planned buy of some two thousand aircraft, it certainly will be the most expensive combat aircraft program in history. Meanwhile, for this program to succeed, it will have to satisfy four demanding customers—the USAF, the USN, the USMC, and the British Royal Navy. To satisfy these customers, the JSF Program Office envisions a family of three closely related but not totally identical airframes.

The USAF sees JSF as a conventional, multi-role strike fighter to replace the F-16. With many foreign air forces planning to retire their F-16 fleets around 2020, there is a huge potential export market for such an aircraft. In addition, the Marine Corps needs some six hundred STOVL (Short Takeoff/Vertical Landing) aircraft to replace both the F/A-18C/D Hornet and the AV-8B Harrier. The similar Royal Navy requirement is for just sixty STOVL aircraft to replace the FRS.2 Sea Harriers embarked on their small *Invincible*-class (R 05) aircraft carriers. In December of 1995, the United Kingdom signed a memorandum of understanding as a collaborative partner in developing the aircraft with the United States, and is contributing $200 million toward the program. The Royal Navy plans to replace the aging V/STOL Sea Harrier with a short-takeoff-and-vertical-landing version of the JSF.

The U.S. Navy's requirement is for three hundred ''highly survivable'' (meaning "stealthy"), carrier-based strike fighters to replace early-model F/A-18's and the last of the F-14 Tomcats. Its version of the aircraft will have a number of differences with the other variants. For instance, the landing gear will have a longer stroke and higher load capacity than the USAF and USMC versions. To help during low-speed approaches, the Navy version will have a larger wing and larger tail control surfaces than the other JSF variants. The larger wing also means increased range and payload capability for the Navy variant, with almost twice the range of an F-18C on internal fuel.

As you would expect, the internal structure of the Navy variant will be strengthened in order to handle the loads associated with catapult launches and arrested landings. There will be a carrier-suitable tailhook, though this may not have to be as strong as on previous naval aircraft, because the JSF will be powered by the same Pratt & Whitney F119-PW-100 turbofan planned for use on the USAF F-22A Raptor. This engine has a "2-Dimensional" nozzle (it will rotate in the vertical plane), which will allow it to have much lower landing approach speeds than current carrier aircraft, and may allow the next generation of carriers (CVX) to do away with catapults altogether.

The Navy's need for survivability means that the JSF design will have a level of stealth technology comparable with the F-22 or B-2 stealth designs, which are the current gold standard in that area. All ordnance will be internally carried, and plans are for it to carry two 2,000-lb/909.1-kg-class weapons in addition to an internal gun and AAMs

Boeing and Lockheed Martin are scheduled to conduct a fly-off of their competing JSF designs in the year 2000, with a contract award the following year. The Boeing model is known as the X-32, while the Lockheed Martin design has been designated X-35. The winning entry should become operational sometime around 2010, at which time it will begin to replace the remaining F/A-18C/D aircraft in service. This is a make-or-break program for all the armed services of the United States. If it works, then the U.S. and our allies will have the pre-eminent strike fighter of the 21st century at their command.

The Future: Common Support Aircraft

While fighters and strike aircraft are important, the various support aircraft like the S-3 Viking and E-2 Hawkeye play equally vital roles in a CVW. And like fighters, they will someday have to be replaced. While this is not going to happen soon, planning for what will be known as the Common Support Aircraft (CSA) is already underway. This aircraft will take over the AEW, COD, ESM/SIGINT, and perhaps even tanker roles currently handled by no less than three different airframes. As always, funding is a problem. Right now, there is very little money available for the development of a new medium-lift airframe that could be made carrier-capable. In current-year dollars, it would probably cost something like $3 billion just to design and develop the airframe. And the price of the various mission equipment packages for each role is anybody's guess.

An artist's concept of an AGM-84 SLAM-ER cruise missile. The SLAM-ER is headed into production, and will be the long-range strike weapon for naval aviation into the 21st century.

Boeing Missile Systems

One likely way around this dilemma might involve adapting for the Navy the new V-22 Osprey tilt-rotor transport currently entering production for the USMC and USAF. A V-22-based CSA could eliminate much of the airframe development costs and allow the design of state-of-the-art mission-equipment packages. It might even replace the SH-60Rs and CH-60's when they begin to wear out.

The Future: Bombs and Missiles

With the introduction of GPS-guided air-to-ground ordnance and improved versions of a number of older PGM systems, the era of Navy aircraft dropping and firing unguided ordnance is dead.[68] In Operation Deliberate Force in Bosnia, for example, something like 70% of the weapons expended in that short but effective air campaign were PGMs. This percentage is likely to rise in future conflicts. What follows is a quick look at the programs that are important to naval aviators.

AGM-84E SLAM Expanded Response Missile

As mentioned earlier, the engineers at Boeing Missile Systems have been working on an improved version of the AGM-84E SLAM missile, which they call SLAM Expanded Response (SLAM-ER). SLAM-ER is designed to add a new generation of technology to the solid foundation laid by Harpoon and SLAM. This new missile will give the Navy a standoff strike weapon with unprecedented lethal power and accuracy. Improvements to the basic SLAM include a pair of "pop-out" wings (similar to those on the TLAM), which will give it more range (out to 150 nm/278 km) and better maneuverability. A new warhead utilizes the same kind of reactive titanium casing used on the Block III TLAM, while its nose has been modified with a new seeker window to give the seeker a better field-of-view. The guidance system of SLAM-ER incorporates a new software technology developed by

68 For a general primer on airborne ordnance, see *Fighter Wing* (Berkley Books, 1995).

A testing version of the Joint Direct Attack Munition (JDAM) guided bomb. JDAM utilizes GPS technology to guide it within just a few yards/meters of the aimpoint.
Boeing Missile Systems

Boeing and the labs at Naval Weapons Center at China Lake, California. Known as Automatic Target Acquisition (ATA, also known as Direct Attack Munition Affordable Seeker—DAMASK), it allows the SLAM-ER seeker to automatically pick out a target from the background clutter. The seeker then ''locks'' it up and flies the missile to a precise hit (within three meters/ten feet of the planned aimpoint). The SLAM-ER is already in low-rate production and has passed all of its tests with flying colors. In fact, this program has become so successful that the Navy has deleted its funding for the planned Joint Air-to-Surface Standoff Missile (JASSAM), since SLAM-ER completely meets the requirements for that. Current plans have SLAM-ER entering the fleet in 1999.

GBU-29/30/31/32 Joint Direct Attack Munition (JDAM) Guided Bomb Family

One key limitation of the current generation of LGBs and Imaging Infrared (IIR)-guided PGMs is that they do not perform well in poor weather. Water vapor and cloud cover are the enemies of these weapons and targeting systems, and have proven to be significant roadblocks to their employment. What airpower planners need is a family of true, all-weather PGMs. Creating this is the goal of the joint USAF/USN/USMC Joint Direct Attack Munition (JDAM) program, which will go into service in 1999.

Now being developed by Boeing Missile Systems (formerly McDonnell Douglas Missile Systems), JDAM is designed to be a ''strap-on'' guidance kit, compatible with a variety of different bomb warheads. JDAM will be equipped with a GPS guidance system and control fins, which can fit around a conventional Mk. 83 (1,000-lb/454 kg), Mk. 84 (2,000-lb/909-kg), or BLU-109 (2,000 lb/909 kg) bomb. Since the JDAM will take its guidance from the constellation of GPS satellites in orbit around the earth, all you'll need to designate a target will be the sixteen-digit numeric code that represents the target's geographic location on the earth's surface.

As currently planned, there will be four separate versions of the Phase I JDAM family. They include:

An F/A-18C Hornet armed with four AGM-154A Joint Standoff Weapons (JSOWs) during a test flight. JSOW is one of a family of precision-strike weapons guided by the NAVISTAR GPS satellite navigation system.
Raytheon Strike Systems

Designation	**Warhead**	**Weight (lb/kg)**
GBU-29	Mk. 80	250 lb/113.4 kg
GBU-30	Mk. 82/BLU-111	500 lb/226.8 kg
GBU-31	Mk. 83/BLU-110	1,000 lb/453.6 kg
GBU-32	Mk. 84/BLU-109	2,250 lb/1,022.7 kg

The majority of the JDAM acquisition will be composed of kits for the GBU-31 and -32 versions. These are sized to fit around both Mk. 83/84 general-purpose bombs, as well as BLU-109/110 penetration warheads. So far, the program is proceeding well in tests, and has proved to be quite accurate. The specified thirteen-meter/forty-three foot-accuracy (six meters/twenty feet when the new Block IIR GPS satellites are put into service) is regularly being beaten in drop tests, and JDAM should come into service on schedule. At a price of only about $15,000 over the price of the bomb, JDAM is going to be quite a bargain. It needs to be, since current plans have the American military alone buying over 87,000 JDAM kits over the next decade or so. One intriguing question about JDAM is whether or not it will be fitted with an ATA-type seeker to enable it to hit really precise targets. While an ATA seeker would only add another $15,000 to the cost of each kit, the accuracy would narrow to less than three meters/ten feet—as good as the Paveway III LGBs in service today. I would expect that you would see an ATA-based seeker deployed on JDAM by 2003.

AGM-154 Joint Standoff Weapon (JSOW)

Well on its way into active service, the AGM-154 Joint Standoff Weapon (JSOW) is intended to be a munitions "truck" able to carry a variety of weapons and payloads.[69] Designed to glide to a target with guidance from an onboard GPS/INS system, it can deliver its payload with the same accuracy as a JDAM bomb. The initial AGM-154A version is armed with BLU-97 Combined Effect Munitions (CEMs), while the -B model will carry BLU-108 Sensor Fused Weapons (SFWs) for attacking armor and vehicles. There are also plans for a -C model for the Navy, which will have a 500-lb/226.8-kg Mk. 82/BLU-111 unitary warhead as well as a man-in-the-loop data-link system similar to that on SLAM. An ATA-type seeker may also be fitted. This weapon is now officially operational with the fleet, with six -A models forward-deployed on the USS *Nimitz* (CVN-68) prior to the 1997 Iraq crisis, where they almost got their combat introduction.

AIM-9X Sidewinder Air-to-Air Missile

For almost a decade, the fighter pilots of the United States have been flying with a short-range AAM that has been thoroughly outclassed by competing products from Russia, Israel, and France. Despite its past successes, the third-generation AIM-9L/M Sidewinder AAM has been passed by and is now thoroughly outclassed. Help is on the way however, in the form of a new fourth-generation Sidewinder, the AIM-9X. Built by Raytheon-Hughes Missile Systems, it will become operational in 1999. The changes in the AIM-9X start at the seeker head, which will be a "staring" IIR array, able to detect targets at ranges beyond those of the human eye. A new guidance and control section at the rear of the missile will make it the most maneuverable AAM in the world. Reduced drag will also extend its range and "no-escape" zone for enemy target aircraft. Finally, the entire AIM-9X system will be controlled by a new helmet-mounted sighting system, which will first see service in the Super Hornet (but it will also be fitted on the Tomcat and earlier-model Hornets). This new missile will be so maneuverable that an AIM-9X can be fired at enemy aircraft that are *alongside* the launching aircraft!

The *Real* Future: Unmanned Combat Aerial Vehicles

Even as the JSF designs are being finalized and the eventual winner selected, it is important to remember that Lockheed Martin and Boeing can't engineer out the nature of the humans that will fly it. Right now, combat aircraft require their air crews to endure dynamic forces that are nothing less than physical torture. At times these stresses can turn deadly. The rapid onset of G-forces in sharp turns literally drains the blood from pilots' heads, causing a sudden "G-Induced Loss-of-Consciousness," or G-LOC. This means that

69 One of the more interesting possibilities for JSOW is to use it as a resupply system for Special Forces units behind enemy lines. It could even be packed full of MRE ration packages, and used for humanitarian relief in "hot" combat zones.

A flight of Lockheed Martin Unmanned Combat Aerial Vehicle (UCAV) concept aircraft. Such remote-controlled aircraft will likely serve in the mid-21st century.
LOCKHEED MARTIN

there is a limit to the performance engineers can put into new aircraft—the physical limitations of the human pilots.

With this in mind, it is likely that the generation of combat aircraft *after* JSF will be unmanned. Today, in roles like photo-reconnaissance and wide-area surveillance, a great deal is already being done with Unmanned Aerial Vehicles (UAVs). Back in the 1970's there were even trials with armed drones, though the threat to pilot billets put short work to that idea. Even so, they make a lot of sense—if not today, then tomorrow. What will be known as Unmanned Combat Aerial Vehicles, or UCAVs for short, will probably start out as modified existing designs (such as leftover F-16's or F/A-18's) whose cockpits will be filled with sensors and data links back to the operators on the ground. In fact, a modified F/A-18C would make an excellent first-generation UCAV, since it already can conduct automatic carrier landings.

The aircraft would fly and operate conventionally, with the exception that when high-G maneuvers are needed, the 9-G limit in the flight-control software could be disabled and the UCAV flown to the actual structural limits of the design. Since we already have in service AAMs that make thirty-G turns, we could easily produce combat aircraft with performances that would make manned aircraft obsolete overnight. UCAVs would doubtless also be much cheaper than current designs, since so much of the money in a manned aircraft design goes into making it safe for the pilot and crew to operate. Keep an eye on this emerging technology. It will be exciting!

Carrier Battle Group: Putting It All Together

Aircraft Carrier Battle Groups (CVBGs) are the single most useful military force available in time of crisis or conflict. No other military unit, be it an airborne brigade or a wing of strategic bombers, gives the leadership of a nation the options and power that such a force commands. This is because the *real* value of CVBGs goes far beyond the simple existence of the unit and its availability for combat; CVBGs also provide *presence*. America's forward-deployed battle groups in the Middle East and the Western Pacific are the most visible symbol of the nation's global commitments. Because of these battle groups, our nation has a say in the affairs of nations and people who threaten *our* vital national interests. The commander of such a battle group bears an awesome responsibility.

Rear Admiral Jay Yakley was one of those commanders. He's gone from flying fighters in Vietnam to commanding his own aircraft carrier battle group (CVBG), based around the USS *Abraham Lincoln* (CVN-72). Back in the early days of August 1990, he was the one of the point men facing down the forces of Saddam Hussein following the invasion of Kuwait. As commander of Carrier Air Wing Fourteen (CVW-14) aboard the USS *Independence* (CV-61), he was in charge of the first organized combat air unit to reach the region following the invasion. In this capacity, together with roughly ten thousand other Americans of the *Independence* CVBG, he had the job of holding the line until other reinforcements could arrive.

He did not have long to wait. Within days, Allied units began to pour in and form the core of the coalition that eventually liberated Kuwait and defeated Saddam's forces. But for those first few days, Jay Yakley and his roughly ninety airplanes were the only credible aerial force that might have struck at Saddam's armored columns, had they chosen to continue their advance into the oil fields and ports of northern Saudi Arabia. Only Hussein himself knows whether or not the *Independence* group was the deterrent that kept Saddam from invading Saudi Arabia.

However, the ability to quickly move the *Independence* and her battle group from their forward-deployed position near Diego Garcia made it possible to demonstrate American resolve to the Iraqi dictator. *That* is the real point of aircraft carriers: *to be seen.* Once seen, they can cause an aggressor to show common sense and back off. But if the aggressor fails to show common sense, then the CVBG can act to make them back off with force.

It is not just the obvious power of the carriers—or more particularly, of the aircraft that fly off them—that is the source of the options a CVBG provides national leadership. In fact, to look at a CVBG without seeing beyond the carrier is to look at an iceberg without seeing what lies submerged. The *real* power of a CVBG is far more than what the flattop with its air wing can bring to bear. Each CVBG is a carefully balanced mix of ships, aircraft, personnel, and weapons, designed to provide the national command authorities with an optimum mix of firepower and capabilities. That the group can be forward-deployed means that it has a presence wherever it goes, and that American leaders have options when events take a sudden or unpleasant turn on the other side of the planet. The downside is cost. CVBGs are among the most expensive military units to build, operate, train, and maintain; a country can only buy so many. Nevertheless, in the years since the end of the Cold War, CVBGs have demonstrated how very useful they can be on a number of occasions. Operations like Southern Watch (Iraqi no-fly patrols, 1991 to present), Uphold Democracy (Haiti, 1994), and Deliberate Force (Bosnia-Herzegovina, 1995) are only a few of these.

Carrier Battle Group Development

Common sense dictates protecting the most valuable warships in your arsenal when they head into potentially hostile waters. And that—simply—is the reason why aircraft carriers are placed in battle groups. Aircraft carriers are useless unless they are *carrying* aircraft. But it takes more than *just* airplanes to insure the carrier's survival. More important, using the CVW's assets for carrier defense defeats the real strength of sea-based aviation. Unless carrier-based aircraft are flying attack missions or defending other fleet vessels (and aircraft are not in fact able to stay airborne long enough to fully accomplish that job), they are being wasted. In other words, sentinels with more staying power than aircraft must protect the carrier against threats—particularly submarines—that can leave it so much burned and twisted scrap metal on the ocean floor. Any ship, no matter how well built, even a huge ninety-thousand-plus-ton *Nimitz*-class (CVN-68) carrier, can be sunk by *conventional* weapons. Without some sort of escort, a carrier is just a very large opportunity for some enemy officer.

The original configuration that gave birth to CVBG development dates from the early experiments with carriers in the late 1920's. Because of their high speeds and medium-caliber gun armament, the large carriers that emerged from the 1922 Washington Naval Treaty tended to be assigned to the scouting or cruiser forces of navies. They initially were used as "eyes" for the lines of battleships that were then the real measure of seapower. But before long, carrier admirals found ways to operate independently, showing that they could survive without the backing of a line of battleships. By the outbreak of the Second World War, they *were* the battle forces.

In 1939, no nation had more than a half-dozen large-deck carriers, and most CVBGs had only a single flattop, with a handful of cruisers and destroyers as escorts. However, this practice began to change very rapidly with the outbreak of World War II. Early in the war, the British began to add fast battleships and battle cruisers to carrier groups, providing protection against enemy surface units. Then the Japanese grouped their six big-deck carriers into a single unit called the *Kido Butai* (Japanese for "Striking Force"). Its escort included a pair of fast battleships, some cruisers, and over a dozen destroyers—enough to stand up against all but the largest surface fleet. With multiple flight decks and hundreds of fighters and strike aircraft, *Kido Butai* could overwhelm any fleet or air force it encountered. Officially known as the "First Air Fleet," and commanded by Admiral Chichi Nagumo, it was *Kido Butai* that struck Pearl Harbor on December 7th, 1941. For the next six months, Nagumo and *Kido Butai* ranged across half the globe, the most powerful force in Naval history. Only the "miracle at Midway" stopped *Kido Butai*, and returned the initiative in CVBG evolution to the Americans.

By early 1943, the power of American industry began to make itself felt as a stream of new *Essex* (CV-9) and *Independence*-class (CVL-22) fast fleet carriers steamed across the Pacific. Before heading for action, they would stop at Pearl Harbor to conduct training and be integrated with fast, new battleships, cruisers, destroyers, and other support ships, and then formed into Task Groups. (Two or more Task Groups formed a Task Force.) Experience gained during raids on Japanese island outposts in 1943 showed that the optimum size for such groups was three or four carriers, a pair of fast battleships, four cruisers, and twelve to sixteen destroyers. More carriers than that tended to make the groups unwieldy. Task Groups were commanded by a senior naval aviator, who assigned strike missions, refueling assignments, independent raids, and other jobs.

By early 1944, Task Force 34/58 had developed into the most powerful Naval force in history. This force, based around four Task Groups and commanded by Admiral Marc Mitscher, won key battles—in the Philippine Sea, off Formosa, at Leyte Gulf, in the South China Sea, and around Okinawa—that eventually led to Allied victory in the Pacific. Task Force 34/58 never lost a battle, and throughout its two-year life span lost only a single flattop, the light carrier *Princeton* (CVL-23).

The end of World War II brought a number of changes to CVBGs. In fact, the massive force reductions following the war almost spelled their end. Results of the early atomic tests at Bikini showed the need to disperse carrier groups. Thus single-carrier CVBGs again became the norm. On the other hand, new technologies began to make these individual carriers much more effective and powerful. Angled flight decks, steam catapults, jet engines, air-to-air missiles (AAMs), and atomic weapons marked just a few of the new systems that Naval aviators saw arrive in the decade of Elvis and Ike. As new technologies arrived, CVBGs began to change their mixes of aircraft and ships. Piston-engined propeller aircraft were sent to the boneyard, and replaced by supersonic jets and high-performance turboprops. The battleships and big-gun cruisers were also retired, as new guided-missile destroyers and cruisers took over the job of escorting a new generation of flattops. Even without the destructive power of the nuclear weapons they carried, each carrier now had more firepower than an entire World War II Task Group.

At the start of the Vietnam War in the mid-1960's, America had more carriers than the rest of the world combined, allowing the USN to easily station three or four CVBGs in the South China Sea. Each group normally had one attack carrier, as well as a guided-missile destroyer or cruiser to provide surface-to-air missile (SAM) coverage. Known as Task Force 77, the flattops were on station near Vietnam from the torpedo boat attacks of the Gulf of Tonkin Incident in 1964 to the evacuation of Saigon a decade later. By then, the older World War II-era carriers were worn out and had to be retired. Yet, with the defense budget drained by the Vietnam War, one-for-one replacement of ships and aircraft was impossible. Instead, the Navy built a new generation of amphibious ships with flight decks for helicopters (the *Tarawa*-class (LHA-1)), and combined the attack and ASW missions into the air wings (CVWs) on the fifteen newer carriers commissioned since the end of World War II. By adding a squadron each of S-3 Vikings and SH-3 Sea King helicopters to the existing attack carrier wings, the so-called "CV Air Wing" was created in 1975. This remained the basic CVW structure for the rest of the Cold War.

While the Navy was reducing the number of carriers and beefing up their air groups, the new *Nimitz*-class (CVN-68) nuclear supercarriers began to arrive. A new generation of aircraft also began to appear on the decks of American flattops. In 1974, the F-14 Tomcat arrived in the fleet, along with new models of the A-6 Intruder and A-7 Corsair attack bombers, and improved models of the E-2 Hawkeye and EA-6B Prowler electronic aircraft.

By the late 1970's the driving force in CVBG development was no longer American plans or technology. That honor fell to Admiral of the Soviet Navy Sergei Gorshkov. In the generation following the high seas humiliation of his fleet during the Cuban Missile Crisis, Gorshkov had managed to create the largest navy in the world. Though much of the Soviet naval buildup was designed to support and protect its growing fleet of ballistic-missile submarines, a large share of its maritime budget was devoted to the destruction of American CVBGs.

Over a period of two decades, Gorshkov grew his fleet with a focus on the large air-to-surface (ASMs) and surface-to-surface missiles (SSMs) that I discussed in the fifth chapter. Supporting this construction effort was a program of tactical development, exemplified by Gorshkov's concept of "the Battle of the First Salvo." His plan was to win a naval war by crippling enemy CVBGs by means of an early series of missile strikes, some of them pre-emptive. By the late 1970's, the Soviet fleet of ASM-armed bombers, and SSM-armed surface ships and submarines, was thought by some to be ready to take on the USN for global maritime dominance.

None of these Soviet developments went unnoticed, and systems like the F-14A Tomcat, AIM-54 Phoenix AAM, and E-2C Hawkeye were the first responses. Then, with the arrival on the scene of President Ronald Reagan and Secretary of the Navy John Lehman in 1981, the men of America's CVBGs finally got the new ships and equipment that had been needed since the 1960's. After years of being undermanned, underpaid, and short on spares and ordnance, the U.S. Navy was ready to win its share of the Cold War's final victory. To meet the increasingly sophisticated Soviet threat, the Navy bought new Aegis SAM ships, and improved aircraft and weapons. However, the basic structure of the CVBG remained unchanged in the 1980's, and would stay that way until the end of the Cold War and the coming of Desert Storm in the early 1990's. What did change was the strategy by which carrier operations were to be conducted. In Secretary Lehman's vision (called "The Maritime Strategy"), in the event of war with the Soviets, massed groups of three or more CVBGs would advance into the Norwegian Sea or North Pacific to strike military bases on the Soviet mainland. In the event, the collapse of the Soviet Empire put an end to "The Maritime Strategy."

The post-Cold War American military drawdown scaled John Lehman's vision of a "600 Ship Navy" back to just over half that number. In addition, the structure of battle groups and air wings was radically altered. Older classes of ships were rapidly retired, along with the entire fleet of A-6 attack bombers and KA-6 tanker aircraft. The Cold War-era CVW of approximately ninety aircraft shrank to just over seventy. Because the Soviet threat of ASMs launched from bombers and SSMs fired by submarines and surface ships was no longer significant, the need for fleet air defense was greatly reduced and the CVW could become an almost purely offensive force. The "outer air battle" was therefore handed off from the squadrons of F-14's, F/A-18's, and E-2's to the Aegis radars and SM-2 Standard SAMs of the battle group's cruisers and destroyers.

Today, the Tomcats and Hornets have been assigned to carry a variety of air-to-ground ordnance, including precision guided munitions (PGMs) for delivery onto targets ashore. In the current era of "littoral warfare" (as defined in "From the Sea" and "Forward from the Sea"), this is to be the primary function of sea-based naval aviation units. Along with delivering air strikes, the battle groups of the 1990's have been given other powerful offensive capabilities. Now CVBGs have each been teamed with a three-or-four-ship amphibious ready group (ARG) embarking a battalion-sized "Marine Expeditionary Unit-Special Operations Capable" (MEU (SOC)).[70] This means that as the first century of naval aviation comes to a close, the

70 The USMC MEU (SOC) is a multi-purpose/capability unit based around a reinforced Marine battalion and medium-helicopter squadron. For more information on MEU (SOC)s and their operations, see my book *Marine: A Guided Tour of a Marine Expeditionary Unit* (Berkley Books, 1996).

CVBG/CVW team stands as an almost purely offensive targeting and striking force for supporting units and objectives ashore in the littoral zones.

Force Structure: How Many Carriers?

Though the power, flexibility, and mobility of CVBGs make them a critical asset for national leaders, and this is unlikely to change in the 21st century, those same leaders must justify the costs of building, training, operating, and maintaining such forces. The costs of CVBGs are immense. The price tag for the U.S. version probably runs close to $20 billion to build and equip, and another $1 billion a year to operate and maintain—a lot of money! With those mind-numbing numbers in mind, let me put a question to you: How many carriers do we need? The answer is complex.

For starters, there are very few nations in the world with the means to even own flattops. The Royal Navy is committed to maintaining two carriers, as is France. Spain and Italy also plan to build additional flattops to give them each two CVBGs. Russia, Brazil, Thailand, and India will struggle to maintain the single carrier groups they currently possess—largely for reasons of national prestige. And then, standing alone, the United States is currently committed to keeping a dozen carriers in commission—as many flattops as the rest of the world combined. In the 1980's, John Lehman's "600 Ship Fleet" included *fifteen* CVBGs, a total driven by the strategy of simultaneous strikes around the Soviet Union's vast periphery. Launching strikes from the Norwegian Sea, the Mediterranean, and the North Pacific required between six and eight groups ready to get under way at any time. Today, with no monolithic threat on the horizon, the need for a dozen carriers in commission seems less obvious. So is twelve CVBGs overkill? No, not really.

The number of carriers our nation requires is ultimately determined by its commitments in the post-Cold War world. In a world without superpower confrontation, our "enemies" become "rogue states," like North Korea and Iraq, while international terrorists, criminal cartels, and chaotic regional, ethnic, or tribal conflicts now are the key threats to day-to-day peace. In today's world order, America's major overseas commitments and interests lie mainly *outside* the Western Hemisphere. At the same time, our victory in the Cold War has burdened the U.S. with responsibility for peacekeeping and stability in areas that, frankly, most Americans would prefer to ignore. Consider the following list of global flashpoints:

- **North Korea**—On the verge of starvation and collapse, North Korea continues to threaten the South Koreans and other nations in the region. It has recently deployed the Tapo-Dong ballistic missile, and may have one or two nuclear warheads.
- **People's Republic of China (PRC)/Republic of China (ROC/Taiwan)**—Following their confrontation over democratic elections and ballistic-

missile tests/exercises in 1996 (in which two American CVBGs intervened), these two estranged countries continue to face off in a slow simmer.

- **India/Pakistan**—As both countries celebrate their golden anniversaries, they confront each other over disputed borders and ethnic and religious differences. An accelerating nuclear arms race raises threats of regional nuclear war, and the proximity of China only exacerbates the problem.
- **Persian Gulf**—UN-sponsored sanctions and "no-fly" operations against Iraq continue, while Iran increases the size and capability of its military forces, causing concern among other countries in the region. Iran and Iraq once again are disputing border areas in the northern end of the Persian Gulf, and firing into each other's territory.
- **Balkans**—The crisis in the Balkans has continued, despite attempts to implement the 1995 Dayton Peace Accords. Bosnia continues to be a hot spot, requiring continuous monitoring by NATO forces, while old ethnic hostilities are erupting in Kosovo and other areas.
- **Algeria**—A chronic Islamic insurrection faces a repressive military regime, as fanatic groups commit brutal massacres in villages near the country's large cities, killing hundreds of innocent civilians.
- **Central Africa**—Hutu and Tutsi factions wage genocidal war, spilling across national borders and defying international relief efforts.
- **West Africa**—Destitute nations continue to be wracked by coups and civil wars that have been endemic since the end of colonial rule in the 1960's, requiring frequent evacuations of foreign civilians.

Current U.S. national military strategy calls for a force structure sufficient to deal with two "major regional contingencies" (small wars or big crises) plus one "complex humanitarian emergency" (natural disaster, epidemic, famine, refugee migration, etc.). You might think that a dozen CVBGs would be enough to handle all that. Unfortunately, the unforgiving demands of complex machinery and the natural limits of human endurance set boundaries that make a dozen carrier groups just *barely* sufficient to maintain two or three carriers on distant deployment at any one time. Let me explain.

When you build a warship like an aircraft carrier, it is not available for deployment overseas all the time. Warships require regular maintenance and upgrades. Thus, in the forty-five-year planned life of an aircraft carrier, it will spend as much as a fifth of its time in docks and yards being repaired and maintained. For example, for every year the ship is in service, two or three months are spent on minor upgrades and maintenance to keep the ship going between "deep" overhauls (when the warship is brought into dry dock for major work). These major overhauls are done every five years or so, take from eight to twelve months to complete, and include everything from repainting the hull to upgrading the living quarters and combat systems. Additionally, nuclear-powered carriers are periodically out of service for a three-year refueling, an intricate surgery (with meticulous attention to radiation safety) that requires cutting great holes through decks and bulkheads and then welding everything back together. All this means that a warship is only available to sail about three years out of every five.

The crew, also, requires its own ''overhaul,'' for the multitude of combat skills embodied in the battle group's ten thousand sailors, marines, and aviators are ''perishable.'' If skills are not taught, practiced, and tested regularly, the combat potential of a ship or air group rapidly deteriorates, even when deployed into a forward area. So a battle group must be assembled and ''worked up'' for almost six months before each six-month deployment.

Finally, and no less important, today's sailors and marines *demand* and deserve a personal life. People are not robots; they need rest, family relationships, and opportunities to advance personally and professionally. Warship crews need some portion of their service careers at or near their home ports. This human factor is the first casualty when politicians deny pay increases and hardship bonuses, or extend emergency deployments to extreme lengths. Because relatively few of today's national leaders have the personal experience of long military deployments, the sea services have especially suffered. To remedy this problem, Admiral Johnson has instituted a six-month ''portal to portal'' deployment policy. That is, the Navy has promised that sailors will spend 50% of their time in home port.

Navy planners struggle constantly to build schedules that maximize the number of carrier groups available for deployment, while providing the best quality of life for embarked sailors and marines. Given a carrier force level of a dozen units, it works out something like this:

- **Deep Overhaul/Nuclear Refueling**—Two or three units at a time. Currently, there are three shipyards (Bremerton, Washington; Norfolk, Virginia; and Newport News Shipbuilding) capable of doing this intensive job, which essentially ''zeros the mileage'' on a ship.
- **Yearly/Periodic Maintenance**—Two or three additional carriers are usually conducting yearly/periodic maintenance, which is mainly done dockside at the ship's home port.
- **Deployment/Workup Cycle**—The remaining six or seven carriers are on an eighteen-month cycle, broken into the following phases:
  —Leave/Unit Training Period—The first six months of the cycle are devoted to resting crews coming off deployment, with leave and training time. Some individual ship or unit training is also conducted then.
  —Workup Period—The second six months of the cycle are designed to refresh unit combat skills, conduct combine training, and validate the unit's ability to conduct joint operations prior to deployment.

—**Deployment**—Designed to be six months long, this is the period where the combined battle group is packaged and forward-deployed for actual operations.

Assuming that this cycle is not interrupted by a major regional contingency, two or three CVBGs can be forward-deployed at any given time. There is always one from the East Coast, which can be assigned to the 2nd (Atlantic), 5th (Persian Gulf/Indian Ocean), or 6th (Mediterranean) Fleets. The West Coast usually has one or two groups available, which work with the 3rd (Eastern Pacific), 5th (Persian Gulf/Indian Ocean), or 7th (Western Pacific) Fleets.[71]

Yes, it takes a great deal of effort and investment to keep just two or three carrier groups forward-deployed at one time. Yet the lack of forward U.S. bases in areas critical to American national interests makes these mobile air bases critical to the national leadership. If America wants to have a voice in a crisis somewhere on the other side of the world, then we need either a friendly allied host nation[72] or a carrier battle group offshore. And CVBGs have one major advantage: They do not need anyone's permission to sail anywhere in recognized international waters.

The current scheme of carrier group rotations assumes a generous (by past standards) allotment of home-port time for ships and sailors, given the operations tempos (OpTempos) of today. In an emergency, though, the groups working up can be rapidly ''surged'' forward to reinforce groups already in the crisis zone. This is exactly what happened in 1990 and 1991 during Desert Shield and Desert Storm. By the time war broke out in early 1991, six CVBGs were in place for strikes against Iraq. Two other American CVBGs had operated in support of Operation Desert Shield and rotated home, while a British carrier group covered the Eastern Mediterranean to fill NATO commitments. In other words, even if forward-deployed carriers are tasked in a crisis, there is enough ''flex'' in the rotational schedule to allow units at home in the U.S. to ''backfill'' other American commitments.

USS *George Washington* (CVN-73)

Let's take a look at one of these groups ''up close and personal.'' Specifically, the CVBG based around the USS *George Washington* (CVN-73), one of the East Coast carrier groups assigned to 2nd Fleet at Norfolk. ''*GW,*'' as

71 There are more carrier groups in the Pacific because the U.S. still maintains one CVBG/CVW at Japanese bases. The remaining groups are based at the West Coast ports of San Diego, California; Alameda, California; Everett, Washington; and Bremerton, Washington.

72 Given the historic unreliability of the United States in foreign affairs and alliances, very few nations are willing to risk the political fallout it takes to invite in U.S. forces. For example, because of the political and cultural risks, Saudi Arabia denied America access to bases during the 1997/98 Iraqi crisis.

The official emblem of the aircraft carrier USS *George Washington* (CVN-73).
JACK RYAN ENTERPRISES, LTD., BY LAURA DENINNO

USS *George Washington* (CVN-73).
OFFICIAL U.S. NAVY PHOTO

her crew calls her, is an improved *Nimitz*-class (CVN-68) nuclear aircraft carrier. One of the second group of three built during John Lehman's glory years of the 1980's, she was laid down at Newport News Shipbuilding on August 26th, 1986; launched from the dry dock on July 21st, 1990; and commissioned on July 4th, 1992. Manned by over six thousand sailors and Marines, the *GW* has conducted three Mediterranean and Persian Gulf deployments since commissioning, a very heavy OpTempo. During her maiden voyage she was ceremonial flagship for the 50th Anniversary Celebration of the D-Day Landings in Normandy, and has conducted ''no-fly'' operations like Southern Watch (Iraq) and Deny Flight (Bosnia).

As a ''working class'' carrier, *GW* lacks some of the glamor and polish that fleet ''showboats'' enjoy (the carrier *John F. Kennedy*, CV-67, once had this reputation). This is a warship, not some floating palace to impress visiting dignitaries. You'll notice on *GW*'s bridge, for example, the row of ''E'' (Efficiency) awards painted there. These are fleet awards, which are given within each class of ships (aircraft carriers, guided-missile cruisers, etc.) to display the ship's visible accomplishments. Each award reflects a particular specialty, ranging from engineering and weapons to food service and tactical ability. In fact, just before leaving on her 1997 cruise, the *GW* crew got the word that they had been selected to wear the Battle "E" (marking them as the top warship for the entire Atlantic Fleet) for 1997, their third such award in just five years. From the bridge to the pump rooms, the men and women who serve aboard the *GW* know they are expected to be the best in the fleet. They make a pretty good case that they have achieved that goal.

What follows is a "snapshot" of the *GW* team in the late summer and fall of 1997, and should help you appreciate the kind of people who make a carrier battle group work. But be aware that Navy crews, like all military units, are in a state of constant transition. The sailors and aviators that appear here will certainly have changed assignments by the time you read this book. One other quick point. Because of the mixed Navy/Marine Corps personnel base aboard the battle group, it is easy to confuse the ranks of officers. To help straighten these out, refer to the following table for clarification:

MILITARY OFFICER RANK TABLE

Rank	Navy	Army/Air Force/USMC
O-1	Ensign	2nd Lieutenant
O-2	Lieutenant, Junior Grade (JG)	1st Lieutenant
O-3	Lieutenant	Captain
O-4	Lieutenant Commander	Major
O-5	Commander	Lieutenant Colonel
O-6	Captain	Colonel
O-7	Rear Admiral (Lower Half)	Brigadier General
O-8	Rear Admiral (Upper Half)	Major General
O-9	Vice Admiral	Lieutenant General
O-10[73]	Admiral	General

Heading the *GW* command team when we were aboard was Captain Lindell G. "Yank" Rutheford, USN. A graduate of the University of Missouri, "Yank" has spent much of his career as an F-14 Tomcat pilot (he also flew A-4 Skyhawks and F-4 Phantom IIs). He commanded a squadron, VF-142 (the "Ghostriders"), aboard USS *Eisenhower* (CVN-69) in 1988 and 1989. Following a staff tour abroad the *Theodore Roosevelt* (nicknamed

73 Currently, O-10 (Admiral and General) are the highest ranks allowed by Federal law. The O-11 rank is a rare honor, voted by Congress for special personnel and occasions. These are known as Admiral of the Fleet, and General of the Army/Air Force/Marine Corps. The last living recipient of this honor was General of the Army Omar Bradley.

Captain Lindell ''Yank'' Rutheford, the Commanding Office (CO) of the carrier USS *George Washington* (CVN-73).

John D. Gresham

''*TR*,'' CVN-71) during Desert Shield and Desert Storm, he decided to take the carrier command track (described in the third chapter).

Two years later, following nuclear power training and command school, he became the Executive Officer (XO) of *TR* for two deployments to the Mediterranean and Persian Gulf. Captain Rutheford then spent eighteen months as Commanding Officer (CO) of the replenishment ship USS *Seattle* (AOE-3), which qualified him for deep-draft command. While driving *Seattle*, he acquired a reputation around the fleet for superb ship-handling and organizational skills—very useful talents during the next step in his career. Following his relief as CO of the *Seattle* in November of 1996, he joined the *GW* as commanding officer. Along with his partner, the commander of the embarked Carrier Air Wing One (CVW-1), Captain John Stufflebeem (whom we will meet later), he provides the commander of the *GW* battle group with a powerful core of striking capability.

The Navy supports its carrier captains with handpicked subordinates who run the day-to-day activities of the boat and her three-thousand-plus-person crew (the air wing brings along more than 2,500 more). Of these, the most critical job on board is the Executive Officer, or XO. While we were aboard the *GW*, we were fortunate to observe a handover between two XOs, when Captain Michael R. Groothousen (the *GW*'s XO since May 1996) left to take command of *Seattle*, and the new XO, Commander Chuck Smith, arrived to take his place. Captain Groothousen, a longtime F/A-18 Hornet aviator, was on his way to a deep-draft command in preparation for commanding a carrier of his own, while Commander Smith is something else entirely, having served in S-3 Viking ASW/Sea Control squadrons.

A tall, lean professional (he resembles a young Peter O'Toole), Chuck Smith is the kind of aviator you'd want at the controls if your plane had to make a night landing in a storm with one engine out. Carrier captains usually come from fighter and attack aviation backgrounds. The ''right stuff'' mystique and old-boy network of TopGun fighter jockeys make it tough for aviators from electronic warfare, ASW, AEW, and sea control specialties to claw their way to the top of the promotion ladder. However, the increasing

Commander Chuck Smith, the Executive Officer (XO) of the carrier USS *George Washington* (CVN-73).

John D. Gresham

Master Chief Petty Officer Kevin Lavin, the Command Master Chief of the USS *George Washington* (CVN-73). Here he is just minutes away from boarding the *GW* for the 1997/98 cruise to the Persian Gulf.

John D. Gresham

importance and versatility of the S-3 in carrier operations, has enabled a few former Viking drivers to get choice commands: big-deck amphibious ships (like the *Tarawa* (LHA-1) and *Wasp*-class (LHD-1) helicopter carriers), and even some supercarriers. Chuck Smith will make a terrific carrier CO when he "fleets up" in a few years. Commander Smith took over the XO job in late August 1997, while *GW* was steaming into the battle group's final training exercise prior to deploying to the Mediterranean. The change happened quickly and seamlessly. The only sign of it aboard the ship was the few minutes it took for every officer who could fit into flight-deck control to see Captain Groothousen off the ship, en route to his next assignment.

Of the three thousand crewmembers aboard the *GW*, something like 95% are enlisted sailors. Their representative, advocate, and ambassador to the Captain is *GW's* Command Master Chief (CMC) Petty Officer, the senior NCO on board. This is a job of great responsibility. If the food or laundry service in the enlisted spaces is unsatisfactory, it is the CMC who makes sure the Captain knows about it. If a sailor's family member ashore needs assistance, he is the one to coordinate solutions through the Red Cross or other appropriate authority. On *GW*, the job is ably filled by CMC Kevin Lavin. When you meet him he seems more like the vice-president of a start-up computer company than the traditional gruff, tattooed Navy chief (his background is in electronics maintenance). Chief Lavin is Commander Smith's senior enlisted advisor, and when he speaks both the CO and XO listen closely!

Captain Rutherford and Commander Smith manage an organization that seems more like a small city or corporation than a ship. Its various departments are key to keeping the *GW* running smoothly for the six months or more that she may spend deployed, or "on cruise" as her crew calls it. Each department performs specific tasks, which make possible the operation of her men, aircraft, and weapons. The alphabetical breakdown of these departments and their heads in the fall of 1997 is shown below:

- **Administration (ADMIN)**—Headed by Lieutenant Jerry Morrison, this is the primary record-keeping group for the ship, and includes personnel, maintenance, supply, financial, and other files.
- **Aircraft Maintenance Division (AIMD)**—With almost six hundred personnel assigned, AIMD provides the *GW*'s embarked air wing with spare parts, maintenance facilities, and specialized support personnel. Commander Gordon Coward heads this division.
- **Air Department (AIR)**—One of the busiest groups aboard the *GW*, the Air Department controls the operations of the *GW*'s hangar and flight decks, as well as the airspace directly around the ship. The Air Department is led by an officer nicknamed the "Air Boss," and his deputy, the "Mini Boss." While we were aboard *GW*, the Air Boss was Commander John Kindred, while the "Mini" was Commander Carl June. Both are experienced pilots who have the skills and knowledge to control every type of carrier-capable aircraft under all weather and sea conditions. In late 1997, Commander Kindred was planning to move on to his next assignment, while Commander June would stay aboard and "fleet up" as the new *GW* Air Boss. Then it will be his job to train a new "Mini" before he moves on in a year or so.
- **Crew Recreation and Morale Department (CRMD)**—This department deals with the crew's spiritual and moral welfare, and is headed by *GW*'s Command Chaplain, Captain Jim Nichols.
- **Deck Division (DECK)**—Even in a "high-tech"age of networked computers and PGMs, there is still a need on every Naval vessel for sailors who can handle lines, small boats, anchors, and all the paraphernalia of traditional seamanship. The *GW*'s "Deck Division" handles everything from launching the ship's boats to manning the replenishment stations during underway refueling and replenishment (UNREP). Lieutenant Commander Johnnie Draughton, who will retire in late 1997, and will be replaced by Lieutenant Greg Worley, leads the sailors of the Deck Division.
- **Dental Department (DENTAL)**—A community of over six thousand people is bound to have some cavities, broken teeth, and dental emergen-

cies while on cruise. Sending them ashore for treatment to a Navy hospital would be impractical and expensive, so *GW* is equipped with a full dental clinic. Headed by Commander Roger Houk, DDS, the Dental Department has everything necessary for good dental hygiene, not only for the crew of the *GW*, but also for the crews on the other ships of the battle group.

- **Engineering Department (ENG)**—The Engineering Department operates and maintains almost every system aboard *GW* except the two A4W nuclear reactors. These systems include electrical power, air-conditioning, jet and diesel fuel, and sewage transfer systems. Supervising literally hundreds of miles of pipes, ducts, and cable runs, and thousands of valves, pumps, switches, transformers, and gadgets, is *GW* Chief Engineer (CHENG) Lieutenant Commander Pete Petry.
- **Combat Systems Division (CSD)**—Without sensors and electronics, modern weapons systems are about as useful in battle as paperweights. The Combat Systems Division cares for the vast array of controlling hardware and software that makes the *GW* an effective weapons and aviation platform. Heading the CSD is the *GW's* Combat Systems Officer (CSO), Commander Diana Turonis.
- **Legal Department (LEGAL)**—Six thousand sailors, Marines, and their families add up to a lot of legal advice. To support this, the *GW* has a fully staffed Legal Department to ensure that everyone aboard has an up-to-date will and power of attorney before deployment, and to handle any investigations and courts-martial that might arise. As it happens, today's military personnel don't get in trouble with the law nearly as often as previous generations. However, long deployments make for high divorce rates and complex family problems. The ship's Legal Officer (LEGALOFF) is Lieutenant Commander Jim Roth, a sharp and able young lawyer. The Legal Officer is also the Captain's technical advisor on Rules of Engagement (ROE), the intricate and ever-changing documents that specify where, when, and how you can shoot.
- **Medical Department (MEDICAL)**—The *GW* Medical Department is equipped and staffed to handle everything from minor lacerations and sunburns to life-threatening trauma and accident cases. Because smaller ships of the battle group have only modest medical facilities and staff, *GW* acts as a central hospital for the force. Heading the *GW* medical team is the Ship's Medical Officer (SMO), Commander Dean Bailey, MD. He is due for relief in the fall of 1997 by Commander Mike Krentz, MD.
- **Marine Detachment (MARDET)**—Traditionally, the twenty-six-man Marine Detachment aboard supercarriers provided security for the "special" (the euphemism for "nuclear") weapons that used to be carried aboard. Today, nuclear weapons are no longer carried aboard carriers, and in late 1997 the Corps reassigned its MARDETs. But until that happened, the *GW*'s MARDET commander was 1st Lieutenant Grant Goodrich.

- **Navigation Department (NAV)**—The traditional nautical skill of navigation has been revolutionized by GPS, digital charts, and real-time satellite weather updates. But it still takes an experienced navigator to advise the bridge watch about how *exactly* to steer the ship in a narrow channel or a tricky tactical situation. *GW*'s Navigation Department is equipped with every kind of navigational instrument, from sextants to GPS receivers. The Navigational Officer (known as "GATOR" for short) is Commander Ron Raymer. He is expecting to leave the ship in early 1998, and will be relieved by Commander Brian Cosgrove.
- **Operations Department (OPS)**—Everything from eating schedules and flight operations to making a rendezvous with a replenishment ship requires a high degree of skill and coordination. This is the job of the *GW* Operations Department, the group that recommends to the CO and XO how they will actually operate and "fight" the *GW*, should the need arise. The *GW* Operations Officer (OPSO) is Commander Don Hepfer, who is a sly and skillful officer.
- **Reactor Department (REACTOR)**—Of all the departments aboard the *GW*, none is shrouded in tighter security than the "Nukes." On non-nuclear Navy vessels the Engineering Department controls the ship's propulsion. But on the *GW*, a dedicated Reactor Department controls the two mighty AW4 units and other associated machinery. They don't like publicity. They won't talk to you. Don't even ask. The department is controlled by a career nuclear surface officer, Captain Joe Krenzel, who will become the commander of the USS *South Carolina* (CGN-37) when he finishes his tour as Reactor Officer (RO) aboard *GW*.
- **Safety Department (SAFETY)**—*GW* is basically a large steel box filled with jet fuel, explosives and rocket fuel, toxic chemicals and waste, fissile material—and, of course, people! Working hard to keep them under control is the *GW*'s Safety Department. This group is charged with monitoring hazardous materials, inspecting firefighting equipment and sanitation gear, and coordinating damage control with the other ship's departments. The *GW*'s Safety Officer (SAFETYO) is Commander Jack Hassinger, who will be relieved in the fall of 1997 by Commander Dave Hegland.
- **Supply Department (SUPPLY)**—The *GW* utilizes thousands of different items during day-to-day operations at sea. Jet fuel and floppy disks, ground beef and paper towels are all used in quantities that stagger the imagination. Keeping up with the ordering, stowing, and record-keeping required to keep *GW* running is the Supply Department. The *GW*'s Supply Officer or SUPPO is Commander Jim Ellison, who can be either the most or least popular officer aboard!
- **Training Department (TRAINING)**—One of the biggest challenges for the *GW*'s crew is to continue their professional growth and training while embarked on cruise. Doing this requires regular refresher training and

The official emblem of Carrier Air Wing One (CVW-1)

Jack Ryan Enterprises, Ltd., by Laura DeNinno

qualification for various skills and equipment, which is the responsibility of the Training Department, commanded by Lieutenant Matt Hempel. He is due to be relieved in late 1997 by Lieutenant Ann Hollenbeck. This department supports correspondence courses to help personnel qualify for their next promotion, distance-learning classes, and video classes, for new tactics and onboard equipment.

- **Weapons Department (WEPS)**—Though *GW* is not as heavily armed as her battle group escorts, she still packs a considerable self-defense "punch"—including Mk. 29 Sea Sparrow SAM launchers and 20mm Mk. 15 Phalanx CIWS systems. *GW* additionally has a number of M-2 .50-caliber machine guns for defense against swimmers and small boats. These are maintained by the Weapons Department, which is headed by the "Gunner," Commander Lee Price. Actual weapons firing is controlled by the Tactical Officer (TACO) in the ship's Combat Direction Center (CDC).

Carrier Air Wing One (CVW-1): THE SHARP END

That the *GW*'s embarked air wing, CVW-1, is a powerful offensive tool is a surprisingly recent development. During the Cold War, the focus of American carriers and their air wings was not the projection of offensive power but the defense of the carrier groups and other naval forces (convoys, amphibious groups, etc.). In those days, air wing training and weapons were mainly oriented toward war-at-sea missions against the Navy of the former Soviet Union, not toward land targets requiring precision deliveries. This is one of the reasons why the performance of Navy aircraft and units during the 1991 Persian Gulf War was so disappointing. Though Navy aircraft flew almost a third of the attack sorties into Iraq, they lacked the PGMs and sensors necessary to kill precision targets. On the positive side, Navy reconnaissance and electronic-warfare aircraft did really useful work, as did the A-6E Intruder all-weather attack bombers (recall that the weather over Iraq was terrible during much of the air campaign). Incredibly, in the name of cost-saving, the entire fleet of A-6E bombers and KA-6D tankers was retired after the Gulf War!

So, after building their plans and policies around a now-dead threat, and buying and retaining the "wrong" aircraft and weapons for the turn-of-the-millennium world, naval aviation entered the post-Cold War era in disarray. Happily, naval aviators are resourceful people, and during the mid-1990's Navy aviators gradually developed technical "quick fixes" and organizational reforms that will equip the Cold War CVW to tackle the challenges of the next decade. Understanding that it would take years and billions of dollars to develop and build new aircraft and weapons, they concentrated on upgrading existing airframes with new systems and weapons.

These focused on supporting the initiatives presented in high-level policy statements like "From the Sea" and "Forward from the Sea," while grimly defending the huge funding allocation necessary for the next-generation "Super Hornet" (F/A-18E/F). Some of the fixes—like acquiring the improved Nighthawk targeting pod for the F/A-18 Hornet and buying more laser-guided bomb kits—were just a matter of money. Others—like turning the F-14 Tomcat (traditionally an air defense interceptor) into a strike and interdiction aircraft—were a bit more difficult. Still, in just a few years things began to turn around. Operation Deliberate Force (the bombing of Bosnian Serb military facilities in 1995) proved that the Navy and Marines could deliver PGMs and suppress enemy air defenses just as well as their Air Force counterparts. Today, the embarked carrier air wings are just as deadly and efficient as any of their land-based counterparts.

CVW-1 is one of the ten air wings in active-duty service today, a survivor of numerous post-Cold War cuts and drawdowns. The wing spent much of its Cold War career aboard USS *America* (CV-66, which was recently retired), and moved over to the *GW* in 1996. A Navy captain (traditionally called the "CAG" for "commander, air group") commands the air wing; and he is a partner, *not* a subordinate, to the carrier's skipper. They both report to the admiral who commands the battle group (generally a two-star rear admiral), and work together as a team. It's an article of faith that flight operations are always controlled and authorized by trained naval aviators, not some distant senior commander who never sat in a cockpit.

As the *GW* battle group prepared to deploy in the late summer of 1997, the commander of CVW-1 was Captain John D. "Boomer" Stufflebeem, USN. Stufflebeem began his naval career as an enlisted sailor aboard a destroyer. He then went to the Naval Academy, graduating in 1975. After duty aboard a frigate, he learned to fly the F-14 Tomcat, rising to command squadron VF-84 (the "Jolly Rogers"). He then took command of CVW-1 in July 1996. While he has spent most of his career flying Tomcats, he generally flies F/A-18 Hornets today. But like most "CAGs," Captain Stufflebeem is qualified to fly the majority of the aircraft assigned to his CVW.[74] A quiet,

74 "CAG" is a term dating back to before World War II, when the air unit aboard a carrier was known as

Captain John D. "Boomer" Stufflebeem, the CO of Carrier Air Wing One (CVW-1). He is shown here in his Pentagon office following the 1997/98 cruise. There he works as an executive assistant to Admiral Jay Johnson, the CNO.
John D. Gresham

The official emblem of Fighter Squadron 102 (VF-102), the "Diamondbacks."
Jack Ryan Enterprises, Ltd., by Laura DeNinno

modest, focused warrior, he prefers to let his air crews and their results speak for him. Stufflebeem's XO is the Deputy CAG (DCAG), Captain Craig Cunninghame. Together they supervise the CVW staff, which acts as a "shell" for managing the various embarked squadrons, each of which may be flying a different aircraft type or model.

As squadrons are shuffled, merged, disbanded, or re-equipped, the makeup of an air wing may change, but the grouping of squadrons tends to be fairly stable over a period of years. The actual wing staff is quite small, just a few dozen officers and enlisted personnel. Because it is a "downsized" post-Cold War air wing, CVW-1 has given up one squadron of F-14's as well as the A-6/KA-6 Intruder squadron, and picked up a third F/A-18 Hornet strike fighter squadron in return. The CVW-1 squadron assignments looked like this in late 1997:

- **Fighter Squadron 102 (VF-102—The "Diamondbacks")**—One of the old F-14 squadrons that survived the drawdown a few years back,

an "Air Group" instead of an Air Wing, thus the term "Commander, Air Group." For most naval aviators, the honor of being addressed as "CAG" by the officers and enlisted men of their wing is considered to be one of the greatest achievements of their careers.

VF-102 is commanded by Commander Kurt Daill. Based at Naval Air Station (NAS) Oceana, Virginia, and flying F-14Bs (equipped with F-110 engines), the Diamondbacks began the 1997/98 cruise with several new capabilities. They were equipped with the new AAQ-14 LANTIRN/GPS targeting pod, the improved Digital TARPS pod, and the GBU-24 Paveway III laser-guided bomb with BLU-109 penetrator warhead. These new systems make the Tomcat a powerful day and night strike fighter, as well as a potent reconnaissance platform. VF-102 operates fourteen F-14Bs, of which four are wired for the new D/TARPS pod. The squadron is also assigned six of the LANTIRN/GPS targeting pods. Of course, the Diamondbacks are still highly capable fighters, deploying the AIM-9M Sidewinder, AIM-7M Sparrow, and AIM-54C Phoenix air-to-air missiles (AAMs).

CVW-1 has three squadrons of Navy and Marine F/A-18C Hornet strike fighters. Each squadron can conduct strike or fighter missions, dropping Paveway II/III LGBs and other PGMs, firing AGM-88 HARM missiles, as well as AIM-9M and AIM-120 AMRAAM AAMs. Normally, each squadron of twelve aircraft deploys with six Nighthawk FLIR/laser targeting pods and three data-link pods for AGM-84E SLAM missiles. However, there are minor differences in the three units, which I'll describe below:

The official emblem of Strike Fighter Squadron 82 (VFA-82), the "Marauders."
Jack Ryan Enterprises, Ltd., by Laura DeNinno

- **Strike Fighter Squadron 82 (VFA-82—the "Marauders")**—Based at NAS Cecil Field, Florida and led by Commander Steven Callahan, the Marauders fly the new Block 18 version of the F/A-18C, with sharper eyes in the form of the new APG-73 radar. This is the radar that will go into the nose of the F/A-18E/F Super Hornet when it comes into service. The APG-73 will finally give the Navy the ability to perform high-resolution ground mapping in any weather, as well as non-cooperative target recognition (NCTR) against enemy aircraft.

The official emblem of Strike Fighter Squadron 86 (VFA-86), the "Sidewinders."
JACK RYAN ENTERPRISES, LTD., BY LAURA DENINNO

- **Strike Fighter Squadron 86 (VFA-86—the "Sidewinders")**—Also based at NAS Cecil Field, Florida, VFA-86 has long been a sister squadron to VFA-82. VFA-86 flies the Block 10 version of the Hornet, and is commanded by Commander Robert Harrington.

The official emblem of Marine Fighter Attack Squadron 251 (VMFA-251), the "Thunderbolts."
JACK RYAN ENTERPRISES, LTD., BY LAURA DENINNO

- **Marine Strike Fighter Squadron 251 (VFMA-251—the "Thunderbolts")**—The Navy periodically requests the loan of Marine Corps carrier-capable strike-fighter and electronic-warfare squadrons to fill out air wings for deployment. Based out of MCAS Beaufort, South Carolina, the Thunderbolts fly the Block 16 version of the F/A-18C, and are commanded by Lieutenant Colonel Tony Valentino.

Along with strike and fighter "muscle," the air wing includes several support squadrons, which provide specialized services to the battle group.

The official emblem of Carrier Airborne Early Warning Squadron (VAW-123), the "Screwtops."

Jack Ryan Enterprises, Ltd., by Laura DeNinno

- **Airborne Early Warning Squadron 123 (VAW-123—The "Screwtops")**—Flying the E-2C Hawkeye, the Screwtops provide the *GW* battle group with airborne early warning (AEW). Based at NAS Norfolk, Virginia, VAW-123 is one of the Navy's oldest E-2 squadrons. The squadron's nickname is reflected in a bold blue and yellow spiral pattern painted on the radome of each aircraft. They are commanded by Lieutenant Commander Edward Rosenquist, and deploy with four late-model E-2C aircraft.

The official emblem of Sea Control Squadron (VS-32), the "Maulers."

Jack Ryan Enterprises, Ltd., by Laura DeNinno

- **Sea Control Squadron 32 (VS-32—The "Maulers")**—Of all the squadrons assigned to CVW-1, none has seen its role changed and enlarged more than the Maulers of VS-32. Flying eight S-3B Vikings, they provide the battle group with surface and ASW services. But their most valuable job is as aerial tankers for the rest of the wing's thirsty aircraft. Based at NAS Cecil Field, Florida, they are led by Lieutenant Commander John J. Labelle.

The official emblem of Tactical Electronic Warfare Squadron 137 (VAQ-137), the "Rooks."

Jack Ryan Enterprises, Ltd., by Laura DeNinno

- **Tactical Electronic Warfare Squadron 137 (VAQ-137—The "Rooks")**—VAQ-137 is tasked with suppressing and jamming enemy radars and communications. The Rooks fly a quartet of EA-6B Prowler aircraft. VAQ-137 is based out of NAS Whidbey Island, Washington, and is commanded by a "retreaded" A-6 Intruder crewman, Commander Craig Geron.

The official emblem of Helicopter Antisubmarine Squadron 11 (HS-11) the "Dragon Slayers."

Jack Ryan Enterprises, Ltd., by Laura DeNinno

- **Helicopter Antisubmarine Squadron 11 (HS-11—The "Dragon Slayers")**—The Dragon Slayers of HS-11 are another squadron that has seen its capabilities and responsibilities grow in recent years. In addition to the traditional role of providing the battle group with ASW protection in the "inner" defensive zone, they have taken on new roles of special operations support and search and rescue. They fly four SH-60F (ASW) and two HH-60H (SAR/Special Operations) variants of the Seahawk. HS-11 is based at NAS Jacksonville, Florida, and is led by Commander Michael Mulcahy.

The official emblem of Fleet Air Reconnaissance Squadron 6 (VQ-6), the ''Black Ravens.''

Jack Ryan Enterprises, Ltd., by Laura DeNinno

- **Fleet Air Reconnaissance Squadron 6, Detachment ''C'' (VQ-6, Det. C—The ''Black Ravens'')**—Flying one of the newest aircraft in Naval aviation, the Black Ravens Detachment ''C'' provides the battle group with electronic surveillance, intelligence, and targeting. Flying a trio of ES-3 Shadow aircraft, the detachment can, in a pinch, load up with a refueling pod and extra fuel tanks, to provide additional airborne tanking services. The small size of the ES-3 force means that squadron headquarters and schoolhouse remain home-based at NAS Cecil Field, Florida, while small detachments or ''Dets'' deploy with each CVW. In the fall of 1997, VQ-6 was commanded by Commander Robert ''Bob'' Wilson, and Det ''C'' aboard the *GW* is headed by their Officer-in-Charge (OIC), Lieutenant Commander Terry Isley.

The official emblem of Fleet Logistics Squadron 40 (VRC-40), the ''Rawhides.''

Jack Ryan Enterprises, Ltd., by Laura DeNinno

- **Fleet Logistics Support Squadron 40, Detachment 1 (VRC-40, Det. 1—The ''Rawhides'')**—Perhaps the least appreciated task in CVW-1 is the vital task of logistical support. The Rawhides fly a pair of C-2A Trader aircraft for CVW-1 out of their Detachment 1. VRC-40 is based out of NAS Norfolk, VA, and is commanded by Commander Paula Hinger. Lieutenant Commander Steven Faggart led Det. 1 while on cruise.

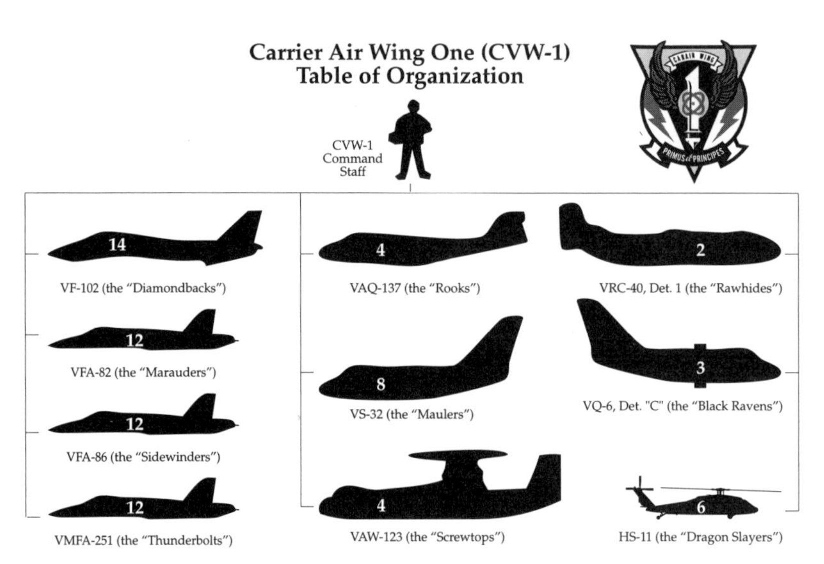

The combat aircraft makeup and organization of Carrier Air Wing One (CVW-1).
Jack Ryan Enterprises, Ltd., by Laura DeNinno

All of these resources give Captain Stufflebeem a compact and powerful air unit, capable of dishing out impressive amounts of firepower upon targets afloat and ashore. If CVW-1 has a vice, it is the lack of depth in its attached units. With just over seventy aircraft assigned, any loss will be felt in both lost resources and unit morale. Nevertheless, it is a well-structured and balanced force, which can hurt an enemy if properly handled.

Cruiser-Destroyer Group Two: Guarding and Guiding "The Boat"

Aircraft carriers and submarines may be the glamorous and expensive superstars of seapower, but the "surface warfare" sailors and their cruisers, destroyers, and frigates are an increasingly visible and vital part of the battle group. These "small" ships (if you can call a ten-thousand-ton cruiser "small") carry an increasing proportion of the Navy's usable "in-your-face" combat power. Systems like the Aegis combat system, SM-2 SAM, Tomahawk cruise missile, and SH-60 LAMPS III helicopter are common aboard surface combatants. In the drawn-down Navy of the 1990's, the Cold War frictions between surface, submarine, and aviation communities have rapidly broken down, making way for a unique kind of "joint" warfare.

To get a view of this synergy, you need to take a look at how today's CVBGs are organized and commanded. Following Desert Storm and the end of the Cold War, a new arrangement was set up to spread CVBG leadership opportunities more fairly. Previously, each CVBG was controlled by a Carrier Group (CARGRU) headed by a Naval aviator rear admiral. The escorts formed a Cruiser-Destroyer Group (CRUDESGRU), and the two groups to-

Rear Admiral Michael Mullen, the commander of the *George Washington* carrier battle group.
John D. Gresham

gether became the battle group. Now, only half the CVBGs are structured this way. The rest (usually three or four at a time) are headed by the CRUDESGRU commander (a rear admiral, lower half, traditionally addressed as "Commodore"), with the carrier and CVW subordinated. Other elements may be attached—the ARG, MEU (SOC), MCM (minesweeping) squadrons, SEAL teams, etc.—and this is the force that deploys for six months at a time.

The *GW* group includes the ships of Cruiser-Destroyer Group Two (CRUDESGRU Two), home-ported at Naval Station Norfolk, Virginia. Commanded by Rear Admiral Michael G. Mullen, USN, it is a well-balanced mix of older and newer ships. Admiral Mullen is a career surface warrior, in the tradition of great surface leaders like Admiral Elmo Zumwalt and Vice Admiral Joseph "Little Joe" Metcalf. He previously commanded the destroyer USS *Goldsborough* (DDG-20) and the Aegis cruiser USS *Yorktown* (CG-48), both top-of-the-line SAM ships. A graduate of the Annapolis Class of 1968 (his notable classmates include Admiral Jay Johnson, the current CNO), he went on to earn a Master's degree from Harvard University. Admiral Mullen represents the new generation of Naval leader, as well educated and savvy as any corporate executive. We'll get to know him better in the next chapter.

Admiral Mullen's "flagship" is a section of the O-3 level "blue tile country" of the *George Washington*, comfortable and nicely furnished, but quite noisy from aircraft operations one deck above. Here he makes his home and office afloat, along with the staff of CRUDESGRU Two. From there he commands the various ships of the force. Let's look at CRUDESGRU Two:[75]

- **USS *Normandy* (CG-64)**—An improved *Ticonderoga*-class (CG-47) Aegis guided-missile cruiser, *Normandy* is commanded by Captain James F. Deppe. *Normandy* has already accumulated an outstanding combat record. In fact, she has fired more BGM-109 Tomahawk cruise missiles in anger (thirteen during Operation Deliberate Force in Bosnia back in 1995) than any other Navy cruiser. The heart of *Normandy*'s combat power is the Aegis combat system, with four SPY-1 phased-array radars to track hun-

75 Much of this information comes from *Combat Fleets of the World* (A.D. Baker, III, U.S. Naval Institute Press), which is the finest such volume in the world. If you want to know more, look it up in *Combat Fleets*.

The Aegis guided-missile cruiser USS *Normandy* (CG-6o), one of the escorts of the *George Washington* battle group.
JOHN D. GRESHAM

dreds of targets at once. Aegis-equipped ships can engage dozens of airborne targets (aircraft and missiles) with SM-2 SAMs, while also controlling the weapons of other ships in the battle group. She carries up to 122 missiles in two Mk. 41 vertical launch systems (VLS), including SM-2 Standard SAMs, BGM-109 Tomahawks, and RUM-139A Vertical Launch Anti-Submarine Rockets. In addition to her twin 5-in/127mm guns and missiles, *Normandy* carries a pair of SH-60B Seahawk helicopters that can deliver ASW torpedoes and AGM-119 Penguin antiship missiles. During 1997 and 1998, *Normandy*'s Seahawks came from Helicopter Squadron—Light Forty-Eight, Detachment Eight (HSL-48, Det. 8), headed by Commander Brent Barrow. For the 1997/1998 cruise, the Detachment 8 OIC was Lieutenant Commander Steve Blaisdell, who commanded two SH-60B aircraft, as well as their flight and maintenance crews.

- **USS *South Carolina* (CGN-37)**—Commanded by Captain David K. Brown, *South Carolina* will be the last U.S. nuclear-powered cruiser in commission. Because of budget pressures, this ship and her sister, *California* (CGN-36), will retire in 1999. Until then, *South Carolina* gives the *GW* battle group excellent service. Thanks to her nuclear power plant, she is the only escort in the group that can stay with the carrier in a long-range, high-speed dash to a distant crisis. *South Carolina* also has the ''New Threat Update'' (NTU), an electronics package that establishes a data link with Aegis ships, and accepts firing commands from their computers. This is especially useful, since *South Carolina*'s missile directors (steerable dish antennas projecting an intense, narrow microwave beam to ''light up'' the target for a missile's seeker head) are considered more effective in coastal areas than those of Aegis ships.
- **USS *Carney* (DDG-64)**—One of the newer ships in the Navy, *Carney* is an *Arleigh Burke*-class (DDG-51) Aegis destroyer. Displacing only 8,300 tons (as compared to almost ten thousand tons for *Ticonderoga*-class cruisers), the *Burkes* are considered the finest surface combat vessels in the world on a ton-for-ton/capability basis. The first U.S. warships designed

with stealth technology to reduce their radar and infrared signatures, they pack the same Aegis combat system and weapons mix as the larger cruisers. They carry fewer weapons (ninety-six missiles in her VLS launchers and a single 5-in/54 mm gun mount), though, and no helicopter hangar is fitted. *Carney* is based at Naval Station Mayport, Florida, and commanded by Commander Mark H. Buzby.

- **USS *John Rogers* (DD-983)**—An updated *Spruance*-class destroyer, *John Rogers* is based at Naval Station Mayport, Florida. Her skipper is Commander James M. Carr, and the two embarked SH-60B LAMPS helicopters are drawn from HSL-46 (headed by Commander Tim Alexander) at Mayport. For the *GW* battle group's 1997/1998 cruise, the HSL-46 Det. 6 OIC was Lieutenant Commander Kenan Shaffer.

Along with the four cruisers and destroyers, CRUDESGRU Two includes a pair of *Oliver Hazard Perry*-class (FFG-7) guided-missile frigates. These smaller (3,660 tons) escort ships are particularly useful for inshore work common in littoral operations. Though limited in gun and SAM capabilities, FFG-7's have good sonars for shallow-water ASW, excellent helicopter facilities, and vast experience in maritime embargo and joint counter-drug operations. Like their namesakes from the age of fighting sail, frigates are fast ships that frequently go in harm's way.[76]

- **USS *Boone* (FFG-28)**—Based at Naval Station Mayport, Florida, *Boone* is commanded by Commander Arthur S. Mobley. Her SH-60B Seahawk helicopter came from HSL-42 (commanded by Commander Robert Presler) during her 1997/1998 cruise with the *GW* group. This unit, HSL-42 Det. 1, is headed by Lieutenant Commander Stuart Howard.
- **USS *Underwood* (FFG-36)**—*Underwood* is based at Naval Station Mayport, Florida, and commanded by Commander Daniel M. Smith. She also has a single SH-60B Seahawk detachment, this one from HSL-44 (headed by Commander Wayne Tunick). The HSL-44 Det. 7 OIC is Lieutenant Commander Jack Shepard.

The last two combat vessels of the *GW* battle group are a pair of improved *Los Angeles*-class (I688) nuclear-attack submarines, which give Admiral Mullen additional capabilities that we can only describe in general terms because of the tight security restrictions that surround and protect the "Silent Service." In addition to hunting down potential enemy submarines inshore, they can fire Tomahawk cruise missiles, track maritime targets,

---

76 Despite their intended "low mix" status in the Cold War structure of the U.S. Navy, the FFG-7's have frequently been in the thick of maritime operations and actions. Two of them, the *Stark* (FFG-31) and *Samuel B. Roberts* (FFG-58), were severely damaged by missiles and mines during operations in the Persian Gulf in the late 1980's, but survived to serve today. FFG-7's were later key assets in the maritime embargoes of Iraq, Bosnia, and Haiti, as well as in combat operations during Desert Storm.

gather electronic intelligence, and covertly deliver and retrieve special operations forces. The assigned SSNs are:

- **USS *Toledo* (SSN-769)**—Based at the submarine base in Groton, Connecticut, *Toledo* is commanded by Commander Bill Burke.
- **USS *Annapolis* (SSN-760)**—Also home-ported at Groton, *Annapolis* is led by Commander Daniel Parson.

Logistics may be the least "sexy" part of Naval operations, but supplies are always the first concern and constant worry of the professional warrior. You don't leave port without carefully planning how the fleet supply train will support your operations at sea over many months and thousands of miles. The mark of a real "blue-water" Navy is the ability to sustain operations at sea indefinitely. The U.S. Navy pioneered at-sea replenishment over six decades ago.

- **USS *Seattle* (AOE-3)**—The great advantage of nuclear-powered warships is that they do not require fuel oil for propulsion. With four nuclear ships (a carrier, cruiser, and both submarines), the *GW* battle group has no need for a flotilla of supporting oil tankers. Only one combat support ship is assigned to the *GW* battle group, but what a ship! With over 53,000 tons displacement fully loaded, USS *Seattle* (of the *Sacramento* class) carries bunker and aviation fuel, fresh food, spare parts, supplies, and ammunition. She can defend herself with a Mk. 29 Sea Sparrow SAM system, a pair of 20mm Mk. 15 Phalanx CIWS, and a full SLQ-32 ESM/ECM system. Captain Stephen Firks commands *Seattle.*

Partners: The *Guam* Amphibious Ready Group

When I wrote about the Marine Corps a few years ago, teaming CVBGs and ARGs/MEU (SOC)s into combined task forces was still a very new concept.[77] Though joining these two powerful and flexible units makes great sense, big technical, institutional, and cultural barriers had to be overcome in order to make it happen. Once the concept was implemented, however, it quickly proved its worth, in the Balkans, the Taiwan Straits, Africa, and the Persian Gulf. The CVBG/ARG/MEU (SOC) teams have held the line for American interests and kept a lid on the chaos of the 1990s.

For their 1997/1998 Mediterranean cruise, the *GW* battle group teamed with an ARG based around Amphibious Squadron Two (PHIBRON-2). These four ships are commanded by an "ARG Commodore," Captain Phillip Sowa, USN, from his flagship, the amphibious helicopter carrier (LPH) USS *Guam*

77 For more on the ARG, MEU (SOC), and their various components and missions, see my book *Marine: A Guided Tour of a Marine Expeditionary Unit* (Berkley Books, 1995).

The amphibious helicopter carrier USS *Guam* (LPH-9), flagship of the amphibious ready group attached to the *George Washington* battle group.
JOHN D. GRESHAM

(LPH-9). Though *Guam* lacks the larger flight deck, well deck, and hangar found on newer *Tarawa* (LHA-1) and *Wasp*-class (LHD-1) assault ships, the ARG has found ways to compensate for these shortcomings. To make up for the shortage of well deck space as well as vehicle and cargo capacity, the ARG has an extra Dock Landing Ship (LSD) to carry the full range of gear needed by an MEU (SOC). Since flight deck parking space on *Guam* is minimal, the AV-8B Harrier II detachment assigned to the MEU (SOC) was reduced to four aircraft. For close air support, the MEU (SOC) can rely on VMFA-251, the Marine Corps F/A-18 Hornet squadron embarked on *GW*. Marines prefer to have their *own* aviators flying cover over a hot beach or landing zone.

- **USS *Guam* (LPH-9)**—*Guam* is among the last of her kind—one of only two of the *Iwo Jima* class (LPH-2) still in service—and is scheduled to be decommissioned after the 1997/1998 cruise. Until then she will be Commodore Sowa's flagship and the headquarters for the embarked MEU (SOC). While she lacks the comforts, as well as some of the advanced communications and electronics, of her younger LHA/LHD cousins, *Guam* is an old warhorse, able to carry up to 1,500 Marines and twenty-four helicopters. Captain William J. Luti commands her.
- **USS *Shreveport* (LPD-12)**—Another "rusty but trusty" veteran of American amphibious operations. An *Austin*-class (LPD-4) Amphibious Transport Dock, *Shreveport* is the "inshore" element of the ARG, with much of the "special" warfare capability of the MEU (SOC) aboard. This includes an organic remotely piloted vehicle (RPV) unit of Pioneer reconnaissance drones, as well as the rubber boats, the force recon element, and a Navy Sea-Air-Land (SEAL) team. Led by Captain Denby Starling II, she can expect at least a decade of service ahead before her final trip to the scrap yard and replacement by a new *San Antonio*-class (LPD-17) amphibious ship.

- **USS *Ashland* (LSD-48)**—Designed to carry a mix of landing craft, vehicles, and cargo, *Ashland* is a *Whidby Island*-class (LSD-41) amphibious ship, capable of carrying up to four Landing Craft, Air Cushioned (LCAC), which are favored in today's amphibious operations. Captained by Commander Timothy R. Hanley, *Ashland* is a modern and very comfortable ship for its crew and embarked Marines.
- **USS *Oak Hill* (LSD-51)**—*Oak Hill* is one of four *Harpers Ferry*-class LSDs, cargo variants of the *Whidby Island* class. By shortening the well deck so that it only has a capacity for two LCACs, *Oak Hill* gains vastly increased stowage for vehicles and cargo. *Oak Hill* is an another state-of-the-art amphibious ship, with Commander Michael A. Durnan at the helm.

Amphibious ships like these are hardly sleek greyhounds of the sea. Sailors joke that the designator "LSD" stands for "large slow duck!" For this reason, Admiral Mullen has teamed several of his escorts for regular operations with the *Guam* ARG, depending upon the mission. For example, during inshore amphibious operations, *South Carolina* and *John Rogers* usually provide cover for the ARG. This includes ASW and gunfire support, as well as a SAM air defense "bubble." Thanks to her NTU system, *South Carolina* can tap into the sensor and automation systems of the Aegis ships, and take firing cues from them.

The ARG's combat power is the cargo it can deliver—Marines and their "stuff." For the *Guam* ARG, that cargo is the 24th MEU (SOC). The 24th was the unit that plucked Air Force Captain Scott O'Grady out of Bosnia back in 1995, after his F-16C Fighting Falcon was shot down by a Bosnian Serb SA-6 SAM. Today the unit is commanded by Colonel Richard Natonski, USMC, and has roughly the same structure as in 1995. The 24th's components include:

- **Battalion Landing Team (BLT) 3-6**—The core of the 24th MEU (SOC) is built around BLT 3-6, a reinforced Marine rifle battalion. Its 1,200 Marines have their own armor, artillery, and transport, providing a capability to launch small coastal raids or spearhead large invasions. During 1997 and 1998, the 24th MEU (SOC) has been commanded by Lieutenant Colonel Richard P. Mills.
- **Medium Marine Helicopter Squadron 263 (HMM-263)**—The air component of the 24th MEU (SOC) is a reinforced squadron of twelve CH-46E Sea Knight medium-transport helicopters. Attached are quartets of CH-53E Super Stallion heavy-lift, UH-1N Iroquois command-and-control, and AH-1W Super Cobra attack helicopters. Normally there is a detachment of six AV-8B Harrier II fighter-bombers assigned, but *Guam*'s limited deck parking space reduced this to just four. Lieutenant Colonel Michael Duva is the commander of HMM-263, and is equipped to operate as a Joint Forces Air Component Commander (JFACC—commanding Army, Navy and Air Force aviation assets on the scene) if necessary.

- **MEU Service Support Group 24 (MSSG-24)**—the 24th MEU (SOC)'s logistical tail is the 24th MSSG. Commanded by Lieutenant Colonel Brian L. Tonnacliff, this compact unit keeps the 24th supplied with everything from floppy disks and ammunition to jet fuel and water.

All of the above components make for one of the best-balanced, most compact fighting forces in the world. What it lacks, like its CVW-1 teammate, is depth—it is only a couple thousand sailors and Marines in a world where dictators command tens of thousands of soldiers. But one of the nice things about being a "gator" sailor or Marine is that if you go in harm's way, a lot more Marines and other American warriors can be on the way, soon to back you up.

"Cats and Dogs": Miscellaneous Attached Units

CVBGs, ARGs, and MEU (SOC)s are just three of the many units "owned," trained, and "packaged" by U.S. Atlantic Command (USACOM) in Norfolk, Virginia.[78] Before a CVBG or ARG/MEU (SOC) goes into combat, it will likely be reinforced with additional support units. While this may not be a complete list, it is representative of what has been regularly used in the last decade or so.

- **Land Based Air Support**—In addition to CVW-1 and HMM-263, the *GW* battle group frequently needs support from land-based aviation to sustain long-term operations in high-threat areas. This was seen clearly during Operations Desert Storm and Joint Endeavor. These aviation units can include:

  —Airborne Tanker Support—Every CAG dreams, hopes, and lusts for more airborne tanker support. Since the retirement of the KA-6D Intruder, the only tankers he actually "owns" are S-3 Viking and ES-3 Shadow aircraft, which can carry under-wing "buddy" refueling pods. Each of these can "give away" about 8,000 lb/3,627 kg of fuel for tanking. This is less than a third of what used to be dispensed by a single KA-6D, so any land-based tanker support is precious. One option is a Marine KC-130 Hercules tanker detachment "on-call" for the 24th MEU (SOC). In addition to Marine tanker support, other tanker assets can make the life of CVW-1 easier. Navy carrier aircraft have refueling probes that fit almost any Air Force, Navy, or NATO tanker aircraft available. Especially valuable are the big Air Force KC-10A Extenders, which can refuel aircraft with either boom or probe refueling systems. One KC-10A carries over ten times the "give-away" fuel load of an S-3/ES-3 tanker, and has

78 In addition to USACOM, there are seven other regional CinCs. These include the Pacific Command (PACOM), Strategic Command (STRATCOM), Southern Command (SOUTHCOM), Central Command (CENTCOM), Special Operations Command (SOCOM), Space Command (SPACECOM), European Command (EUCOM), and Transportation Command (TRANSCOM).

much greater loiter time. A favorite Navy ''trick'' is to use a large land-based tanker, like a KC-135, as a ''milk cow'' to top off S-3/ES-3 tankers, which then distribute the fuel to other carrier aircraft.

—Airborne Early Warning (AEW) Support—Back in 1982, the Royal Navy learned a hard lesson about operating without AEW support in a high-threat environment. Ships were sunk, sailors died, and the expedition to retake the Falkland Islands was seriously jeopardized. While VAW-123's four E-2C Hawkeyes provide excellent AEW capability, Admiral Mullen is happy to have additional ''eyes in the sky.'' If support from Air Force E-3 Sentry AWACS aircraft is available, count on him using it with gusto. Data links on Navy ships can talk to computers on Air Force and NATO Sentries. Secure voice radios are also compatible, assuming everyone has correctly loaded the proper codes (this is a common foul-up in exercises, and it's been known to happen in combat).

—J-STARS Ground Surveillance—The *GW* CVBG has powerful sensors to monitor the sky, but its ability to detect and track surface and ground-based targets is more limited. The S-3Bs of VS-32 can fly effective day and night maritime surface surveillance missions, but this competes with their other missions as tankers, ASW aircraft, and sea control platforms. Real-time tracking of ground targets ashore is even more difficult. If you need to monitor a ground threat, have your CinC ask the Air Force for the loan of few E-8 Joint Surveillance, Tracking, and Reconnaissance System (J-STARS) aircraft. Based on a commercial Boeing 707 airframe, the E-8's carry side-looking synthetic-aperture radar (SAR) that can detect both stationary and moving targets on the earth's surface. The J-STARS system also has special modes for detecting and tracking maritime targets, then passing the data to the battle group via a data link. Because J-STARS is designed for ''wide area'' surveillance, it can monitor thousands of square miles of territory at a time.

—Air Force Wings—CVW-1 is trained and equipped to work jointly with air units of other services, or even of allied countries. This reflects a profound transformation in Navy thinking since 1990, and even as late as 1995 the capability was marginal. But now the CVBG finally has its own capability to generate and use Air Tasking Orders to synchronize and ''de-conflict'' all kinds of air operations (strikes, sweeps, recon, cargo delivery, airmobile assault, cruise-missile salvos, search and rescue—you name it!). Someday we may see ''expeditionary'' Air Force wings (like the 366th, a composite unit of fighters, bombers, and tankers) directly supporting carrier operations or providing cover for a CVBG or ARG.

—Marine Aviation Support—The Marine Corps can quickly deploy squadron of two-seat F/A-18D Hornet all-weather strike fighters to support CVBG, ARG, and MEU (SOC) operations. Equipped with Nighthawk laser-targeting pods for LGBs, as well as AIM-120 AMRAAM and AGM-65 Maverick missiles, the F/A-18Ds are highly capable strike fighters.

Marines also fly land-based EA-6B Prowler electronic warfare (EW) and jammer aircraft to augment those already aboard the carrier. Also, Marine Hornets and Prowlers can easily operate from carriers if necessary, since they too have tailhooks!

- **Reconnaissance Support**—To plan air strikes effectively you need high-quality, up-to-the-minute imagery of potential targets, and intelligence analysts who understand how to interpret these images. The main reconnaissance asset on the carrier is a quartet of TARPS-equipped F-14's. Older TARPS pods bring back reels of film that has to be developed in an onboard photographic lab. New Digital TARPS pods have a data link that can return a stream of pictures to the carrier before the F-14 lands. Other sources for imagery tend to sound like a bowl of alphabet soup: Satellite imagery will come from the National Reconnaissance Office (NRO) in Chantilly, Virginia, with its fleet or orbital imaging and radar satellites. UAVs and other airborne imaging systems can also be tasked for the CVBG staff. To process and distribute imagery and other products, the National Imaging and Mapping Agency (NIMA) was created in 1996. NIMA will combine the services of NRO, the Central Imaging Office (CIO), the National Photographic Center (NPIC), and the Defense Mapping Agency (DMA), all under a single roof. Digital and paper maps, annotated photography, and customized target graphics will be "pulled" on demand by regional joint intelligence centers and "pushed" down to the wing and squadron intelligence officers who will need it.

The big NRO satellites produce a huge volume of high-quality "close look" and "wide area" imagery, but many urgent demands compete for limited time slots on these precious national assets. Smaller and less expensive collection systems will come on line in the early 21st century. NRO is trying to improve the timeliness of the images they deliver, and to "downgrade" the Super-Secret classification of final products, so that more people and organizations can see them. An alternative that will become available in 1998 is commercial satellite imagery at one-meter resolution (good enough to distinguish tanks from trucks, but not to identify specific models). In the long run, the military will probably be one of the biggest users of commercial imagery, since it will be cheap, timely, and best of all, unclassified!

UAV systems continue to make steady progress toward the goal of long-duration, stealthy, unmanned airborne reconnaissance. The Pioneer UAV continues in service with the Marine Corps. Also, the first Air Force Predator unit (the 11th Reconnaissance Squadron), at Nellis AFB, Nevada, formed several years ago. Predator is derived from the Gnat 750 UAVs, which were combat-tested by the CIA in Bosnia back in 1994; and Predators were used during a recent workup of the *Carl Vinson* (CVN-70) CVBG off the southern California coast. The Navy today can receive the data feed from Predators, and control them from carriers. Trials have even tested controlling the big

A Predator Unmanned Aerial Vehicle (UAV) flies over the USS *Carl Vinson* (CVN-70) during an exercise off the southern California coast. Within a few years, such UAVs will be a common asset supporting carrier groups.
OFFICIAL U.S. NAVY PHOTO

UAVs from nuclear submarines! On the downside, there are no facilities for carrier takeoffs or landings, and Predators must be launched from a land site. But the multi-day endurance of the Predator makes this a minor limitation in most areas.

In addition to the Predator and Pioneer programs, progress is being made on long-endurance surveillance UAVs, like Dark Star, produced by the Lockheed Martin ''Skunk Works.'' The even-longer-range Teledyne Ryan Global Hawk is also on track, as well as the data links and common control stations needed to make the UAV available to users. Traditional manned reconnaissance systems include the RF-18D Hornet introduced in 1997, equipped with a new Advanced Tactical Reconnaissance System (ATARS).

- **Intelligence Support**—In addition to imagery and mapping support, the *GW* battle group can also make use of many of the other products generated by the various ''spook'' agencies. Some of these include:

  —National Security Agency (NSA)—The NSA, which controls all electronic and signals intelligence collection, is a significant supporting agency for an amphibious unit like the *GW* battle group. A cramped, high-security compartment called the ''Ships Signals Exploitation Space'' (SSES) lets battle group, ARG, and MEU (SOC) commanders tap into a wealth of electronic intelligence sources including RC-135 Rivet Joint and EP-3 Orion electronic intelligence aircraft, as well as communications intelligence satellites. Ship-based sensors (like the Classic Outboard ESM system) can intercept and analyze electronic signals, from SAM and air traffic radar to cellular phones and television signals.

  —U.S. Space Command (US SPACECOM)—Based at Falcon AFB, Colorado, US SPACECOM provides space-based services to support combat operations. Key assets include weather, GPS navigation and communications satellites, ballistic missile warnings, and in the future, theater missile defense command and control.

—Cable News Network (CNN)—Intelligence analysts, after a few drinks, will usually agree that CNN is the finest real-time intelligence-gathering service in the world. In fact, one of the greatest benefits of the Challenge Athena system was to give commanders and staffs access to networks like CNN, Skynet, and MSNBC. A side benefit is improved crew morale when up-to-the-minute news and sports from home are available.

Battle Group Operations: The Concept

When the Army guys talk about ''doctrine,'' Navy guys talk about CONOPS. ''Doctrine'' or CONOPS is how we think about the way we fight. For American Naval commanders, tradition, more than technology, dominates the way they look at CONOPS. Carrier-based air units tend to be more ''brittle'' than land-based wings and squadrons. When you have very few aircraft, and reinforcements may be weeks away, the loss of each plane and crew hurts—a lot. That means that battle group and air wing commanders have to be very cautious when they commit their limited resources, yet very bold in employing them to make the effort effective. Balancing such conflicting objectives takes a special gift—even a kind of operational and tactical genius. The capacity for such judgment is rare.

Aviators are an incredibly select group to begin with, and only the best of the best ever rise to command carriers, air wings, and battle groups. Perhaps this explains why naval aviators seem to be so successful, not only in the top ranks of the military, but also in government and industry. They don't just have ''the right stuff.'' They have the right CONOPS.

A trained, well-led battle group is an ideal tool for many different missions. These missions include:

- **Presence**—Global naval presence is the primary mission of every peacetime CVBG deployment. Just having a carrier group in the neighborhood encourages regional bullies and opponents to back off, as seems to have happened in the Straits of Taiwan in 1996 when a pair of CVBGs deterred aggression by the People's Republic of China against Taiwan. It's really very simple: If you attack even a *small* American ship, you've got yourself an international incident. If you attack an American carrier, you've got yourself a war against the people of the United States. *You lose.* Any questions?
- **Show of Force**—This might be best described as a one-time application of military power for the purpose of sending a message. It is a punitive

military action designed to apply a measured amount of force against a specific target. Operation Desert Strike against Iraq in 1996 and the 1986 raid on Libya are prime examples.

- **Maritime Embargo**—In the days of sailing ships, this mission was called a blockade. Though modern interpretations of international law have made the historic concept of blockade obsolete, maritime embargo and inspection operations are a staple of CVBG operations today. At one point in 1994, the U.S. and its allies were simultaneously running embargo operations against Iraq, Haiti, and in the Balkans.
- **Freedom of Navigation Exercise**—"Freedom of the Seas" is a concept the U.S. Navy inherits from the Pax Britannica—the age of British global empire, when the Royal Navy enforced the "right of innocent passage" at gunpoint, anywhere in the world. To tyrannical regimes, the notion of free navigation is as incomprehensible as most other freedoms. Dictators are often tempted to simply close their coastal waters, applying the surfer punk's rule of "my beach, my wave!" As a matter of policy, the United States will challenge any attempt to prevent or restrict free navigation in recognized international waters. Our operations on the Libyan "Line of Death" in the Gulf of Sidra in 1981 and 1986 are classic examples of this mission.
- **Maritime Escort**—High-value commercial vessels, like tankers and container ships, or military transports (like Maritime Prepositioning Squadrons) sometimes have to be escorted past hostile shores and through dangerous waters. CVBGs provide powerful force for escort missions, with their array of air, surface, and subsurface capabilities. Our escort of American-flagged Kuwaiti-owned tankers during 1988 and 1989 in the Persian Gulf demonstrates this mission.
- **Expeditionary Support**—"Expeditionary" is a fancy word for invasion, the ultimate exercise of force. Though rare, invasions still happen. Recent examples are the 1983 invasion of Grenada, our landings in Beirut during the same period, and the liberation of Haiti in 1994. CVBGs and ARGs were the key units in prosecuting each of these actions.
- **Power Projection**—Sometimes, you just have to fly a lot of sorties over some beach for a long time to support a long-term national commitment. Desert Storm and operations in the Gulf of Tonkin during the Vietnam War are examples. This kind of operation usually involves two or more CVBGs sharing the burden of operations over a period of days or weeks. The key to this kind of warfare is pacing the air crews and maintaining an adequate supply of munitions for delivery onto the targets.

Each of these missions depends on having all the elements of the CVBG team in place, so that the battle group commander has a full range of options to block or defeat any hostile action. A battle group commander can only do

A Non-Combatant Evacuation Operation (NEO) being run by U.S. Marines. These operations have become almost commonplace since the end of the Cold War.

Official U.S. Navy Photo

his job, however, if he has political support from the National Command Authorities, necessary freedom of action from his regional CinC, and Rules of Engagement (ROE) that allow him to accomplish the mission.

So how would a CVBG commanded like Admiral Mullen use the tools at his command to prosecute an actual crisis? Consider the problem facing many American travelers in some of the world's rougher neighborhoods: walking into the middle of a civil war. Since the end of the Cold War in the early 1990's, literally dozens of national, tribal, ethnic, or religious conflicts have sprung up. They drag on for years without one side or the other gaining a decisive advantage. During particularly chaotic periods, immediate evacuation of American citizens and other non-combatants from the war zone is required to prevent massacres or hostage situations from developing, as they did in Iran back in 1979.

In a ''typical'' NEO (Non-Combatant Evacuation Operation), Marines from the MEU (SOC) fly into a capital city to reinforce the guard in the American embassy, and then safeguard the helicopter evacuation of non-combatants to the waiting ships offshore. The U.S. and our allies have conducted literally dozens of NEOs in the last few years, particularly in West Africa and the Balkans. Usually an NEO is accomplished by detaching and dispatching the ARG/MEU (SOC) team and a few escorts to conduct the mission independently. During Desert Shield (1990), simultaneous NEOs were required in Somalia and Liberia, without taking forces away from the buildup in the Gulf. Most combatants in civil wars these days have the good sense to let us run our NEOs without getting in the way. However, some truly fanatical terrorists or tribal warriors just can't resist the urge to test their prowess against the Marines.

A ''worst case'' NEO in our time might involve rebel forces closing in on a city; not just ragtag guerrillas with AK-47's and RPG-7's, but well-equipped and trained forces with artillery, armor, helicopters, and fighter-bombers. Let's say that they have taken a hard stand against American intervention; not just nasty rhetoric but closing off normal evacuation routes. Overland roads, seaports, and airports have been closed, and several thousand civilians are trapped inside embassies in the surrounded city. To emphasize their dislike of Westerners, rebels have taken shots at embassy guards and killed some news crews, inciting outrage from the world media.

Since the situation is clearly getting out of hand, an entire CVBG/ARG/MEU (SOC) team is dispatched to bring the civilians out with a minimum of losses. With the orders and ROE from the National Command Authorities in hand, the battle group commander brings the force to the offshore waters of the embattled country, then calls his unit commanders together for a planning conference aboard the flagship. Meanwhile, intelligence agencies will be working overtime to gather, generate, and deliver the necessary imagery, maps, and data on potential threats.

As the battle group arrives offshore, rebel and loyalist forces are fighting in the streets around the embassies, and the rebel air force is bombing the capital city. The embarked CVW is tasked to take out the air threat to the Marine helicopters, and escort and protect the evacuees as they are flown back to the ships of the ARG. The CVW sets up a combat air patrol (CAP) station of F-14 Tomcats backed up by an E-2C Hawkeye AEW aircraft. At the same time, the State and Defense Department spokesmen back in Washington, D.C., are making the intentions of the force clear to the rebels and the rest of the world: Americans will fire on anyone interfering with the evacuation.

Initially, the show of force deters any rebel response, which is exactly the intent. Meanwhile, additional air units are already being prepared for operations. Each operational mission is known as an "event" to the air boss on the carrier, who manages and controls the local airspace to ensure that no midair collisions or enemy incursions interfere with operations. Events are normally scheduled about one hour apart to allow a little slack time for flight and deck crews to "respot" aircraft (shuffle them around on the deck) and take a few breaks. These "cyclic ops" can be maintained indefinitely if required.

Meanwhile, an Aegis ship moves closer inshore, with one of the HH-60G SAR/special operations helicopters aboard standing by for Tactical Recovery of Aircraft and Personnel (TRAP) in case a plane goes down. The escort provides inshore radar coverage of the airspace, and establishes a forward SAM "bubble" to protect the ships of the ARG should the rebels try an air strike against them. About this time, the first wave of Marine reinforcements aboard its helicopter transports arrives at the embassy and prepares to take out the first of the evacuees. Within a few hours of the initial Marine guard reinforcement, the helicopters loaded with evacuees will be shuttling out to the ARG ships on a regular schedule. Since evacuations can involve rescuing hundreds and sometimes thousands of civilians, and it can take hours to get them all clear of the war zone, trouble can easily occur during this period.

A few hours into the evacuation cycle, the rebels have carefully noted the timing and routes of the evacuation helicopters, with a view to teaching the Americans a lesson. Under cover of darkness a mobile battery of radar-controlled SAMs moves into the evacuation corridor. And just as dawn is breaking, a flight of MiG fighters attempts to intercept one of the helicopter groups. These threats do not go unnoticed. A few miles offshore, systems operators in the backseats of an ES-3 Shadow ESM aircraft intercept radio messages between the SAM battery and rebel headquarters, as well as communications traffic going out to the MiG base outside the city. To deal with this threat, the battle group commander reinforces the next rescue group with additional CAP escorts, and orders alternative route planning for the helicopters.

Suddenly an escorting EA-6B Prowler begins to detect the telltale signals of the SAM surveillance radar. The Prowler fires a pair of AGM-88 HARM anti-radiation missiles at the radar vehicle, destroying it in seconds. Moments later, a pair of F/A-18 Hornet fighter-bombers dive into the area, destroying the SAM launch vehicles with LGBs and AGM-65 Maverick missiles.

As the strike aircraft finish their deadly work, word goes out over the secure data links that the AEW Hawkeye and ES-3 have detected several flights of MiG fighters being vectored into the area. Rapidly, the CVW commander maneuvers his fighters into position to stop them. He orders his most capable fighters to intercept (these aircraft have Low Probability of Intercept—LPI—radars equipped with NCTR operating modes). As the American fighters move into place, they acquire the rebel fighters on radar, and fire their AIM-120 AMRAAM AAMs at maximum range. Over half of the rebel MiGs are vaporized in clusters of oily fireballs. The rest flee back to their base. Now that the immediate danger is past, the evacuation continues without a break until the last of the civilians and embassy staff have safely reached the ships of the ARG.

As the Marine guard and reinforcement force is recovered, the National Command Authorities decide to punish the rebel violations of international law prior to withdrawal. With the Marines and helicopters back aboard, the ARG heads back out to sea, and one last strike is planned, briefed, and assembled aboard the carrier. Throughout the NEO operations, national intelligence agencies and the CVBG's own TARPS F-14's have been trying to locate and identify critical rebel command posts and heavy-weapons sites. Because of the earlier attacks on U.S. forces, a one-time show of force against the rebels is authorized and rapidly executed.

Close inshore, the Aegis ship launches several dozen BGM-109 Tomahawk cruise missiles to attack fixed radars and command posts. Behind them come a wave of fighter-bombers and support aircraft, which attack the rebel MiG base, as well as the headquarters of the insurgent forces. To minimize collateral damage, only PGMs are used to hit bunkers and aircraft shelters. Once their ordnance is expended, all aircraft safely return to the "boat" for a well-deserved rest.

As the battle group withdraws, Aegis ships and one CAP section provide a "rear guard" until the force exits the threat area. A few days later, the evacuees safely disembark; and weapons, fuel, and supplies are replenished. Then the battle group moves on to its next destination, the cycles of normal operations are reestablished, and the crews begin to think about their next port call and the exercises that will follow. While this scenario is much simplified, it illustrates how CVBGs can rapidly adapt to a fast-breaking situation. Getting a battle group into such a high state of readiness is, of course, no simple matter. The next chapter explains how Admiral Mullen, Captain Rutheford, and CAG Stufflebeem spent the hot summer of 1997 preparing their people, ships, and aircraft for the challenge of an actual deployment. Join me, and I'll show you how they spent their vacation!

Final Examination: JTFEX 97-3

"This is 4.5 acres of sovereign U.S. territory"

Rear Admiral Michael Mullen, Commander, *George Washington* Battle Group

In the fall of 1997 trouble was once again brewing in the Persian Gulf. Once again, Iraq was defying the authority of the United Nations Security Council, trying to hide from the world the weapons of mass destruction Saddam Hussein had spent so much to produce. As usual, the Iraqi dictator railed against UN weapons inspectors' attempts to detect his research and production centers for chemical, biological, and nuclear arms. And once again, the world went to the brink of war.

As in previous years, this crisis required a U.S. response that was both rapid and clear. Quickly, units of the Army's XVIII Airborne Corps were put on alert; and the U.S. Air Force dispatched reinforcements to the aerial task force (based at Prince Sultan Airbase in Saudi Arabia) already enforcing the southern Iraqi "no-fly" zone. But this time there was a complication. For the first time since August of 1990, our Persian Gulf allies denied us the use of bases on their territory. Though we still do not know whether this action resulted from pent-up frustration over our failure to form a clear policy toward Iraq, or from fear of the reaction of their own Islamic fundamentalist factions, this much was clear. If America were to react to this crisis, then the response would have to come from U.S. ships sailing in international waters.

To this end, the newly installed Chairman of the Joint Chiefs of Staff, General Henry Shelton, sent the word down the chain of command: "Send in the carriers." Within days, the carrier battle groups (CVBGs) based around the aircraft carriers *Nimitz* (CVN-68) and *George Washington* (CVN-73) were sailing for the Persian Gulf, where they could quickly mount air and cruise-missile strikes against Iraqi targets should these be required. As the CVBGs rattled sabers, UN Secretary General Kofi Annan carefully constructed a diplomatic effort to persuade Saddam that further intransigence would lead to

The carriers *Nimitz* (CVN-68) and *George Washington* (CVN-73) in the Persian Gulf during fall 1997. These two vessels and their battle groups comprised the bulk of the striking power that stood down Saddam Hussein during the arms inspection crisis.
OFFICAL U.S. NAVY PHOTO

falling bombs. The persuasion—eventually—worked, and the inspectors were able to return to their jobs.

Meanwhile, the two battle groups spent almost six months on station in the Gulf, until they were relieved of their vigil in the spring of 1998 by two more CVBGs, centered around the carriers *Independence* (CV-62) and *John C. Stennis* (CVN-74). The U.S. kept two carrier groups in the Persian Gulf until late May 1998, by which time tensions in the region had relaxed. Back home in America, most of us gave little thought to the thousands of men and women on these ships. Even though we may have worried a great deal about the Iraqi crisis itself, they were out there, doing a vital and dangerous job for us, and generally making it look easy. This last is a significant point: Making it look easy is hard work. It takes practice, training, intense education, constant drilling.

The process of preparing a CVBG for an overseas deployment begins months before it deploys, and it takes the efforts of every person assigned to the group, as well as thousands of others who do not leave American waters. Let's look at part of that effort, as the *GW* (*George Washington*) group ratcheted up its combat skills in the summer of 1997.

Getting the Group Ready: Joint Training

You fight like you train!

Commander Randy "Duke" Cunningham, USN (Ret.)
First U.S. Air-to-Air "Ace" of the Vietnam War

This statement dates from the spring of 1972, soon after then-Lieutenant Cunningham and his valiant backseater, Lieutenant, J.G. "Willie" Driscoll, shot down five North Vietnamese MiG fighters and became America's first confirmed Vietnam fighter aces. Cunningham and Driscoll's success did not come out of the blue; their generation of naval aviators had been the first to be given a new kind of pre-combat schooling, called "force-on-force" training. Simply put, force-on-force training involves training units and personnel against role-players who simulate enemy units at the peak of their game. The first of these programs was the famous "TOPGUN" school, then located at NAS Miramar near San Diego, California. While the tools and curriculum were rudimentary by today's standards, the results were spectacular. The Navy's air-to-air kill ratio in Vietnam, the measure of aerial fighter performance, improved by an amazing 650%. Not surprisingly, the other services took notice.

Today, every branch of the U.S. military has multiple force-on-force training programs and facilities, and each of these has been validated by the outstanding combat performance of their graduates.[79] CVBGs, like fighter pilots, do best when they have been tuned up by means of intense force-on-force training—a tune-up that's considerably complicated by the variety and multiplicity of roles a CVBG might be required to undertake. Today's CVBG is more than a group of ships designed to protect the flattop. When properly deployed and utilized by the National Command Authorities (NCAs), a CVBG's mission can range from "cooling off" a crisis to spearheading the initial phases of a major invasion or intervention.

Meanwhile, preparing a war machine as large and complex as a CVBG for a six-month overseas cruise is a huge undertaking. In fact, the various components of the group spend twice as much time recovering from the last cruise and getting ready for the next as they actually spend out on deployment. And all of this has grown more complicated in the last decade as a result of the changes in the NCA command structure stemming from the Goldwater-Nichols Defense Reorganization Act. Back in the 1980s, and before Goldwater-Nichols, the Navy was the sole owner and trainer of carrier groups before they were sent overseas. Today, that ownership has moved to another organization, the U.S. Atlantic Command (USACOM) based in Norfolk, Virginia.

Led by Admiral Harold W. Gehman, Jr., USACOM is a mammoth organization—in fact, the most powerful military organization in the world today. USACOM essentially "owns" every Army, Navy, Air Force, and Marine unit based in the continental United States. Its job is to organize, train, package, and deliver military forces for the commanders of the other unified Commanders in Chiefs (CINCs)—the heads of the various regional commands responsible for conducting military operations around the globe. Whenever the NCAs need to send an American military force somewhere in the world, the phone usually rings first in Norfolk at USACOM headquarters.

Goldwater-Nichols has also brought practical changes to the U.S. military. For instance, CVBGs now no longer operate independently of other units—or indeed of other services. So an air strike from a carrier may receive aerial tanking and fighter protection from U.S. Air Force units, and electronic

79 Though there are literally dozens of such programs (ranging from staff-level exercises to war games involving tens of thousands of participants), the best known are at the Army's National Training Center at Fort Irwin, California, and at the Air Force's Operation Red Flag at Nellis AFB, Nevada.

warfare support from a U.S. Marine Corps EA-6B Prowler squadron, and have the target located and designated by an Army Special Forces team. This, in essence, is what is meant by ''joint'' warfare, and it's far removed from Cold War practices that gave the Navy few responsibilities other than the killing of the ships, aircraft, and submarines of the former Soviet Union. Needless to say, joint war fighting skills don't just happen. They must be taught and practiced before a crisis breaks out. The CVBG must practice not only ''Naval'' skills, but also ''joint'' skills with other services and nations.

This job falls to the joint training office (J-7) of USACOM, which lays out the training regimes for units being ''packaged'' for missions in what are normally known as JTFs or Joint Task Forces. Getting a particular unit ready for duty in a JTF is a three-phased program, which is supervised by individual groups of subject-matter experts. For example, on each coast a Carrier Group (CARGRU) composed of a rear admiral and a full training staff is assigned to prepare CVBGs for deployment. On the Pacific Coast, this is done by CARGRU One, while CARGRU Four does the same job for the Atlantic Fleet. The training CARGRUs supervise the various elements of the CVBGs through their three-phase workups. These break down this way:

- **Category I Training**—Service-specific/mandated training that focuses on the tactical unit level. Examples include everything from carrier qualifications to missile and ACM training at the ship/squadron/CVW level.
- **Category II Training**—This is joint field training, in which the various pieces of the CVBG come together in what are known as Capabilities Exercises (CAPEXs) and Joint Task Force Exercises (JTFEXs).
- **Category III Training**—This is a purely academic training phase, which takes place just prior to the JTF deploying. Composed of a series of seminars, briefings, and computer war games, it is designed to give the unit commanders a maximum of up-to-the-minute information about the areas where they will likely be operating and the possible contingencies that they may face.

These exercises provide a multi-level training regime for *every* member of the battle group, from the sailors in the laundries to the CVBG commander and his staff. And most participants will tell you that the pre-workup training is usually tougher than the actual overseas deployment. The old saying that sweat in training is cheaper than blood in combat remains true. In a world as uncertain as today's, we as a nation *owe* the men and women of our armed forces the very toughest training we can provide for them.[80] All of this brings us down to the men and women of the *GW* group in the summer of 1997,

80 Most veterans of Desert Storm will tell you that the extensive force-on-force training prior to their deployment to the Persian Gulf was tougher than anything the Iraqis threw at them. This is hardly a surprise, since the Opposing Force units they trained against are usually the best-trained and motivated units in the U.S. military.

facing a terribly real experience, designed to test the limits of their endurance and skills.

Getting the Group Ready: Part I

The countdown to *GW*'s deployment in the fall of 1997 actually began in February of 1996. That is when the battle group based around the USS *America* (CV-66) returned from their own six-month cruise to the Mediterranean Sea.[81] Since *America* had been scheduled for decommissioning and eventual scrapping, this was her final cruise. The *GW* would replace her. Other ships in this combined CVBG/Amphibious Ready Group (ARG) were scheduled for deep maintenance as soon as they arrived back home. Thus the *Wasp* (LHD-1) and the *Whidbey Island* (LSD-41) were headed into dry dock for almost a year of overhaul. Replacing them would be the amphibious ships *Guam* (LPH-9), *Ashland* (LSD-48), and *Oak Hill* (LSD-51). At the same time, a number of the escorts and submarines were swapped out, as the personnel at Atlantic Fleet Headquarters and USACOM packaged the new CVBG. Even though the CVBG would make just one cruise in this form, the plans are to reconstitute it again in a more permanent form for its 1998/1999 cruise.[82]

In February of 1996, while the thoughts of most of the group's personnel were on their upcoming leave periods and visiting with their families and friends, at the USACOM and Atlantic/2nd Fleet headquarters planning for the CVBG's training and deployment in 1997 had already begun. For starters, there was the scheduling of minor overhauls for the ships assigned to the CVGB that would deploy in 1997, as well as managing the usual flow of personnel coming and going to new assignments. These months of relative quiet offered a time for getting the new folks up to speed, and a chance for those remaining in the group's units to attend technical and service schools or to take some leave.

By the fall of 1996, the various pieces of the battle group were ready to begin their Category I training. So, for example, the *Guam* ARG and the 24th Marine Expeditionary Unit—Special Operations Capable (MEU (SOC)) were beginning their own workups, supervised by teams of USACOM training mentors. Meanwhile, even as CARGRU Four personnel were deep into the training of the *John F. Kennedy* (CV-67) CVBG (which would proceed the *GW* group to the Mediterranean in the spring of 1997), the CARGRU staff had already begun to assign personnel to the *GW* group to start the workup process. At the same time, the various squadrons of Carrier Air Wing One (CVW-1) began to come to life at Naval Air Stations (NASs) from Whidbey Island, Washington, to Jacksonville, Florida. About half of these

81 This is the same group that I highlighted in my 1996 book *Marine: A Guided Tour of a Marine Expeditionary Unit* (Berkley Books, 1996).

82 The *GW*, *Normandy* (CG-64), *South Carolina* (CGN-37), and *Guam* were all scheduled either for deep overhaul or scrapping at the end of the cruise in 1998.

squadrons would also be breaking in a new commanding officer, normally a freshly frocked commander (O-5) who had just ''fleeted up'' from the executive officer's job in the unit. Along with the command changes came in-squadron training. A *lot* of it!

Getting a squadron ready to deploy starts with refresher/upgrade training for all the flying personnel in the unit. This brings everyone in the unit up to a common level of proficiency across a range of skills and missions. As they do this, the squadron maintenance chiefs begin to bring the squadron's aircraft up to standard. This is not to say that the aircraft have been allowed to go to seed. But since the squadron is not a ''deployed'' unit, and personnel were away on leave and at service schools, keeping every aircraft fully mission-capable has not been a priority. Deployed units get the pick of the ''good'' airplanes, as well as first priority on training ammunition and spare parts.

While the air units were starting on their road to deployment, so too were the crews of the ships of the battle group. And the officers and enlisted personnel were re-learning the details of their trade on short training cruises out of their home ports. During these cruises, the crews powered up all the ship's systems to find out the new capabilities and liabilities the yard workers had installed. Also, during these cruises the new crew members began the bonding process with their shipmates. This is especially important in the escorts (known as ''small boys''), which will do so much of the work supporting and protecting the carrier and ARG.

For the men and women of the *GW* battle group, their final run to deployment started in May of 1997, with the departure of the *John F. Kennedy* battle group. Now that that group was on its way, the CARGRU Four staff could devote their full attention to making the *GW* group ready for their early October deployment. Several key training events, whose dates had been previously been set by the USACOM J-7 staff, began to have immediate importance. These included:

- **Naval Strike and Air Warfare Center Rotation**—Since ''Boomer'' Stufflebeem, the commander of CVW-1, had units spread over five bases in four different states on two coasts, the rare opportunities for getting his units together were more precious than gold. One of these golden opportunities happened at the Naval Strike and Air Warfare Center (NSAWC) at NAS Fallon near Reno, Nevada. For over three decades, it has been Navy policy that *every* CVW preparing to deploy do a rotation at the NSAWC. Over a period of three and a half weeks, the various CVW components are based at NAS Fallon, where they can practice the art of composite warfare together. While there, they undertake a series of air strikes against the target arrays up on the Fallon bombing range. Supervised by the Center Staff, and assisted by aggressor aircraft and ground units acting as surrogate enemies, the wing works up through a number of phases, culminating with the three-day Advanced Training Phase

(ATP).[83] During ATP, the wing must conduct a series of large ''Alpha'' strikes (usually involving between two and three dozen aircraft) against targets up on the Fallon ranges. All of this training brings the entire wing, from pilots and planners to maintenance personnel and photo interpreters, up to combat standards. With the air wing now molded into a unified fighting unit, it was time to merge it with the *GW* and her battle group. The USACOM Category I training now completed, the *GW* group was ready to move onto the challenges of Category II.

- **Capabilities Exercise (CAPEX)**—In mid-June of 1997, after CVW-1 had returned from NAS Fallon, the ships of the *GW* battle group met off the Virginia capes to conduct what is called a CAPEX. This exercise, which was run over two weeks, was designed to integrate CVW-1 into the rest of the battle group's operations. This meant doing a number of things in a very short time. Once the battle group had assembled, the CVW-1 aircraft and crews flew aboard from staging bases along the Atlantic coast. What followed were several days of carrier requalification for everyone in the wing, including Captain Stufflebeem. With qualifications completed, the wing and battle group began a series of training exercises, designed to show the CARGRU Four staff that they could safely and effectively conduct strike operations. During this time, the rest of the battle group practiced various skills, such as simulated Tomahawk cruise-missile strikes, and combat search-and-rescue (CSAR) training. The CAPEX was a test of skill and endurance, designed to stress everyone in the *GW* battle group from Admiral Mullen down to the chiefs of the various ships' laundries. Again, all went well, and its end had melded the assorted parts of the battle group into a fighting unit. Next came the final part of the *GW* group's Category II training, the JTFEX.
- **Joint Task Force Exercise (JTFEX) 97-3**—Run over three weeks in late August and early September of 1997, JTFEX 97-3 was a ''final exam'' for the combined *GW* CVBG/CVW/ARG/MEU (SOC) team. JTFEXs—the crown jewels of USACOM exercises—are the largest and most complex series of exercises regularly run by USACOM. Even as the sea services are using them as benchmark exercises for Navy groups, the other services are utilizing them in the same way: to test their own fast-reaction units (such as the 82nd Airborne Division at Fort Bragg, North Carolina, or the 2nd Bombardment Wing based at Barksdale AFB, Louisiana).

With the Category II training completed, the ships and aircraft of the battle group headed home for a final leave period. During this time, the Category III training and briefings for the battle group staff and leaders took place around Washington, D.C. While their actual sequence and locations are

83 This similar to the Red/Green Flag exercises conducted by the USAF at Nellis AFB to the south of Fallon. Although somewhat smaller than the Nellis exercises, there is a greater emphasis on live-fire and electronic-warfare issues.

classified, the briefings and war games were conducted by a variety of military and intelligence agencies, with the goal of sharpening the minds of the CVBG/CVW/ARG/MEU (SOC) leadership. When these exercises were finished, the CARGRU Four staff started preparing for the next group, which was based around the new *Nimitz*-class carrier *John Stennis* (CVN-74).

JTFEX 97-3

In the confusing (maybe anarchic is a better word) post-Cold War world of joint and coalition warfare, the USACOM staff must package and deliver to the unified/regional CinCs units that are ready to ''plug in'' to a joint/multinational JTF. The JTF must start combat operations on almost no notice, and function in an environment where the ROE can change on a moment's notice. That means the units assigned to the JTF must be trained with an eye to functioning in a variety of scenarios that were unimaginable as recently as a decade ago. Some of these may even involve situations where conflict may be avoided (if a show of force is sufficiently effective), or where conflict may not be an option (in what are called Operations Short of War).

Training units for situations like these requires more than the simple force-on-force training that was good enough for the military services during the Cold War. Exercises like Red Flag (at Nellis AFB, Nevada) and those at training facilities like the Army's National Training Center (at Fort Irwin, California) were always based upon assumptions that a ''hot'' war was already happening. Because of this, the engaged forces' only requirement was to fight that conflict in the most effective manner possible. While the services teach combat skills quite well, teaching ''short-of-war'' training is a much more complicated and difficult undertaking. Only in the last few years (after high-cost lessons learned in Haiti, Somalia, and Bosnia) has progress been made on this daunting training challenge.

So far, the leader in this new kind of ''real world'' force-on-force training has been the Army's Joint Readiness Training Center (JRTC) at Fort Polk, Louisiana.[84] The JRTC staff, for example, was among the first to insert into traditional force-on-force training what the military calls ''friction'' elements and non-traditional ideas like ''neutral'' role-players on the simulated battlefield, and to include a greater emphasis on logistics and casualty evacuation. JRTC's focus on these kinds of layered issues have made it a model for other joint training operations run by USACOM (such as the JTFEX-series exercises, which are run approximately six times a year—three on each coast).

The result of all this thinking has been a gradual evolution in the scenarios presented to participants in the JTFEXs. As little as three years ago, every JTFEX was essentially a forced-entry scenario into an occupied country

---

84 For a closer look at the outstanding JRTC program, see my book *Airborne: A Guided Tour of an Airborne Task Force* (Berkley Books, 1997).

that looked a lot like Kuwait, and the opposing forces were structured much like the Iraqis. The critics who were complaining that USACOM was preparing to "fight the last war" were making a good point. Today there'd be no justice in that criticism. Now, each JTFEX is made a bit different from the last one, or for that matter from any other. For one thing, USACOM has gotten into the habit of making the JTFEXs truly "joint," by spreading out the command responsibilities. By way of example, a JTF headquarters based at 8th Air Force headquarters at Barksdale AFB, Louisiana, controlled JTFEX 97-2 (run in the spring of 1997), while the first of the FY-1998 JTFEXs will be an Army-run exercise, controlled by XVIII Airborne Corps at Fort Bragg, North Carolina. Now that each of the services has opportunities each year to be the JTFEX "top dog," the scenarios have tended to become not only more fresh and innovative, but also more fair in the distribution of training responsibilities and opportunities.

The quality of JTFEX exercises has also been improved by means of what is called a "flexible" training scenario—that is, a scenario without highly structured schedules and situations. In more structured scenarios, for example, participants knew exactly when and how the exercise would transition to "hot war" status. In current JTFEXs, there is much more uncertainty. Furthermore, the actions of the participants can affect the "flexible" elements of the scenario, and these actions can be scored positively or negatively. It is even possible that participants might contain a JTFEX "crisis situation" so well that a transition to a "hot" war situation might *never* occur. But creative work by the USACOM J-7 staff makes this unlikely. Thus when a commander or unit does well, "friction" and challenges are added so no participant gets a chance to "break" the scenario. On the other hand, if a unit has itself been "broken" by the situations it faces, the exercise staff may choose to give it additional support or opportunities to "get well enough to go back into the game," as it were. You have to remember that exercises like the JTFEXs are designed to build units up, not break them down.

For the *GW* group, the focus in the late summer of 1997 was getting ready for their particular "final exam," JTFEX 97-3 (the third East Coast JTFEX of FY-97). With their deployment date scheduled for early October 1997, every person in the battle group was eager to get through the exercise and move on to the Mediterranean. But the USACOM J-7 training staff wasn't going to make that easy. To that end, several new elements were being added to the scenario in anticipation of new capabilities soon coming on-line. Within a couple of years, for example, the entire force of *Ticonderoga*-class (CG-47) cruisers and *Arleigh Burke*-class (DDG-51) destroyers will be receiving software and new Standard SAMs capable of providing the first theater-wide defense against ballistic missiles. Thus in JTFEX 97-3, the opposing forces were assumed to have a small force of SCUD-type theater ballistic missiles, some possibly armed with chemical warheads. The U.S. forces were not only expected to hunt these down, but

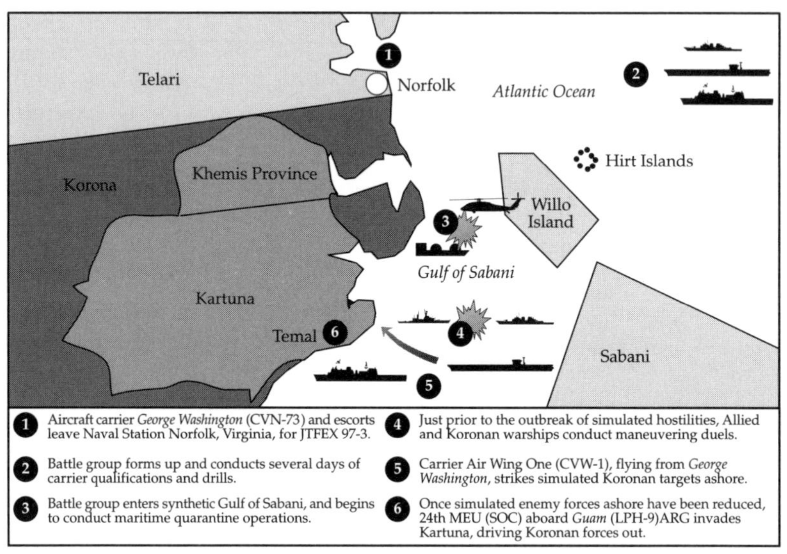

The activities of JTFEX 97-3 in August and September 1997.

Jack Ryan Enterprises, Ltd., by Laura DeNinno

to "shoot" them down with Patriot SAMs or with the Aegis systems on board several of the escorting vessels. The group's abilities in this area would be closely watched by USACOM.

In addition, CVW-1 was testing procedures for generating more sorties from the *GW*. This effort was based on a demonstration called a "SURGEX"—or Surge Exercise—run the previous July off the Pacific coast by the *Nimitz* battle group. SURGEX attempted to discover how many sorties a single carrier/air wing team could generate over a four-day period. By augmenting the air wing and ship's company with additional air crews and flight deck/maintenance personnel, and by adding the services of a number of land-based USAF tankers to support the effort, the *Nimitz* and her embarked air wing were able to generate 1,025 sorties in just ninety-six hours. This was almost 50% better than had been planned (though flight and deck crews wore out rather quickly). By the late summer of 1997, the *GW*/CVW-1 team was already implementing these lessons. Though they wouldn't have the additional flight personnel used by *Nimitz*, the flying day would be extended, USAF tankers would be made available, and some new procedures for monitoring crew fatigue would be tried. By doing things smarter, it was hoped that the average of around one hundred sorties a day might be increased by as much as half.

JTFEX 97-3: Players, Places, and Plans

The scenario for this JTFEX 97-3 was inspired by the 1990 invasion of Kuwait by Iraq, with the coastal waters of Virginia, North Carolina, and

South Carolina providing the primary battle arena. But thanks to the magic of today's electronics and GPS satellite technology, USACOM has been able to dispense with the actual geography of these littoral spaces and invent "synthetic" terrain for this and other recent exercises. Specifically, the USACOM staff created a series of "no-sail" zones off the East Coast that formed a simulated battlespace that looks a *great* deal like the Persian Gulf or Red Sea—long and narrow, with only a limited number of entrances and exits. It was into this arena that the *GW* battle group would sail during JTFEX 97-3.

For JTFEX 97-3 the opposing players would be known as Koronans, and the neutral victims of Koronan aggression would be Kartunans. Kartuna's homeland would be the center of the scenario's crisis. The Koronan forces would be played by various elements of the 2nd Marine Division at Camp Lejeune, North Carolina, 2nd Fleet at Norfolk, Virginia, and some Marine aviation units from the Marine Corps Air Stations at Cherry Point and New River, North Carolina and Beaufort, South Carolina.

While the simulated Koronan military would be nowhere as large as that of Iraq prior to the invasion of Kuwait in 1990, it would nevertheless have some distinct similarities. For example, Marine F/A-18 Hornet fighter/bombers from MCAS Beaufort would simulate Mirage F-1Cs armed with Exocet anti-shipping missiles and MiG-29 Fulcrums equipped with advanced air-to-air missiles (AAMs). Several *Spruance*-class (DD-963) destroyers and *Oliver Hazard Perry*-class (FFG-7) frigates would simulate Russian *Kashin*-class missile destroyers and Chinese missile corvettes. Finally, Marines based at Camp Lejeune would play Kartunan ground units, while Lejeune itself would play the part of the Kartunan homeland.

The forces of the Allied coalition would, of course, be played by the *GW* battle group and CVW-1, as well as their attached *Guam* ARG and the embarked 24th MEU (SOC). Though a battalion of the Army's 82nd Airborne Division and a number of USAF KC-135 aerial tanker aircraft based at Langley AFB, Virginia, would also play, the focus of this particular exercise was naval and expeditionary. This meant that if combat units could not fly or float into the JTFEX 97-3 scenario, they would *not* participate.

One of the largest (and most interesting) of the participants was Standing Naval Force Atlantic (STANAFORLANT). STANAFORLANT, established in 1967, was the Cold War equivalent of a World War II Hunter-Killer (HUK) ASW group—but with a unique twist. Each NATO nation involved was to assign a destroyer or frigate from their Navy to STANAFORLANT, and then the total force was placed under a single joint NATO commander. This arrangement has the advantage of providing the STANAFORLANT commander with an eclectic mix of weapons and sensors, and with personnel whose training, talents, and experiences are widely varied. Ships from Canada, Germany, the Netherlands, the United Kingdom, and the United States of America form the permanent membership of STANAFORLANT (there are usually a half-dozen ships operating in it at any given time); but they are joined periodically by Naval units from Belgium, Denmark, Norway, and Portugal.

STANAFORLANT carries out a program of exercises, maneuvers, and port visits, and can be rapidly deployed to a threatened area in times of crisis or tension. Throughout the Cold War, STANAFORLANT provided a rapid-response escort group for NATO naval commanders, in case of a sudden "surge" by the submarine and naval forces of the former Soviet Union. Today, STANAFORLANT's mission has broadened from this Cold War mission. Now STANAFORLANT is one of several NATO ready-reserve naval units that provide sea control services to the alliance (another of these units is in the Mediterranean supporting operations around Bosnia); and it could easily be found enforcing a maritime embargo or providing disaster/humanitarian relief. During JTFEX 97-3, it would practice all of these missions, and some others that would have been hard to imagine as little as ten years ago.

While STANAFORLANT would not technically be part of the *GW* battle group, it would nevertheless be attached to it. Since ships constantly cycle in and out of STANAFORLANT, there is no such thing as a "standard" mix of ships and weapons. However, the STANAFORLANT group that participated in JTFEX 97-3 is representative. Let's look at it:

STANAFORLANT Ships Participating in JTFEX 97-3

Ship Name	Ship Class	Country	Displacement (Tons)	Speed (Knots)	Armament	Helicopters
HMLMS *Witte De With* (F 813)	*Jacob Van Heemskerck* (F 812) FFG	Netherlands	3,750	30	Mk. 14 (1) with SM-1 SAM (40), Mk. 29 (1) with RIM-7 Sea Sparrow SAM (8 plus Reloads), Mk. 141 (2) with RGM-84 Harpoon SSM (2 Each), 30mm Goalkeeper CIWS (1), and Torpedo Launchers (2) with Mk. 46 ASW Torpedoes (2 Each)	N/A
HMS *London* (F 95)	Boxer (Type 22 Batch 2) FFG	Great Britain	4,800	30	MM-38 Exocet SSM (4), GWS.25 (2) with Sea Wolf SAM (6 Each with Reloads), 30mm Cannon (4), and STWS.1 Torpedo Launchers (2) with NST 75 Stingray ASW Torpedoes (3 Each)	HAS.3 Lynx (2) with NST 75 Stingray ASW Torpedoes, Mk. 11 Depth Charges, and CL 834 Sea Skua ASMs

Ship Name	Ship Class	Country	Displacement (Tons)	Speed (Knots)	Armament	Helicopters
FGS *Köln* (F 211)	*Bremen* (Type 122) FFG	Germany	3,800	30	76mm OTO Melara Cannon, Mk. 29 (1) with RIM-7 Sea Sparrow SAM (8 plus Reloads), Mk. 141 (2) with RGM-84 Harpoon SSM (4 Each), and Mk. 42 Torpedo Launcher (1) with Mk. 46 Torpedoes (4 plus 4 Reloads)	Lynx Mk. 88 (2) with Mk. 46 ASW Torpedoes
HMCS *Iroquois* (DDH 280)	*Iroquois* (DDH 280) DDG	Canada	5,100	30	127mm OTO Melara Cannon, VLS (2) with VL RIM-7 Sea Sparrow (8 Each with 24 Reloads), Mk. 10 LIMBO ASW Mortar (1 with 3 Barrels), and Mk. 32 Torpedo Launchers (2) with Mk. 46 Torpedoes (3 Each)	Sea King (2) with Mk. 46 ASW Torpedoes
NRP *Vasco da Gama* (F 330)	MEKO 200 FFG	Portugal	3,200	31.8	100mm (1), 1 Mk. 15 20mm CIWS, Mk. 29 with RIM-7 Sea Sparrow (8 plus 16 Reloads), Mk. 141 (2) with RGM-84 Harpoon SSM (4 Each), and Mk. 32 Torpedo Launchers (2) with Mk. 46 Torpedoes (3 Each)	Sea Lynx (1) with Mk. 46 ASW Torpedoes
USS *Peterson* (DD-969)	*Spruance* (DD-963)	United States of America	8,050	32.5	5"/54 Mk. 45 (2), 20mm Mk. 16 CIWS (2), Mk. 41 (1) VLS with BGM-109 Tomahawk SSM & ASROC (61 Total), Mk. 141 (2) with RGM-84 Harpoon SSM (4 Each), Mk. 29 with RIM-7 Sea Sparrow (8 plus 16 Reloads), and Mk. 32 Torpedo Launchers (2) with Mk. 46 Torpedoes (3 Each)	SH-60B Seahawk LAMPS III (2) with Mk. 46 ASW Torpedoes and AGM-119 Penguin ASMs

As you can see, the STANAFORLANT group assigned to participate in JTFEX 97-3 was a compact, powerful surface action group (SAG) that could bring a wide variety of weapons and systems to bear on a particular threat or mission.

For this exercise, the command of STANAFORLANT fell onto Rear Admiral Peter van der Graaf of the Netherlands, a tall blond bear of a sailor, who was based aboard his flagship, HMLMS *Witte De With* (F 813). With his vast smile and hearty laugh, Admiral van der Graaf quickly became a favorite of the *GW* battle group. In fact, he would hoist his flag aboard the *George Washington* at one point during the exercise. That he was a superb ship handler and leader only made STANAFORLANT's presence in JTFEX 97-3 the more telling.

One other small, but useful, naval force took part in JTFEX 97-3: a special mine warfare component. This unit would test a number of new ships, systems, and technologies designed to counter what most professional Naval analysts agree are the single greatest threat to naval littoral operations. These deadly "weapons that wait" are extremely cost-effective. They are not only relatively cheap to make, but they don't have to be high-tech to do the job. In fact, many current mines have basic technologies going back decades. At the same time, like their land-based counterparts, sea-based mines can make areas of ocean uninhabitable for years at a time. The fact that four of the five U.S. ships suffering combat damage in the last two decades have been hit by mines (the supertanker SS *Bridgeton*, frigate *Samuel B. Roberts* (FFG-58), Aegis cruiser *Princeton* (CG-59), and helicopter carrier *Tripoli* (LPH-10)) only highlights the threat. I should note that three of these ships were damaged by mines whose Russian design actually predates the First World War.

Despite the obvious threat presented by mines, over the last two decades mine warfare has been allocated less than one percent of the USN budget. The problem: Mine warfare is not glamorous. Compared with commanding a sleek destroyer or submarine, or flying a combat aircraft, it is considered decidedly "un-sexy" by most naval professionals. Much like infantry combat in urban areas, it is a nasty, dangerous business. Clearing mines takes a lot of time, it's filled with headaches, it generates casualties, and failure is easy to come by—not a smart career move. All the same, if the U.S. sea services are to become a littoral-capable force, mine warfare will have to become an equal partner with surface, subsurface, and air components of the fleet. The Navy has been taking serious action to make this intention a reality.

For JTFEX 97-3, a "rainbow" mine warfare force of ships, helicopters, and personnel was assembled from units along the coast of the Gulf of Mexico. These units represent the state-of-the-art of USN mine warfare technology and doctrine. Under the command of Captain Bruce Van Velle (who would act as the unit's Commodore and the mine warfare component commander), the unit was composed of the units shown below:

JTFEX 97-3 Mine Warfare Task Force

Ship Name	Ship Class	Commanding Officer	Displacement (Tons)	Speed (Knots)	Armament	Helicopters
USS *Inchon* (MCS-12)*	Ex-*Iwo Jima* (LPH-2)	Captain Matt Tuohy	18,798	23	20mm Mk. 15 Phalanx CIWS (2), 25mm Mk. 38 (2), and M2 .50 caliber HMG (8)	HM-15 (CO, Commander John Brown) with 8 MH-53E Sea Dragon helicopters
USS *Sentry* (MCM-3)	*Avenger* (MCM-1)	Lieutenant Commander J.J. Olsen	1,312	13.5	M2 .50-caliber HMG (2)	N/A
USS *Champion* (MCM-4)	*Avenger* (MCM-1)	Lieutenant Commander Chris Nygard	1,312	13.5	M2 .50-caliber HMG (2)	N/A
USS *Heron* (MHC-52)	*Osprey* (MHC-51)	Lieutenant Commander Rob Fink	918	12	M2 .50-caliber HMG (2)	N/A
USS *Oriole* (MHC-55)	*Osprey* (MHC-51)	Lieutenant Commander Bob Hospodar	918	12	M2 .50-caliber HMG (2)	N/A

*** Flagship of Mine Warfare Component, Captain Bruce Van Velle Commanding**

JTFEX 97-3 was the first occasion that the Navy's new emphasis on mine warfare was included in a major Atlantic Fleet joint exercise. The core of the mine warfare component was the converted helicopter carrier *Inchon* (MCS-12). Designed to act as a command ship for the mine countermeasures force of mine-hunting helicopters and minesweepers, the *Inchon* is the largest, most capable ship to ever take on this task. For JTFEX 97-3, she would act as a mobile air base for eight RH-53E Sea Dragon mine-countermeasures helicopters and four hundred personnel from Mine Countermeasures Squadron Fifteen (HM-15, the "Blackhawks"). Based at Moffet Field near Sunnyvale, California, and commanded by Commander John Brown, the Blackhawks are a mix of active and reserve personnel who fly one of the most interesting aircraft in the Navy inventory. Their MH-53E Sea Dragon, a modified version of the Marine CH-53 Super Stallion heavy transport, tows mine-countermeasures "sleds" and other equipment from a few hundred feet above the sea, and is one of the most effective means of clearing lanes through mine-infested shallow waters. Looking much like their Marine CH-53E brethren, the Sea Dragon is easily distinguishable by the large side sponsons filled with extra fuel, which the MH-53E guzzles at low altitude.

Along with the *Inchon* and her mine-countermeasures helicopters, four new *Avenger* (MCM-1) and *Osprey*-class (MHC-51) mine-hunters with their reserve crews would participate in the exercise in order to demonstrate new

RH-53E Sea Dragon helicopters of HM-15 aboard the flagship of the Mine Countermeasures Task Force during JTFEX 97-3, the USS *Inchon* (MCS-12).

OFFICAL U.S. NAVY PHOTO

ideas and technologies. These included an autonomous mine-detection vehicle, along with an explosive mine-clearing system to clear lanes for landing craft in the surf zone of an invasion beach. This increased emphasis on mine warfare is long overdue, and the efforts being pursued during JTFEX 97-3 are just the first of what will be many badly needed steps.

JTFEX 97-3 was under the control of the 2nd Fleet commander, Admiral Paul Reason, who would watch over the exercise from the fleet flagship, the USS *Mount Whitney* (LCC-20). Aboard the *Mount Whitney* would be the various warfare component commanders (air, naval, ground, special operations, etc.) that would run JTFEX 97-3, as well as many of the exercise observation personnel. JTFEXs and other large-scale exercises require significant numbers of people to manage and record what is going on. Thus, JTFEX 97-3 required the efforts of several thousand military and civilian personnel to observe, document, and analyze all that went on over the millions of cubic miles of battlespace off the Atlantic coast. These included observation teams from the Center for Naval Analysis (CNA—a U.S. Navy-funded "think tank") and members of the Senior Officers Observer Team (SOOT). The SOOT team is made up of ship, squadron, and other commanders who are temporarily detached from their own commands and assigned to observe and evaluate the actions of their counterparts.

It was against this backdrop of objectives, plans, technologies, ships, aircraft, and personnel that John Gresham (my researcher for this series) and I traveled south to the Virginia Tidewater to take in the events of JTFEX 97-3 in late August and early September 1997—almost three weeks. Partly because space doesn't permit, but even more because much of what I saw concerned operationally sensitive issues, I cannot begin to tell you about all of what went on there. Nevertheless, I'll show you some of the high points, as well as some of what life is like aboard U.S. warships.

Sunday, August 17th, 1997

August of 1997 was hot and humid in the mid-Atlantic. As if the challenges of the coming JTFEX were not enough, the weather gods were going to make the sailors and Marines suffer. On the afternoon before the group sailed, it was hot. Really hot! So hot that a new high-temperature record of 104° F/40° C had been set that afternoon at the Norfolk Naval Base. In this kind of heat, officers and NCOs had to watch closely for signs of heat stroke and exhaustion in their enlisted personnel as they labored to finish loading supplies and equipment, while ships and equipment had become so heat-soaked that they would stay hot for days to come. Even the heavy-duty air-conditioning of ships like the *GW* and *South Carolina* was having a hard time keeping up with it.

To my good fortune, I managed to miss much of the heat wave, since I would fly into the exercise several days later. But for John Gresham, the heat and humidity would become part of his permanent memory of JTFEX 97-3. John arrived late in the afternoon, thus missing the worst of the midday heat. As he pulled his car up to the long row of carrier docks, he could see all manner of ships. Two other big *Nimitz*-class carriers, the *Theodore Roosevelt* (CVN-71) and *John Stennis* (CVN-75), were tied up in the deep-water berths next to the *GW*. The "*TR*" was about to head into Dry Dock 12 across the river at Newport News Shipbuilding for her first major overhaul, while the "*Johnny Reb*" was working up for her first overseas deployment in 1998.

Hauling his bags up the long brow to the hangar-deck-level entrance, together with what seemed like thousands of other officers and men, John felt very small and very diminished. And well he might. For in fact, the *GW* was *full*. Every single bunk, stateroom, and cot was filled; some were actually being *shared* by contractor, observation-team, and training personnel who were augmenting the normal ship's crew. Though Lieutenant Joe Navritril, the capable young Public Affairs Officer (PAO) for the ship and battle group, had managed to find a stateroom for me up on the 02 level, John had to make do in somewhat less comfortable accommodations lower in the ship. Meanwhile, as John waited for Lieutenant Navritril to escort him below, he was able to meet a number of the ship's officers, including Captain Rutheford and the newly appointed Executive Officer (XO), Commander Chuck Smith.

After the young PAO arrived, he and John headed into the interior of the ship and went below. Once they reached the third deck, they headed aft to a small fifteen-man bunk room that was being used for overflow berthing during the coming exercise—hardly what you would call "plush" accommodations. Yet a quick look showed that nobody under the rank of lieutenant commander, or less than a contractor's technical representative, was getting even these berths. So John, counting himself lucky, wedged his gear and large frame into a center bunk to get some rest.

Even after nightfall, the entire ship was still like a sauna bath from the brutal pounding of the day's sun. Despite the best efforts of the air conditioners, some spaces would not cool down during the coming voyage. Unfortunately, John's bunk room was going to be one of these (it was located directly over one of the reactor/machinery spaces). Stoically accepting what couldn't be changed, John and his shipmates settled down for the night, quietly sweating away the hours until the carrier sailed the following morning.

Monday, August 18th, 1997

> JTFEX 97-3—Day 1: The situation in the Gulf of Sabani continues to grow worse, with the forces of Korona continuing to mass along the Kartunan boarders. Responding to requests from the Kartunan government, Allied coalition naval forces are being assembled to move to the Gulf to protect Allied interests, and to be prepared for possible contingencies such as evacuations of civilians and other endangered personnel.

After 0600 reveille, John rolled out of his rack to start the first of what would be many long, hot days at sea. Despite his cramped and steamy accommodations on the third deck, his location offered some advantages. For one thing, he was close to a nearby officers' head (bathroom), complete with shower facilities; and the main officers' dining room was located just a few steps forward of his berthing room. "Wardroom 3," the largest of the officers' dining areas aboard the *GW*, with room for over a hundred personnel to sit at once, is the social center of the ship. Service comes in two ways there. You can either pass through a cafeteria-style serving line, or you can have a mess specialist take your order.

This morning, since he wanted to have a good view of the sailing, and Captain Rutheford had announced that the ship would slip moorings promptly at 0700, John hurried through his breakfast. Afterward, he headed up to "Vultures Row" on the island and found a place outboard to watch the proceedings. "Yank" Rutheford was prompt, and the lines were slipped at the top of the hour. Overhead, an HS-11 HH-60G helicopter kept watch for signs of trouble, as well as to feed the traffic situation in the channel down to Captain Rutheford. Gently putting a few turns of reverse power on the screws, he eased the big flattop far enough away from the pier for the three tugs to take a hold on the carrier's hull.

All of these actions are done with exquisite precision and patience. On the one hand, the berths in this part of the harbor give a *Nimitz*-class carrier only about ten feet/three meters of clearance from the muddy bottom. On the other, the huge propellers tend to stir up the mud and sand, which can clog the delicate seawater inlets and condensers, and thus they are used carefully until the ship is in the middle of the main channel.

After backing carefully away from the pier, Captain Rutheford conned the ship in a wide reverse "Y" turn, leaving the *GW* aimed directly down the main channel leading to Hampton Roads. Calling, "All ahead one third," he now began the run down the channel to the right-hand turn that leads to Hampton Roads and the mouth of the Chesapeake Bay.

Meanwhile, other ships of the battle group and ARG were pulling out of their berths behind the *GW*, preparing to follow her down the channel. These included the cruisers *Normandy* (CG-60) and *South Carolina* (CGN-36), the *Seattle* (AOE-3), and the *Guam*. The cruisers would act as guards for the *GW* until the other ships of the battle group arrived to assist with that job. The *Guam* was on her way to meet up with the other three ships of her ARG, which were based down the Bay at Little Creek. From there, they would head south to Moorehead City, North Carolina, to pick up the personnel and equipment of the 24th MEU (SOC).

In fact, all along the eastern seaboard of the U.S., warships were leaving port to join up with the *GW* for the coming exercise. At the submarine base at Groton, Connecticut, the nuclear attack submarines *Toledo* (SSN-769) and *Annapolis* (SSN-760) cleared the Thames River channel and Long Island, heading south to join up with the rest of the group. Similarly, down at Mayport, Florida, the destroyers *Carney* (DDG-64) and *John Rodgers* (DD-983) and guided-missile frigates *Boone* (FFG-28) and *Underwood* (FFG-36) were clearing the mouth of the Saint John's River and heading north to the rendezvous point off of the Carolina coast. Finally, STANAFORLANT was finishing its trans-Atlantic run, planning to arrive several days hence.

While all of this activity was going on, the various elements of CVW-1 were finishing their movement to airfields in the mid-Atlantic region and preparing to conduct their ''fly-on'' to the *GW* the following day. For most of the squadrons, things had gone quite well. Most of the air wing would fly aboard the following afternoon, after which they would conduct several days of carrier qualifications prior to the actual start of the JTFEX scenario.

As they proceeded toward the open sea, the crew of the *GW* concentrated on getting things squared away after several months of berthing in Norfolk. All over the ship, electrical lines, hoses, and other pieces of equipment were being coiled, stowed, and put away. Even before the ship passed over the Hampton Roads Tunnel, there was a buzz of activity all over *GW*. Captain Rutheford had several training evolutions to run before the carrier exited the mouth of the Chesapeake Bay—anchoring drills in the forecastle and tests of the various firefighting and damage-control systems. After *GW* passed the outer bay, Captain Rutheford ordered course 090° (due east), and headed for the Virginia capes, where they met the two cruisers, and began a run south to pick up the air wing.

Tuesday, August 19th, 1997

JTFEX 97-3—Day 2: The war of words between Korona and Kartuna continues, with additional Kartunan military units being brought to a heightened state of readiness. In addition, Allied National Intelligence Sources have begun to track suspected Kartunan SCUD ballistic missile units to their training and test ranges.

That morning the *GW* and her cruiser escorts were operating in clear sunshine about 125 nm/230 km southeast of MCAS Cherry Point, North Carolina. Earlier they had broken into an area of high pressure and lower humidity, allowing the air-conditioning to get a start on removing the heat soak from the ships. Meanwhile, the *GW* flight deck crews were getting ready to take aboard their first jet aircraft of this cruise—not CVW-1 jets, but a group of USMC AV-8B Harrier II jump jets from Cherry Point. About 1100, the Harriers arrived, attracting lots of attention up on Vultures Row. Many of the young sailors had never seen a Harrier perform a vertical landing, a wondrous sight to folks used to the normal arrested landings of conventional carrier airplanes. After a short break for lunch, the Marines were off, so that the deck would be clear for the aircraft of CVW-1 arriving that afternoon.

1300 found the *GW* and her escorts steaming into the gentle southwest wind, preparing to take aboard the first of the aircraft from CVW-1. This was a particularly dangerous time for the pilots and crews of the air wing, since landing skills are easily lost without practice and it had been several months since their last ''traps'' aboard the *GW* during the CAPEX. For that reason, everyone on board the ship and in the air was being extra careful. Across the deck public-address system came the booming voices of the ''Air Boss,'' Commander John Kindred, and his assistant, the ''Mini Boss,'' Commander Carl June. Neither was risking ruining their perfect safety record during their assignments on board the *GW*. After making sure that the Landing Signals Officers (LSOs) were on their platform, the deck/safety crews were ready, the plane guard helicopter from HS-11 was overhead, and the cruisers were at their stations, the Air Boss and Mini Boss turned on the lights of the landing system and began to bring the air wing aboard.

One of the first to land was Captain ''Boomer'' Stufflebeem, flying an F/A-18C Hornet. Behind him came the approximately seventy aircraft of CVW-1. As soon as each plane was safely down and the arresting wire was cleared of the tailhook, the pilot was directed forward to a parking area ahead of the island. There the aircraft were either chained down or directed two at a time to one of the elevators, where they would be struck below to the hangar deck. There they would be carefully parked, sometimes with only inches between each bird. The crews then headed below to their squadron ready rooms down on the O-2 level, where they would check their landing scores from the LSOs. These scores were important to the flight crews, since each of them would have to requalifiy to land aboard the *GW* before being allowed to fly operationally in the coming JTFEX. For the next several days in fact, carrier qualifications would be the major activity of *GW* and CVW-1. Until each squadron's entire complement of air crew had successfully completed their required day and night ''traps,'' the *GW* would be nothing more than a training base. All of this took several days of nearly round-the-clock flight operations, and was to be the first real test of endurance for the *GW* and her crew.

Wednesday, August 20th, 1997

JTFEX 97-3—Day 3: At the direction of the U.S. Department of State, all U.S. citizens in Korona and Kartuna have been ordered to evacuate due to the threat of armed conflict. In particular, due to the threat of Koronan armed intervention, the National Command Authorities of the United States have ordered the 24th MEU (SOC) to conduct a Non-Combatant Evacuation Operation (NEO). This will be composed of U.S. citizens and other at-risk personnel from the Kartunan capital as soon as they can take up station in the Gulf of Sabani. In addition, elements of the *George Washington* Battle Group, the *Guam* Amphibious Ready Group, STANAFORLANT, and other units will stand by to support the NEO and other operations as required.

My own expedition to join JTFEX 97-2 began at midday, when I boarded a VRC-40 C-2A Greyhound Carrier Onboard Delivery (COD) aircraft of VRC-40 at NAS Norfolk. As I had been warned, the flight was fully booked and every seat filled. Everyone traveling out to the battle group had only the three COD flights to and from the *GW* as available transport.

Flying as a passenger on board a C-2 is unlike any other flying experience I've known. For starters, you sit in mildly uncomfortable ''bucket'' seats, facing aft in rows of four. Since payload and range are more important than creature comforts, the Greyhound has no sound-deadening material, and the air-conditioning system is decidedly crude, though robust. In the killing heat and humidity of the NAS Norfolk ramp, the vents spewed a chilling fog, which did not let up until we climbed to cruise altitude.

To help protect the passengers against the noise of the twin turboprop engines, we were each issued a ''Mickey Mouse'' cranial helmet with ear protectors. In addition, we each wore a ''float coat'' life preserver, just in case we had to ditch during the flight. When all of us were strapped in, the two crew chiefs gave us a safety briefing, then raised the rear cargo ramp, and the flight crew started the engines. As soon as these had warmed up, the aircraft was taxied to the end of the runway, and we were off.

Once the C-2A reached 10,000 feet/3,048 meters, the ride became more comfortable. The fog from the air-conditioning vents had become a flow of fresh air, and except for the constant rumble of the twin turboprops, everything was pleasant. As the aircraft turned southwest, we crossed over the coast and went ''feet wet.'' From my small window, I could see the four amphibious ships of the *Guam* ARG loading up the elements of the 24th MEU (SOC). When they finished this task, they would join up with the rest of the battle group, now operating approximately 200 nm/370 km offshore.

It took us just under an hour to reach the battle group's operating area, at which point we were put into a wide port turn to hold for landing. Since qualifying pilots is considered more important than landing VIPs, we circled the battle group for almost a half hour before the command came from the *GW*'s air traffic control center to get into the landing pattern. Soon after the Greyhound broke into the landing pattern, the flight crew gunned the engines and headed onto the final approach to the carrier.

Back in the passenger/cargo compartment, the crew chiefs ordered all of us to brace ourselves. After a big ''thump'' when the wheels touched down, I was jammed back into my seat as the tailhook snagged one of the arresting wires. Once the aircraft was stopped, the deck crews quickly disengaged the hook and began to fold the wings. The flight crew then taxied forward to a parking space ahead of the island, where the deck crews immediately began to chain the bird down.

Moments later, I was following the other passengers to a hatch in the island structure, and then down a ladder to the *GW*'s Air Transport Office (ATO) on the O-2 level. There we checked in with the ATO watch officer, handed in our survival gear, and picked up our bags. By this time, John Gresham and Lieutenant Navritril had arrived to escort me to my stateroom on the O-2 level. After a shower and a quiet dinner in Wardroom 3 with John and Lieutenant Navritril, I went to bed. Even the pounding noise of aircraft launching and landing one deck above did not keep me from sleeping.

Thursday, August 21st, 1997

> JTFEX 97-3—Day 4: The Koronan government has continued to threaten their Kartunan neighbors. They are claiming that the Northern Kartunan province of Khemis is legally part of the Koronan homeland, and are demanding its concession to avoid hostilities. Meanwhile, the various personnel to be evacuated during the planned NEO have begun to assemble at the American embassy in the Kartunan capital city of Temal.

I awoke at 0600, showered, and headed down the ladders to Wardroom 3 for breakfast. There I met John and Lieutenant Navritril. As we finished breakfast, Lieutenant Navritril informed us that we would be meeting with Admiral Mullen later that morning to discuss his plans for the upcoming exercise and his philosophy of running a carrier battle group.

Promptly at 1000, we arrived in the flag officers quarters in ''Blue Tile'' land on the O-2 level, and soon after that we headed into the admiral's sitting room. There Rear Admiral Mike Mullen, the commander of Cruiser-Destroyer Group Two (CRUDESGRU 2) and the *GW* battle group, warmly greeted us. Admiral Mullen is a surface line officer, one of the new generation of battle group commanders now sharing command opportunities with naval aviators. He is a handsome man, cool and intellectual; it's no surprise that he has a Harvard master's degree. At the moment, he was clearly preoccupied with getting the battle group set up for the coming JTFEX 97-3 exercise.

Spacing his remarks between the metallic roars of Catapult Number One (located directly above his stateroom), he discussed his vision of CVBG operations. Back in the Cold War, he explained, the Navy feared that the Soviet Union would try to eliminate the U.S. naval presence by sending multiple regiments of Tu-22M Backfire and Tu-16 Badger bombers armed with huge air-to-surface missiles (ASMs). The idea was to kill the CVBGs, after which Soviet submarines and surface groups would clean up the survivors with their own SSMs. It was against this threat that systems like the F-14 Tomcat fighter and Aegis were developed; their function was to shield the CVBG from waves of incoming missiles.

Today the world has greatly changed. With the bomber regiments of the Soviet Union a thing of the past, the threat of air attack on a CVBG has been so reduced that it is no longer necessary to maintain standing combat air patrols (CAPs) of airborne fighters. At the same time, the once-formidable fleet of Soviet submarines and missile ships is now either rusting at anchor, or else has been cut up into scrap metal. At this point the new CVBG tactics now practiced by Admiral Mullen and his contemporaries take the stage.

Today's CVBG tactics revolve around the reality that in the post-Cold War world very little threatens U.S. naval forces. The only navies in any way capable of hurting us are not likely to do that, since they are already friends like our NATO allies and Japan. Even if the friendship were somehow to cease, we could probably beat all of them together in a fair fight. All of this means practically that we can greatly reduce our commitment of aircraft and vessels to self-protection, and their resources can now be dedicated to the projection of offensive power. Now that they are relatively unconcerned about the military threat from other nations, CVBG commanders like Admiral Mullen want to *be* the military threat in a theater of operations. Obviously, Admiral Mullen does not plan to ignore enemy threats. To do that would be both foolish and irresponsible. Rather, he plans to leverage his best systems so as to defend the fleet with the greatest possible efficiency and effectiveness.

The practical consequences of all this doctrinal thinking is that the "Outer Air Battle" concept that drove fleet air defense tactics in the 1980's is dead and gone. Instead of massed CAPs of F-14 Tomcats and F/A-18 Hornet fighters, backed by a wall of SAM-armed escorts directed by the Aegis cruisers and destroyers, a more modest defense plan has been adopted. To that end Admiral Mullen now planned to depend on the SM-2 Standard SAMs aboard the *Normandy*, *South Carolina*, and *Carney* to project an air defense "bubble" over the battle group and ARG, while continuing their other jobs of keeping naval and submarine threats at bay. This would allow him to reserve the sorties of his Tomcats and Hornets for the job of delivering ordnance onto land and naval targets. In other words, the escorts would be doing a kind of "double duty" so as to allow the rest of the force to project offensive power into the littoral regions that would be their operating areas.

Tomahawk and Harpoon cruise-missile strikes would be used wherever possible on fixed or naval targets, so that manned aircraft sorties would be saved for more mobile or more difficult missions.

For JTFEX 97-3, he would take the fullest possible advantage of both STANAFORLANT and the USAF KC-135 tankers he was being supplied with. Also, wherever possible, he would make use of special operations forces (from his own onboard SEAL (Sea-Air-Land) team and the 24th MEU (SOC) as force-multipliers and reconnaissance assets. Finally, though he was discreet in his references to them, Admiral Mullen planned to make full use of the two available SSNs, the *Toledo* and *Annapolis*. Both represented immensely capable platforms for a number of missions (such as intelligence-gathering, ASW, and Tomahawk cruise-missile attacks), and he clearly had big ideas for their use.

Just before our visit came to an end and we headed back down to Wardroom 3 for lunch, Admiral Mullen told us what he would stress the most in the coming exercise: safety. He had good reason to be concerned. There had been numerous deaths during the previous three East Coast JTFEXs. In JTFEX 96-2, for example, which was run in May of 1996 by XVIII Airborne Corps, thirteen deaths had resulted from the collision and crash of two USMC helicopters at Camp Lejeune, North Carolina. For the JTFEX about to begin, Admiral Mullen had just one simple objective: to bring every one of the participants home alive with all the important parts attached and in working order! He planned to accomplish this objective by a variety of means, ranging from "buddy system" checks of deck crews for fatigue, to regular drills for damage control and battle stations. As John and I left, we could only pray that these plans would work.

Friday, August 22nd, 1997

> JTFEX 97-3—Day 5: As the Koronans continued their threat, the United Nations Security Council voted the first of a number of sanctions against them, in the form of a maritime embargo on weapons and petroleum products. This embargo was to go into effect immediately, and would be enforced by the military forces of the U.S. and her coalition allies. To support this operation, the coalition naval forces will move into the Gulf of Sabani immediately. Also, the *Guam* ARG and 24th MEU (SOC), supported by elements of the *GW* CVBG and STANAFORLANT, will begin to conduct the planned NEO from the American embassy compound in Temal.

This morning John and I were extremely busy, as we were about to move from the *GW* to the Aegis cruiser *Normandy*, where we would spend time with the "small boy" sailors of the *GW* battle group. As luck would have it, our rushing around turned out to be unnecessary. A heavy squall line had moved through overnight, leaving behind rough seas, high winds, and heavy patches of rain—and delaying our departure. Meanwhile, the high summer heat continued, with peak temperatures over 90° F/32 C°. All of this meant that flight operations around the battle group and ARG were extremely dicey.

Before our departure, we had a scheduled meeting with three of the CVW-1 squadron commanders down in the air wing ready room. Joining us for coffee and a chat in the comfortable, leather-covered ready room chairs were Commander Curt Daill of VF-102 (flying F-14B upgrades), Commander Robert M. Harrington of VFA-86 (flying the Block 10 F/A-18C Hornet), and Commander Michael Mulcahy of HS-11 (flying SH-60F and HH-60H Seahawk). The three men's comments about the aircraft they flew and the units they commanded turned out to be both candid and informative.

Curt Daill is every inch the classic F-14 Tomcat driver, with all the ego and ambition that go with the job. As commander of VF-102, he headed a squadron that was rapidly acquiring new and useful capabilities. Already flying one of the most powerful warplanes in the world with its F-110 engines and AWG-9/AIM-54 Phoenix weapons system, the "Diamondbacks" had recently added two new systems to their aircraft. These are the new Digital Tactical Airborne Reconnaissance Pod System (D/TARPS) and the AAQ-14 Low Altitude Navigation (LANTIRN) pod. The four D/TARPS pods assigned to VF-102 allowed them to take and transmit near-"real-time" targeting images while still over the target. This capability would allow Admiral Mullen to plan a strike on target just minutes after the D/TARPS-equipped F-14 locates it. The AAQ-14 LANTIRN pod (which has a built-in GPS/INS system) gives the F-14 community the ability to conduct day and night precision strikes with Paveway LGBs, as well as accomplishing wide-area reconnaissance with GPS positional accuracy. Both of these new capabilities made VF-102 one of the most desirable air units that a CinC might be assigned in a time of crisis.

Commander Bob Harrington, a quiet, intense man, who lets his squadron's actions speak for him, heads VFA-86. Another long-time Naval aviator who has seen his chosen flying community move in surprising directions, Commander Harrington has gone from operating A-7 Corsairs armed with "iron" bombs to taking up the F/A-18C armed with the most advanced PGMs.

Our third squadron leader, Commander Michael Mulcahy of HS-11, arrived just a little late. Though he didn't tell us then, we later learned that the skipper of the "Dragon Slayers" had just flown one of the first missions of the UN-mandated maritime embargo of the Koronan forces. Flying an HH-60H loaded with a SEAL team, he had swooped down on the fleet oiler *Merrimack* (AO-179), which was being used by USACOM to simulate a merchant ship transporting concealed arms and other sensitive cargo to Korona. Hovering over the oiler's deck, the SEAL team had "fast roped" down to the ship and conducted a simulated "takedown" of the suspected weapons cache that intelligence sources had reported there. After seizing the ship, the SEAL team had called for a prize crew from one the battle group escorts and turned the ship over to them.

"I haven't really been up to anything important this morning," Commander Mulcahy remarked as he coolly joined us in the ready room.

HS-11 is one of the few carrier aviation units that fly two different aircraft. In addition to the SH-60F variant used for submarine hunting, they also fly the HH-60H search and rescue (SAR)/special operations version. This means that in addition to helping protect the *GW* from submarines that might penetrate the so-called "inner zone," inside the protective ring of escort vessels, they also provide the battle group with the ability both to rescue downed air crews and to deliver and retrieve special operations teams. This is a wide range of roles and missions for a unit with only six aircraft (four SH-60Fs and two HH-60Hs), and it means that they almost always have a bird or two in the air somewhere.

Following our chat with the squadron leaders, John and I returned to our quarters to pick up our bags, and then we headed up to the ATO office with Lieutenant Navritril. There we checked in with the ATO watch officer, gathered our float coats and cranial helmets, and tagged our bags. Once we had taken care of these details, Lieutenant Navritril introduced us to Captain James F. Deppe, the CO of the *Normandy*. Jim Deppe, a tall, slim, handsome, native Texan, is a 1974 Naval Academy graduate who has spent his career in the surface warfare community. After serving most of his sea time on frigates (he commanded the USS *Kauffman* (FFG-59) from 1992 to 1994), he was selected to take over command of the *Normandy* in early 1997.

As we began talking with him, the ATO watch officer announced that it was time to head up to the flight deck and board an HS-11 SH-60F for the flight over to the *Normandy*. Grabbing our bags and other gear, we followed a yellow-shirted flight deck handler up a ladder, exited the island, and walked into a full-blown squall, complete with forty knot winds over the bow, blasting horizontal raindrops (heated to over 80° F/27° C by the local weather) into our faces! Leaning into the storm, we struggled across the deck between other aircraft preparing to take off. The Seahawk was parked on a spot over one of the waist catapults with its engines already turning.

Soon after we had crowded aboard and were strapped in, the crew got ready to take off. But as the pilot ran through his checklist and throttled up, he got a warning light indicating a problem in one of the T700 engines. Quickly, both power plants were shut down, and we were asked to leave the aircraft and head back over to the island. By this time thoroughly soaked, we descended back to the O-2 level and the ATO office, while flight deck crews cleared the broken bird from the deck and started up the next flight event. Within minutes, the voice of Air Boss John Kindred boomed over the flight deck PA system, soon followed by the roars of jet engines and the screech of catapults.

As we stripped off our soaked survival gear, the ATO personnel handed us dry towels and cold drinks. Then we sat down to wait. Fifteen minutes later, we were told that the *Normandy* would launch one of her own SH-60B Seahawks, which would collect us following the flight event currently under way. The bad news was that it would take at least three hours before they could land aboard the *GW*. We had a long wait ahead of us. The good news was that this would give us a chance to talk with Captain Deppe, and get some feel about how he and his ship were being used by Admiral Mullen.

As CO of one of the most capable Anti-Air Warfare (AAW) platform in the fleet, Deppe had been assigned the job of AAW coordinator for the entire force. Since most of the other warfare functions coordinators (ASW, Anti-Surface Warfare (ASUW), etc.) were based aboard the *GW*, and the *Normandy* had nothing like the secure, wide-bandwidth satellite communications systems that would allow secure teleconferencing, he had to make the commute over to the *GW* almost daily. This was necessary in order to attend secure conferences among the officers responsible for the battle group's defense. Add to this the relative novelty of the battle group tactics being practiced by Admiral Mullen, and you have Jim Deppe spending several hours in the air each day going back and forth between *Normandy* and "Blue Tile Land" in *GW*. This new way of running a CVBG is an extremely "hands on" way of doing business, and until new wide-bandwidth satellite telecommunications systems become more common in the fleet, you're going to see a lot of ship COs flying back and forth between ships.

It was almost 1500 (3 P.M.) by the time the last of the CVW-1 aircraft were brought aboard, and the waist helicopter landing spots cleared. The HSL-48 Seahawk had circled the *GW* for almost an hour, and the crew was clearly in a hurry to get back home, approximately 100 miles/161 kilometers away. By this time, the squall had cleared enough for us to cross the flight deck without getting soaked. This time, the preflight checks all went well, and within minutes, the crew was cleared to launch. After we lifted off, we headed east to rendezvous with the *Normandy*. Flying at around 1,500 feet/457 meters altitude, we stayed below the cloud base and ran flat out to the east. About halfway to the cruiser, I looked out a window and saw below a dirty brown streak in the water spreading out for miles. When I asked the crew chief about it, he frowned. "Pollution," he said. Some ship had passed through and pumped its bilge into the blue of the Atlantic. It occurred to me just then that an antiship missile might come in handy—*pour encourager les autres*.

Soon our new home, the Aegis cruiser USS *Normandy*, came into view. Steaming into the wind, she was making ready to take us aboard. The deck crews were making quick work of it. After just a single circle of the cruiser, the pilot ran up the wake of the ship, matched his speed to the ship's, and hovered over the helicopter deck. At this point, the crew chief winched down

Captain Jim Deppe, the CO of USS *Normandy* (CG-60), cons his ship while refueling under way from the USS *Seattle* (AOE-3).
JOHN D. GRESHAM

a small line with a ''messenger'' attachment at the end. When it reached the deck below us, a deck crewman scampered across to the messenger and inserted it into the clamp of the ship's Recovery, Assist, Secure, and Traversing (RAST) system—a system of mechanical tracks in the deck of the ship's helicopter pad. The clamp, which runs on the tracks, is designed to hold the messenger at the end of the line. The helicopter can then be winched down safely and securely onto the deck, even in heavy seas. Soon, we found ourselves on deck, and Captain Deppe was rushing up to the bridge.

The reason for his hurry was quickly evident. The huge bulk of the USS *Seattle*, the *GW* battle group's fleet replenishment ship, was showing on the horizon. We had arrived just in time for him to take over the delicate and sometimes difficult job of conning the ship while replenishing under way. After leaving our bags for the deck crews to take to our quarters, we followed him to the bridge—not an easy undertaking. To reach the top of the cruiser's massive deckhouse requires climbing some seven ladders. The effort was worth it, though, for up there we had a splendid view of one of the most beautiful dances performed by U.S. Navy ships.

I've always believed that the skill that separates great Navies from the also-rans is the ability to sustain a fleet at sea with underway replenishment (UNREP). Something of an American invention prior to World War II, UNREP is a little like an elephant ballet. The dynamics of conning a ship in close proximity to another are completely different from any other kind of ship handling, and Captain Deppe was about to give us a textbook lesson in the art.

Initially, he allowed Captain Stephen Firks, CO of the *Seattle*, to come up on *Normandy* and position his ship on the cruiser's port (left) side. Once this was done, the *Seattle* began to shoot messenger lines across the gap to the deck crews of the *Normandy*. After these were recovered, the deck crewmen pulled larger lines across and began to rig the refueling lines. For this UNREP, only two refueling lines would be set, since only JP-5 jet fuel for the *Normandy*'s gas turbine engines and helicopters was being transferred, so there would not be any ''high lines'' for moving cargo or other supplies. There would also be no use of the *Seattle*'s UH-46 Sea Knight Vertical Replenishment (VERTREP) helicopters, as the *Normandy* was still well stocked with food and other consumables.

Within ten minutes, the lines were rigged, and the refueling hoses were pulled across the hundred feet/thirty meters or so of space between the two ships. Each hose has a ''male'' probe, which locks into a ''female'' receptacle on the receiving ship. These can be rapidly disconnected in the event of an emergency, what the Navy calls a ''breakaway.'' When properly set and pressurized, each hose can move several thousand gallons a minute of distilled petroleum products. As soon as the refueling probes were secured into their receptacles, the *Seattle* began to pump JP-5 over to the cruiser. Gradually, the pressure was built up, and the flow increased.

While all of this was going on, the two ship captains were carefully conning their vessels, making sure that the spacing and alignment remained constant. This can be difficult with ships of different sizes. Since the larger one wants to ''suck'' the smaller vessel into its side, maintaining station during UNREPs is a delicate business measured in an additional rpm or two of shaft power, or a twitch of propeller pitch. This afternoon all went exceedingly smoothly, and Captains Deppe and Firks (of *Seattle*) put on a show of ship handling that one could only admire.

Part of the beauty of this operation is that it is done virtually without radio or other electronic signals. To keep things simple and quiet, only lights and flags are used. After about thirty minutes of refueling, the call came up from engineering that the *Normandy*'s fuel bunkers were full and the UNREP completed. As they uncoupled the hoses, the crews of both ships were careful to limit JP-5 spills into the sea, to minimize pollution. Not many of us realize how tough pollution-control rules are on the military, and how hard they work to be ''green.'' Once the hoses were retracted back to the *Seattle*, the deck crews began to strike their lines and drop them over the side to be retrieved by the oiler's personnel. Now came one last ticklish operation.

Captain Deppe ordered all ahead two thirds (about twenty knots/thirty-seven kilometers an hour), and then began a gradual turn to starboard, a maneuver designed to make the breakaway from the 53,000-ton oiler as smooth and easy as possible. Deppe ran the cruiser through a full 360° turn and almost 10,000 yards/9,144 meters of separation from the *Seattle* before he felt free to maneuver again. At the completion of this turn, he ordered the cruiser to head west to join up with some other ships of the *GW* battle group. After that, we all adjourned below to freshen up for dinner.

I was escorted to quarters usually reserved for an embarked flag officer—very luxurious after the cramped quarters of the *GW*. With only around 350 personnel, the *Normandy* is much more intimate and pleasant than the carrier. People can actually find privacy here and there on *Normandy* if they want it. Another nice thing about being on one of the ''small boys'' was the absence of the hundreds of extra VIPs, observers, media personnel, and contractors now on the carrier, making space and comfort more plentiful than aboard the *GW*. Perhaps the only thing I missed was the live video feeds from CNN and other networks provided by the onboard Challenge Athena system.

As we gathered in the *Normandy*'s wardroom for dinner, I was struck by the youth of Captain's Deppe's officers. While the department heads were mostly lieutenant commanders, most of the others were lieutenants with less than five years service. Escort duty is a young person's profession, and around the table the majority of the faces were under thirty. Aboard the "small boys" of the cruiser/destroyer/frigate force, the officers' wardroom is the center of their social world. The wardroom table is a place of open expression, with rank and position holding little sway. Here problems are discussed, assignments made, and professional experience passed along to young officers. There is *very* little formality. The only real rule is that everyone stands for the captain, and waits for him to serve himself before everyone else does so. As for the food, it's as good as any you will find in the fleet. From the *Normandy*'s small galley came a mountain of edibles, including a fine salad bar and excellent baked chicken and rice. The only problem you'll find is dealing with the roll of the ship. And therein lies a story.

The *Ticonderoga*-class (CG-47) Aegis cruisers were built upon hulls originally designed for the *Spruance*-class (DD-963) general-purpose destroyers. They share a common structural hull power plant and many other systems. However, the extra load of weapons and other equipment associated with the Aegis combat system has definitely "maxed out" the original *Spruance* design. The "*Ticos*," as they are known, displace fully 15% more than a *Spruance*, much of which is located in the tall deckhouses that mount the four big SPY-1 phased-array radars that are the heart of the Aegis system. What this all means is that the *Ticos* are top-heavy. Not enough to make them unstable or prone to capsizing, mind you; but enough to make them less than comfortable for those who don't enjoy pitching, swaying, and rolling. In fact, they handle the seas quite well and maneuver like a small Italian sports car in the hands of a professional. However, they do roll a *lot*! In a heavy sea or sharp turn, they can heel up to 40° from the vertical. It is not particularly uncomfortable, and does not tend to cause motion sickness. However, it does make activities like eating meals potentially exciting. And for us that evening, more than once the ship took rolls steep enough to force us to grab hold of plates and serving dishes.

After dinner, we were given a tour of the engineering departments and combat center. While *Normandy* is almost ten years old (she was commissioned in 1989) and coming to the end of her second five-year operating period, she is in terrific shape. In fact, I was amazed how well her crew has maintained her. Everything was spotless, even the deck corners; and all the sensor and combat systems were "up" and ready for action.

Normandy is representative of the "Baseline 3" *Ticos*, with improved lightweight SPY-1B radars (each Aegis ship has four of these) and new computers. Following the 1997/98 cruise, she will head into the yard for a major overhaul, which will completely update her Aegis combat system to the latest version. When she comes out of the yard sometime in 1999, she will be equipped with the new SM-2 Block 4 SAM, which will give her an ability to engage and destroy theater ballistic missiles (TBMs). Eventually, the entire fleet of Aegis cruisers and destroyers will have this capability, which will greatly reduce the risks from enemy TBMs to our forward-deployed forces. Today, the crew of the *Normandy* and the Aegis destroyer *Carney* were simulating some of the engagement techniques that will be part of that future capability.

After the tour, I headed down to the commodore's stateroom and sleep. John and I were scheduled to return to the *GW* in the morning, as we had been hearing rumors that the "hot war" part of the JTFEX scenario might start within a day or two. I had wanted to be aboard the *GW* when that happened in order to have the best possible view of the start of the hostilities. As it happened, things didn't work out according to schedule—to our great good fortune, for we ended up experiencing the most interesting day of the exercise.

Saturday, August 23rd, 1997

> JTFEX 97-3—Day 6: The Koronan government today continued to pressure Kartuna by test firing several SCUD ballistic missiles on their test range. This is seen as a sign that they are bringing their theater ballistic missile combat units to a high state of combat readiness. In addition, the Koronan fleet has been surged out of their ports, and is currently moving into position to track and trail the Coalition Naval forces massing in the Gulf of Sabani. Meanwhile, elements of the 24th MEU (SOC) and *Guam* ARG have commenced their NEO of the American embassy compound in Temal. It is expected that this operation will be completed early on the morning of August 24th.

By Saturday morning, much had happened in JTFEX 97-3. Overnight, the *Normandy* and the other escorts had rejoined the *GW*, and the combined battle group had entered the northern end of the Gulf of Sabani. Passing by the (imaginary) Willo and Hirt Islands, the group turned south into the Gulf to support the *Guam* ARG/24th MEU (SOC) in their NEO of endangered personnel from Kartuna.

Meanwhile, the USACOM J-7 exercise leaders were working hard on the "flex" part of the scenario, trying to bait Admiral Mullen and his commanders into actions that would cause hostilities to break out immediately. For the admiral and his staff, their job was to keep a "lid" on the scenario for as long as possible—important in the light of the NEO the 24th MEU (SOC) which had begun in the predawn hours. Here was to be the "eyeball-to-eyeball" phase of the exercise, simulating the "short-of-war" realities that our commanders would face in an actual crisis. Even though this was a training exercise, you could feel the tension of the emerging situation. Everyone in the battle group knew that they were being evaluated for their readiness to go into a potential combat situation during JTFEX 97-3, and nobody wanted to let the rest of the force down.

All around the battle group, ships from the Atlantic Fleet were being used to simulate Koronan Naval vessels in an "aggressor" role. And numerous other ships were simulating neutral shipping traffic, trying to get clear of the emerging fracas, or to get one more cargo run in before the "war" started. The final proof that the "hot" phase of the exercise was about to begin arrived on a UH-46 transport helicopter's morning run in the form of the *Normandy*'s SOOT team representative. This was Captain James W. Phillips, the CO of the Aegis cruiser *Vella Gulf* (CG-72), who had come aboard to observe the proceedings and to evaluate the performance of Captain Deppe and his crew during the exercise. Captain Phillips is a courtly gentleman who quickly attached himself to Jim Deppe, and they were soon chatting away like two old friends working out the best place to catch a prize bass. But you only had to look out a porthole of the *Normandy*'s wardroom during breakfast to see that the game afoot in this pond beat the hell out of any fishing you might find ashore.

Things were about to get very interesting in this little patch of the Gulf of Sabani. About 1,000 yards/914 meters off the starboard beam, a *Normandy* whaleboat was taking a maritime inspection team to the frigate *Samuel Elliot Morrison* (FFG-13), which was currently playing the part of a neutral merchant ship. Breakfast was hardly finished when the word came over that the frigate had a real casualty who needed to be evacuated back to the mainland, an action that caused a problem for John and myself. The diagnosis was hepatitis, and the patient was being transported over in the whaleboat with a corpsman.

With only a single HS-11 sortie scheduled to fly from *Normandy* to *GW* that day, this meant that the casualty and corpsman would take our places on the Seahawk, and we would have to wait another day or two to return to the carrier. Captain Deppe made it clear that he would do his best to get us back as soon as possible. And besides, he went on to say, there was plenty of room for us aboard, and since it was Saturday in the "real" world, it would be pizza night on the *Normandy*. Since *Normandy* had one of the best galleys in the Atlantic Fleet, this sounded like making the best of a bad situation.

After the HS-11 Seahawk arrived and collected the casualty and corpsman, the ship passed into a comfortable high-pressure zone, which had the effect of dropping the temperature to a refreshing 80° F/27° C, and drying out the air to a sparkling clarity. Visibility became almost unlimited, with line-of-sight ranges running to almost 30,000 yards/27,400 meters. It soon became the most beautiful day I'd seen in months, with a flat calm sea and almost no wind. Meanwhile, the "bubble" of visible space around us had become crowded with ships.

Later that afternoon, around 1600 (4 P.M.), as I stood on the helicopter platform aft, I noticed something strange. One of the nearby ships suddenly closed from astern to around 2,000 yards/1,828 meters, and tried to move around us, much as a car tries to pass a truck on an interstate highway. A moment later, I felt the deck shudder underneath my feet, and heard the sharp whine of the *Normandy*'s four LM-2500 gas turbines going to full power. In just seconds the cruiser jumped from twelve to thirty knots, and Captain Deppe radically cut in front of the other ship, blocking the pass. Somewhat dazzled by this maneuver, I looked aft at the other vessel, a *Spruance*-class destroyer that I initially expected to be the USS *John Rodgers* (DD-983) from our battle group. But then I noticed that this *Spruance* did not have the ASROC launcher of the *John Rodgers*, and a quick glance at her pennant number confirmed my suspicions. It was the USS *Nicholson* (DD-982)—a VLS-equipped *Spruance* simulating a Koronan *Kashin*-class guided-missile destroyer. Clearly the JTFEX 97-3 scenario was growing hotter. John and I headed forward to the bridge at a dead run to find out what was going on.

As we arrived on the port bridge wing, I saw the *Nicholson* trying to slip up our beam. Over at the edge of the bridge were Captains Deppe and Phillips, watching intently as the destroyer maneuvered. At the same time, the TBS (Talk Between Ship) radio circuit came alive with traffic from all around the battle group. Two frigates simulating Koronan guided-missile gunboats were maneuvering aggressively. Looking to one of the young lieutenants, I asked, ''What the hell is going on?''

''They're playing chicken,'' he said, ''like the Russians.'' The remark was like a trip through time for me.

Back in the Cold War, the ships and submarines of the Soviet Navy used to trail our CVBGs the way *Nicholson* was doing. This was a favorite tactic of the late Admiral Sergei Gorshkov (the longtime chief of the Soviet Navy), and took advantage of the ''freedom of navigation'' rules accorded ships on the high seas. The idea was to maneuver for a clear line of sight to the carrier the way they'd do just before the outbreak of a real conflict. In the ''first salvo'' of that war, the ships and subs would fire their missiles, torpedoes, and guns and attempt to put the flattop out of action. The only way to defeat this threat was for our own escort ships to maneuver aggressively, physically placing themselves between the enemy ships and the carrier. At times, vessels of both sides would actually ''bump.'' Such aggressive maneuvering now and then increased tensions between the superpowers.[85] We used to call it ''Cowboys and Russians,'' and I had thought that it was a thing of the past. I was clearly wrong.

Though it's not publicized by the U.S. Navy, the tactic of interposing an escort ship between an opponent and the carrier is still practiced; it re-

85 It was these kinds of ''bumping'' incidents that caused the creation of the ''Incidents at Sea'' treaties between the United States, the Soviet Union/Russia, and a number of other nations.

The destroyer USS *Nicholson* (DD-982), during her maneuvering duel with the USS *Normandy* (CG-60).
JOHN D. GRESHAM

sembles the ''hassling'' that fighter pilots engage in to keep themselves sharp. But ''dogfighting'' with billion-dollar cruisers and destroyers is riskier. Clearly the USACOM training staff wanted to stress Admiral Mullen and his staff into a situation where the Koronan forces could claim a provocation and initiate hostilities while the 24th MEU (SOC) was still conducting their NEO in Temal. The challenge was clear. If a Koronan ship was able to draw a line-of-sight bead on the *GW*, then the escorts would be required to ''fire'' on the offending vessel to keep the flattop safe. At the same time, because *GW* was conducting flight operations, there was very little Captain Rutheford could do to help combat the intruders.

For the next few hours, it would be up to the ''small boys'' of the battle group to keep the Koronan missile ships at bay. Clearly, the *Normandy*'s Saturday night pizza tradition was about to go on hold for a while. Captain Deppe, immediately grasping the challenge, went to the task with a grin on his face. Opportunities were rare to maneuver his ship to its limits against a fellow skipper in an almost perfectly matched ship. This was just such a chance. Although there are clear exercise rules about how close opposing combatants are allowed to approach, these rules were about to be bent. In fact, the only rule seemed to be: Don't actually *touch* the other guy!

The next few hours went by very quickly, as we parried and thrusted with the *Nicholson*. The captain of the *Nicholson* (Commander Craig E. Langman) was extremely aggressive, doing everything he could to get past us. He never succeeded. Captain Deppe maneuvered the *Normandy* like a Formula I racing car, keeping the destroyer solidly away from the flattop. At times we raced ahead at over thirty knots, only to crash-stop within a ship length or two. Then we might sit for ten or fifteen minutes, with just a thousand yards or so separating the two vessels. Suddenly, the *Nicholson* would jam on the speed, and the maneuvering would begin again. Each time, Captain Deppe would match his counterpart move-for-move. At times the *Normandy* would heel as much as 40°, and you could hear the sounds of pizza pans and crockery hitting the deck back in the galley. Other times, it would be a race to see if the *Nicholson* could inch ahead just a little, followed by a radical turn to try to gain position.

It wasn't until sometime after 2000 (8 P.M.) that the *Nicholson* and the other two Koronan intruders finally turned away, and the jousting was over. As Captain Deppe ordered the engines throttled back and began to con the *Normandy* to her assigned position in the defense screen, Admiral Mullen's voice came up on the TBS circuit. For several minutes, the admiral commented on the performance of each ship in the screen, after which he paid a glowing compliment to the skippers of the three escorts that had fended off the Koronan warships. After his hearty "Well done," you could feel the tension ease around the ship. Though we did not know it at the time, the *GW* battle group had passed a significant test; they had bought two more days of "peace" for the Kartunans and their coalition allies.

Aboard the *Normandy*, life began to settle back to normal. Down in the galleys, the mess specialists salvaged what they could of the pizzas they would serve at mid-rats. Though the 2300 (11 P.M.) feeding was heavy that night, many of the officers and crew chose to just hit their racks and grab some sleep instead. These were the veterans, who knew that what they had seen today was only the beginning of what could be another two weeks of "combat." Those with less experience and more adrenaline munched on thick-crust pan pizza, and chatted about the terrific ship-handling Captain Deppe had shown the entire battle group that day. As I lingered over a piece of the baked pie, I answered a question that had been in my mind for some time: Since the end of the Cold War, the surface forces of the USN have not had a serious enemy. Such a condition can breed complacency and lead to "sloppy" habits in commanders and crews. Jim Deppe's performance on the bridge of the *Normandy* this Saturday evening convinced me that *our* surface Navy still has "the right stuff."

Sunday, August 24th, 1997

JTFEX 97-3—Day 7: The 24th MEU (SOC) completed their NEO early today, and is evacuating the civilians to a neutral location. The aggressive actions of Koronan Naval forces yesterday have been reported to the UN Security Council, which has issued an additional resolution allowing expanded use of force in the event of further harassment. The only Koronan government response has been additional mobilization of their military forces.

The morning after the game of "Cowboys and Russians" dawned humid, overcast, and stormy. I awoke to a knock on my door from a chief petty officer at 0600 (6 A.M.). He informed me that the captain had arranged for a UH-46 VERTREP helicopter to pick up us and shuttle us over to the *GW*.

Quickly showering and packing up my bag, I met John in the wardroom for breakfast, and we discussed our plans for returning to the carrier. Since the helicopter was due overhead at 1000 hours (10 A.M.), I took the time to go up to the bridge and thank Captain Deppe for his hospitality. Afterward, on my way down, I ran into Captain Phillips, who confirmed my own thoughts about the previous night's proceedings. He had noted *Normandy*'s impressive performance in his report to the SOOT team leader aboard the command ship *Mount Whitney*. ''Keep an eye out for things to break tomorrow,'' he added slyly. Armed with this information, John and I collected our bags, and then headed aft to the helicopter hangar to await our ride back to the *GW*.

At the hangar, a chief handed us float coats and cranial helmets, and gave us a quick safety briefing on the Sea Knight. And then at the appointed time, the UH-46 set down gently on the *Normandy*'s helicopter pad. The big twin-rotor Sea Knight was a tight fit on the small landing platform, and you could see the deck personnel carefully watching the clearance between the rotor blades and the superstructure. We quickly boarded the bird and strapped into our seats. Two minutes later, the crew buttoned up the UH-46 and lifted off into the overcast. The ride back to the *GW* took about fifteen minutes.

In the ATO office, Lieutenant Navritril had good news for John. Since many of the VIPs, contractors, and other extra ship riders had flown home, he would now get to occupy a two-man stateroom up on the O-2 level near mine. He also let us know that the Challenge Athena link was working well, which meant that we could expect to see one of the opening-day NFL football games the following Sunday. ''So take it easy,'' he told us, ''and relax the rest of the day.'' Both of us gratefully took him up on this suggestion, and retired to our staterooms for a little ''down'' time. If things got ''hot'' on Monday, I wanted to be ready.

Monday, August 25th, 1997

> JTFEX 97-3—Day 8: At dawn this morning, the armed forces of Korona began a general invasion of the Kartunan homeland. Elements of every branch of the Koronan military are involved, and have been identified, and are rapidly overrunning the country. The UN Security Council, the U.S. government, and the government of all coalition allies have condemned this action. Meanwhile, the UN Security Council has voted a number of resolutions, including one which encouraged ''use of all necessary and appropriate force'' to halt the aggression.

As soon as word of the invasion reached him, Admiral Mullen initiated a revised ROE, and put into effect the attack plans that he and his staff had been working on since we had sailed. One of his first acts was to activate Captain Deppe's fleet air defense plan. With Deppe designated as "Alpha Whiskey" (AW—the fleet AAW commander), the three SAM ships were spread through the area to fully cover all the high-value units. The *Normandy* would stay close to the *GW*, while the *South Carolina* would move closer to the *Guam* ARG (the superior over-land performance of her missile radar directors gave her better inshore characteristics than those of the Aegis ships). The *Carney* would act as a "missile trap," and work as the AAW "utility infielder" for the fleet. She would stay "up threat" of the main fleet, and do her best to break up any air attacks from Koronan air units.

This day would see the opening of the air campaign (which would follow the model set forth during Desert Storm). Today's air and missile strikes were designed to eliminate the Koronan ability to hurt the coalition fleet; CVW-1 would destroy the Koronan air defense system, air force, and navy, while Tomahawk cruise-missile strikes from the *Normandy*, the *Carney*, and the submarines would decapitate the Koronan command and control network. It was a good plan. Still, the key to making a plan work is to keep it flexible enough to respond to any countermeasures that an enemy might respond with. This meant getting the TARPS F-14's of VF-102 into the air to sweep the Gulf of Sabani, Kartuna, and Korona for targets worthy of CVW-1's attentions. With only four TARPS-capable F-14's, and whatever satellite imagery that could be downloaded from the Challenge Athena system, the battle group intelligence would be half-blind. Luckily, they would also have the services of the three VQ-6 ES-3's, giving them "ears" to supplement their eyes.

This day launched the entire group into wartime operating conditions; they would stay that way until the End Exercise (ENDEX) time, sometime the following week.

Tuesday, August 26th, 1997

> JTFEX 97-3—Day 9: The Koronan military forces, continuing their invasion of Kartuna, claim to have taken control of more than half the country, and have flown numerous missions against the coalition air and Naval forces in the Gulf of Sabani (with results that are currently not known). Meanwhile, the coalition forces, based around the carrier USS *George Washington* (CVN-73) and her battle group have begun counterattacks against the Koronan invaders.

One of the first things you get used to aboard an aircraft carrier is you never find total quiet. Down below, you hear the machinery noises that are the heart and lungs of the ship. As you rise through the decks, the noises of the flight deck begin to make themselves heard, until you reach the O-2 level,

A young Navy maintenance technician works on an HS-11 helicopter in the hangar bay of the USS *George Washington* (CVN-73).
JOHN D. GRESHAM

where the "airport" is on your roof. Surprisingly, you can even sleep through all the noises of the catapults firing, arresting wires straining, the tailhooks and landing gear slamming into the deck, and the jet noise coming through the armored steel deck over your head. After a while the noises blend into one another and you just sleep in spite of it all.

On this second day of the "war," I wandered around the ship to get a sense of how the young men and women who were doing most of the work were handling both their work and what leisure was available to them. Down on the hangar deck, for example, I witnessed some amazing mechanical and technical exploits. Jet engines weighing five tons were changed with less than a yard's clearance between aircraft. Kids who don't look old enough to own a "boom box" back home handled million-dollar "black boxes." Sweat, oil, jet fuel, hydraulic fluid, metal shavings, and salt air all mixed into a pungent smell that says only one thing: You're in an aircraft carrier hangar bay. This is a land ruled not so much by the ship's officers, as by those mythic people who hold the naval service together—the chiefs.

In the Navy, there is a saying that officers make decisions and the chiefs make things happen. It's true. Here on the hangar deck, the bulk of the maintenance and repair work is done by senior enlisted personnel and non-commissioned officers (NCOs), who spend their days (and frequently nights) putting back into working order the aircraft that officers go out and break. Any machine, no matter how robust and well built, will eventually break or fail if used long enough. It therefore falls to these unsung heroes of naval aviation to do the dirty and not very well rewarded work of keeping the airplanes flying. How do the taxpayers of the United States reward these dedicated young people? While the pay of enlisted/NCO personnel has slipped a bit in the last few years (by comparison with what the average civilian earns), it is still light-years ahead of the near-poverty level of the 1970's. In fact, the Congress has recently voted a small pay raise, and it should be in pay envelopes by the time you read this.

As for accommodations, well, as we've already seen, don't expect a four-star hotel. With 90% of the crew made up of enlisted/NCO personnel,

The officers' mess in Wardroom 3 aboard USS *George Washington* (CVN-73)

JOHN D. GRESHAM

so-called ''personal space'' for non-officers is almost absurdly lacking. Most enlisted and NCO berthing is made up of six-man bunk/stowage units, with an attacked locker unit. Each person has an individual bunk, bunk pan, and locker. Each bunk has a reading light, privacy curtain, and fresh-air duct, all packed into a space about the size of a good-sized coffin. The six-man modules are grouped into berthing spaces, which share a communal head/shower, as well as a small open area equipped with a television, table, and chairs. Normally, when you walk through these spaces, red battle lamps (to preserve night vision) illuminate the area and allow those off their work shifts to get some sleep. In the common areas there's usually a television going and someone is probably ironing their clothes.

The Navy, recognizing the necessary shortcomings of the personal accommodations, does what it can to make up for that by giving naval personnel the finest food money can buy. It's not fancy, tending toward good, basic chow, but the mess specialists work hard to throw in favorites like pizza, stir-fry, or Mexican dishes several times a week. In addition, the dietitians try to keep food relatively low in fat by offering fresh vegetables and salads whenever possible. For the enlisted sailors, meals are usually served cafeteria-style in the large serving area forward of Wardroom 3. One of the largest open spaces in the ship, this is the central focus of the enlisted personnel aboard ship. Here they can eat, talk, attend a class, play a video game, and perhaps escape the routine for a little while. There are also other diversions.

Workout facilities are located here and there throughout the ship. These have become extremely popular in recent years, as the ''hardbody'' culture has become fashionable. For more serious fitness enthusiasts, there are exercise and aerobic classes held on the hangar deck several times a day, as well as a jogging group that makes the circuit of the flight deck, weather and flight operations permitting. The ship's cable television system normally broadcasts over six channels from a small studio on the O-1 level under the island. Run by a technical team under Lieutenant Joe Navritril, it shows movies, news, ship's bulletins, and other programming. There is also a small cable radio station, which broadcasts an ''eclectic'' mix of rock and roll, blues, and jazz. A four-page newspaper, *The Guardian*, comes out every day at lunch. It is a delightful mix of news from "the world," as well as more topical pieces relating to daily life aboard the *GW*. Finally, movies (complete with bags of popcorn) and VCRs can be rented for off-duty video parties back in enlisted berthing areas.

An innovation made possible by the Challenge Athena system is personal E-mail over the Internet for everyone on board. This is handled through the ship's own onboard Intranet, which feeds into a central file server. Each person is assigned an E-mail account and address (aboard the *GW*, this ends with the suffix *@washington.navy.mil*). The messages are then routed through the server and Challenge Athena system to and from the Atlantic Fleet communications center in Norfolk, Virginia. This means that everyone on the ship with access to a computer (some are in common areas in kiosks for those who do not have personal laptops or office machines) can receive E-mail messages from home. Already, it is changing the face of shipboard life.

For example, the three thousand sailors and Marines aboard the amphibious ship *Peleliu* (part of the *Nimitz* battle group, which deployed from the West Coast a month before the *GW* CVBG) sent over fifty thousand E-mail messages in just their first month under way! The effect on crew morale has been astounding. The arrival of Naval E-mail has come none too soon for our sailors, since the old Navy draw—"Join the Navy and See the World"—has become all but obsolete. Over the last decade, the ships of our battle groups have made less than half of the port calls on deployment that they used to make. This means that seeing foreign countries, long a recruiting attraction, has been almost eliminated. Ever since the 1979 Iran Crisis, long (ninety-plus days) line periods have become the norm for CVBGs, and this has been tough on crew morale.

Wednesday, August 27th, 1997

> JTFEX 97-3—Day 10: The military forces of Korona have today completed their occupation of Kartuna, including the capital city of Temal. The last elements of the Kartunan government evacuated to the country of Telari, which today announced its joining of the Allied coalition. Meanwhile, the coalition forces have been stepping up their attacks on Koronan military targets, reportedly inflicting heavy damage. The battle continues. . . .

This morning found the *GW* battle group continuing to dish out punishment on the Koronans. The objective was to destroy enough of their armed forces to allow the Marines of the 24th MEU (SOC) to make a landing near the Kartunan capital city of Temal (in actuality, Camp Lejeune, North Carolina) sometime the following week. A battalion from the 82nd Airborne Division assaulting a nearby airfield would support this landing. This would allow follow-on forces to be landed from the sea and air.

Before this could be done, the Koronan forces would have to be reduced in size and power, and this was the job of the ships, missiles, and aircraft of the *GW* group. Already, significant progress had been made toward this goal. Though Koronan air and Naval forces had aggressively launched attacks on the naval units of the allied coalition, Admiral Mullen's detailed plans for protecting the ships of the force had been working to near perfection. Throughout the battle group, the various warfare commanders had been working hard to eliminate the specialized threats they were responsible for.

Captain Deppe on the *Normandy* (the group AW commander) had been especially busy in dozens of AAW engagements between his SAM ships and the planes of the Koronan Air Force. Deppe's disposition of his SAM ships had worked particularly well, with the *Normandy* absorbing most of the attacks aimed at the *GW*. The *Carney* and *South Carolina* also shot down their share of enemy intruders, with the result that the USACOM exercise controllers rapidly had to strengthen the Koronan Air Force, lest it be completely destroyed before the shooting phase of the exercise was only three days old. Once again, the J-7 controllers from USACOM were being forced to "ratchet up" the threat level of the exercise, just to keep it challenging for the *GW* group.

"Give me a fast ship for I intend to go into harm's way!' "

Captain John Paul Jones, Continental Navy

The Koronan Navy was put out of action equally fast. Because safe distances had to be maintained between the Koronan KILO-class diesel boats (being played by borrowed USN nuclear submarines) and the ships and subs of the *GW* group, exercise rules tended to make them sitting ducks. The Koronan surface ships died a little harder, though they did die quickly. Within hours of the outbreak of hostilities, every one of the Koronan missile destroyers and patrol boats had been hunted down and dispatched by the allied forces. Sometimes, their elimination came at the hands of aircraft firing standoff missiles like AGM-65 Mavericks and AGM-84 Harpoons. Particularly effective against the missile patrol boats were SH-60B LAMPS III helicopters from the escorts armed with AGM-119 Penguin air-to-surface missiles (ASMs). Using these little helicopters as perimeter security guards proved to be an efficient way of keeping the Koronan patrol boats at arm's length, without requiring a mission by an F/A-18 or S-3B to kill them.

There were also a number of surface engagements by ships of the *GW* escort and STANAFORLANT—not all going in favor of the allied coalition. In just a single day of surface combat, hits by simulated Koronan missiles (assumed to be Chinese-built C 802's) damaged the *Carney*, *Samuel Elliot Morrison*, and *Seattle*, putting them out of action (and the exercise) for various lengths of time. In addition, the *Boone* was assessed to have been hit by Naval gunfire. In return, the *Underwood* and HMS *London* were assessed to have sunk a Koronan missile patrol boat with RGM-84 Harpoon SSMs. Littoral Naval warfare is like knife fighting: close and bloody. However, by setting his units up to fight this way, Admiral Mullen was able to maximize the number of attack sorties that could be generated by CVW-1 off the *GW*. Though he risked his surface ships, he got the desired results on the beach.

Thursday, August 28th, 1997

> JTFEX 97-3—Day 11: There have been reports today of various atrocities by Koronan military forces against the population of Temal, the capital city of occupied Kartuna. In addition, it appears that the Koronan forces are digging in to protect their gains against a possible counter-invasion by Allied amphibious and airborne forces.

Ever since our sailing, Captain Rutheford had made a point of exercising his crew with a series of battle drills—a deadly serious business aboard any warship, and particularly on an aircraft carrier. Most of the damage suffered by flattops in combat has come as a result of fire. It is the worst nightmare of carrier sailors, whose home is basically a big metal box full of jet fuel, explosives, and other combustible materials. Until a carrier like the *GW* is fully "buttoned up" (that is, put in a condition where it is most survivable), a fire can rage through it much like those that devastated the *Oriskiney* (CVA-34), *Forrestal* (CV-59), and *Enterprise* (CVN-65) back in the 1960's. Buttoning up usually comes when the ship goes to "General Quarters" (GQ) or Condition "Zebra." Since it takes time for sailors to learn to live and work at GQ, Captain Rutheford makes a point of practicing it regularly. Every Tuesday and Thursday evening at 2000 Hours (8 P.M.) while under way, the *GW* goes to GQ for several hours of combat and damage-control drills. It is at GQ when a warship truly becomes a living organism, with the personnel aboard acting as nerves, muscles, and immune systems, making it capable and strong.

It takes just a few minutes for the ship to get fully buttoned up and ready to take whatever punishment an enemy might care to dish out. Every person on the *GW* (even John and I) had an action station, where they are expected to be during GQ. So at 2000, we were manning our action station—a couple of desks inside Lieutenant Joe Navritril's small public affairs office on the O-1 level. From there we could sit, sweat a little (it's warm with all the computer and television gear), and listen to the drills around the ship.

This evening, a firefighting training drill was going on several levels above us on the island. At the same time, weapons drills were being run with the Mk. 29 Sea Sparrow launchers. All around us, you could feel the crew bonding with the big ship, becoming as much a part of it as the nuclear reactors, plumbing, and catapults. It also is a time of great stress and concern, even during training. This is because the ship must still function while buttoned up. Moving from one compartment to another becomes difficult, as heavy hatches and watertight doors must be opened, and then redogged. There are chances for mistakes to be made, and this evening there was one.

One of the important jobs that must be done daily aboard ship is the testing of the various petroleum systems to make sure that their contents are pure and free of contamination like water or dirt. This evening, a young sailor was carrying several glass jars of samples down from the flight deck to the metrology lab for testing when he dropped one up on the O-2 level above us. Unfortunately, in the darkened compartment he lost track of the spill, and wound up slipping and falling in the slippery puddle. Almost instantly, there was an alarm over the 1MC system of "MAN DOWN!" and a call for a medical team. Within seconds the young sailor had a corpsman at his side, and a hazardous-materials team on the way to clean up the dangerous spill. I was struck by the way that the others in the compartment with John and me stopped what they were doing to wait for word on the young man, almost like waiting for a player to stand up after an injury at a football game. It was quiet for a few minutes, until Captain Rutheford came back on the 1MC to tell us that the sailor's injuries were slight (an injured wrist was all), and that the response teams had done a great job of taking care of him. As I stepped out of the public affairs office for a drink from the water fountain in the passageway, I saw the young man being carried down the ladder from above on a Stokes litter, not unlike a dozen eggs being cradled by a housewife on her way home.

A few minutes after the GQ alert was lifted, I headed back to my stateroom two levels up. I had to admire the way that the GQ had been handled. It was just like combat. It was at this moment that I knew the *real* truth about this ship. The *GW* and her crew were *ready* for whatever the coming deployment would bring, and God help the enemy foolish enough to try to hurt them. It would not be a fair fight. You can always tell a military unit that is functioning well: When it is stressed, you cannot even see them sweat!

Friday, August 29th, 1997

JTFEX 97-3—Day 12: Press reports from the Allied Coalition report the air and naval forces of Korona have been heavily damaged, and rendered effectively harmless. In addition, air units flying from the USS *George Washington* (CVN-73), USS *Guam* (LPH-9), and

Commanders John Kindred (the Air Boss, left) and Carl June (the Mini Boss, right) in Primary Flight Control (Pri-Fly) aboard the USS *George Washington* (CVN-73).

JOHN D. GRESHAM

other naval vessels have been flying over 100 attack sorties every day since the outbreak of hostilities.

The payoff for all the efforts of the "little boys" of the *GW* battle group and STANAFORLANT was the ability of CVW-1 to concentrate on their real job—attack sorties against Koronan military targets ashore. Did they destroy the occasional air or naval target? Absolutely. And they did so with a ruthless efficiency when the targets were available. But an old saying explains what Admiral Mullen had in mind for his flyers:

"Fighter pilots make movies. Bomber pilots make history!"

Unknown Navy Attack Pilot

The aircraft and crews of CVW-1 were *really* earning their keep only when they were delivering ordinance onto targets of value ashore. This meant that the fifty F-14 Tomcats and F/A-18 Hornets aboard the *GW* were flying morning, noon, and night to hit as many high-value targets as possible. In particular, they would give special attention to enemy units and systems that could threaten the Marines of the 24th MEU (SOC) and the airborne troopers of the 82nd Airborne Division when they came into play in a few days. These included targets like mobile antiship missile sites along the coast (which could hit the amphibious ships of the *Guam* ARG), mobile SCUD ballistic-missile launchers, and SAM/AAA sites in the planned invasion area. Along with these high-value Koronan targets, there would be attacks on the fielded forces of the Koronan military in and around Kartuna. Because it is the air crews who fly the planes and drop the weapons who give naval airpower its worth, let's take a closer look at how their dangerous job gets done on the *GW*.

If you want the best view of a carrier's air operations, there is only one place to go: Primary Flight Control—or "Pri-Fly" as it is known. This is the domain of Commanders John Kindred (the Air Boss) and Carl June (the Mini Boss). Kindred and June are the lords and masters of the *GW*'s flight deck and the airspace around the ship. The Navy has for generations made it a practice to hand responsibility to highly qualified naval aviators for those jobs aboard carriers that relate directly to flying—jobs like catapult and landing signals officers (LSOs). These jobs have to be done *right.* People who do them properly are promoted. Those who don't can look forward to new civilian careers. Of these jobs, the captain's, of course, carries the greatest responsibility. However, right after the captain comes the Air and Mini Bosses. No other pair of individuals has so much influence on the core services (flying aircraft in support of Naval/Marine operations) the ship was designed to deliver. These two officers control virtually every aspect of the boat's air operations, from the pace and number of missions flown to how the aircraft are parked and serviced. This means, practically, that when the ship is flying aircraft, there is no margin for error despite massive stress, a thankless work schedule, and very little sleep. Clearly you need special people to be Bosses.

Since a good long look at Pri-Fly seemed essential to the total experience of carrier operations, I asked to spend a day there with Kindred and June. After climbing the five ladders from my stateroom to the O-7 level of the island, I joined the crowded and busy team in Pri-Fly. Along the port side overlooking the flight deck are three chairs, much like Captain Rutheford's chair on the bridge one level below. Here is where Kindred and June spend their days and nights. Soon after I entered, they very graciously invited me to sit in the center chair between them. It was an impressive view. On command from Commander Kindred, a chief petty officer behind me passed forward a steaming cup of coffee and the Air Tasking Order (ATO) Flow Sheet—or Air Plan for short—the document that explains and controls the day's air operations. Printed double-sided on a single legal sheet of paper, it is the daily bible for the flight deck. On one side is a set of time lines, with a line for each squadron or air unit participating that day. These time lines are then broken down into individual "events," each of which represents a particular planned launch/landing cycle on the flight deck. The flip side shows detailed notes about the flight schedule and the schedule of tanker aircraft, and is personally signed (they have to review it daily) by the *GW*'s Air, Strike, and Operations officers.

As I read the Air Plan, I was struck by the number of flight "events" on the schedule. All told, there were nine of them, which was normal for this phase of JTFEX 97-3, Commander June informed me. Because of the Navy's recent effort to increase the number of daily air sorties, the two Air Department officers were trying to implement some of the lessons learned during the recent SURGEX by the *Nimitz* group. To support their SURGEX, *Nimitz* and CVW-9 had been heavily reinforced with additional air crews and deck personnel, allowing them to run over two hundred sorties a day. *GW* and CVW-1 had no such augmentation. Even so, there was still room for increasing the number of events and sorties over the fleet norms. More efficient use of personnel and resources (such as better organization of the hangar and flight deck crews) and enforced rest and eating periods between events had allowed Kindred and June to safely expand the five or six flight events of a "normal" day to as many ten or twelve. In this way, CVW-1 could easily run over 150 sorties a day for an indefinite period of time, should it be called upon to do so.

After taking in all that I could of the Air Plan, I lifted my head to watch as the two men took aboard a dozen or so aircraft from the day's second air event. Included in this gaggle was CAG "Boomer" Stufflebeem flying a VMFA-251 Hornet, who bagged a perfect "OK-3" trap. Meanwhile, another strike was getting ready up forward to head out on the noon mission (Event-3), which would concentrate on hunting enemy SAMs and mobile missile batteries. Most of the aircraft for this mission were on the bow, and would have to taxi aft once the area around the deck angle was clear.

As soon as the last of the Event-2 aircraft were aboard, the Air Boss called for the LSOs to stand down for a while and the landing light system to be shut down (the longer it is lit, the sooner it will wear out). Moments later, Commander June pointed out several helicopters in holding patterns. There would be just enough time to bring them aboard before the next event, he explained. During a window of less than fifteen minutes, two SH-60's from HS-11 and an SH-3 Sea King carrying VIPs from the *Mount Whitney* (LCC-20, the fleet command ship) arrived, and then were either parked, towed away, or flown off for the next event. Once the helicopters were taken care of, the flight deck went relatively quiet, while hundreds of people with colored jerseys swarmed about, doing their various jobs.

Up in Pri-Fly, the pace had hardly slackened. To my rear, the chiefs and representatives from the various CVW-1 squadrons were exchanging information and making sure that everyone was in sync. If anyone had a question, he would come forward and wait respectfully until Kindred or June took notice. After a short discussion, a decision would be made. As these continued, I turned my attention back to the flight deck, where—as always in my experience—I saw great energy and purpose, and no wasted movements, no unnecessary actions. It is the world's most dangerous dance—a dance made even more risky in light of the necessity to run the deck crews until the late hours of the evening because of the expanded number of air events. But for now, Kindred and June were working hard to get Event-3 into the air.

As I watched, I was amused to see then that some things never change—and shouldn't. In spite of the array of new computers and other available high-tech tools, Kindred and June still use many of the same tools and procedures that their predecessors in World War II might have used. For example, each man has a set of colored grease pencils to make notes on the thick windows in front of them to remind them about which aircraft are aloft and the state of their fuel.

"Why do you do that?" I asked.

"Some things computers and software will *never* improve on," they explained. "Computers and electronics might fail; an explosion might put them out of action; but grease-pencil marks and paper hardcopies will work

The busy flight deck of the USS *George Washington* (CVN-73), filled with CVW-1 aircraft during JTFEX 97-3.

John D. Gresham

as long as they still exist. Perhaps not as well or as quickly, but they will *always* work.''

About 1150 hours, Commander Kindred picked up his headset and announced to the flight deck that it was time to start engines for the Event-3 launch cycle. Carefully, the aircraft handlers directed the air crews to taxi their aircraft aft for their launch from Catapults 3 and 4. Moments later, after everyone was in the proper place and Captain Rutheford had given his approval, Kindred gave the catapult officer permission to conduct the launch. Already overhead was the plane guard HH-60G from HS-11, as well as a USAF KC-135 airborne tanker flying from Langley AFB (in the simulated country of Telari).

First off the *GW*'s deck were a pair of VS-32 S-3B Vikings configured as tankers, followed by a replacement VAW-123 E-2C Hawkeye AEW aircraft. These were followed by a quartet of VF-102 F-14's, two configured with D/TARPS pods for the noon reconnaissance run, while the others had LANTIRN pods for laser bombing. Following the Tomcats was a VAQ-137 EA-6B Prowler loaded for Suppression of Enemy Air Defense (SEAD) operations. Last off were a dozen F/A-18 Hornets, with a variety of loads—from HARM missile and LGB simulators, to live iron bombs for training runs on a local range. This was a large launch, with almost two dozen aircraft involved. More than that would make the ninety-minute Air Event cycle impossible to sustain, and would actually reduce the total number of sorties that CVW-1 could generate per day. It also made optimum use of the flight deck space, which even on a ship the size of the *GW* is limited.

Once the Event-3 aircraft had launched (it only took about fifteen minutes), the angle was reset for recovery of the HS-11 HH-60G on plane guard duty and the launch of its replacement. After this, activities on the flight deck slowed down, and there was time for a few bites of sandwiches brought up from below by one of the chiefs. By 1300 hours (1 P.M.), the F/A-18's were back in the landing pattern, preparing to come aboard. The relatively short range of the Hornets means that they normally operate within a single air event, while the Tomcats and other aircraft would come back at the completion of Event 4. For now, the deck aft was cleared, the arresting wires checked, and the landing light system turned on. The landing cycle took about fifteen minutes, after which the Hornets were rapidly taxied forward to the bow, where they could be refueled and rearmed to take part in other strikes later in the day. Also coming aboard was one of the three VRC-40 C-2 COD aircraft, bringing its load of mail, personnel, and spare parts from NAS Norfolk. It would return to Norfolk with a load of nonessential folks. The ship had been packed since sailing, but now as Labor Day weekend approached, the many VIPs, news media types, and technicians were finding reasons to head back to the beach.

As for me, it was time to leave the two Air Bosses to their grueling task. They had six more launch/recovery cycles ahead before they could grab some sleep and get ready for the first launch the following morning.

Saturday, August 30th, 1997

> JTFEX 97-3—Day 13: There are reports that Allied forces are preparing to invade the occupied country of Kartuna to evict the Koronan military forces. At this time, Allied public affairs officers will only report that operations against the Koronan forces continue, with no new information on results.

While most of America was getting started on their Labor Day holiday weekend, the participants of JTFEX 97-3 were just beginning to hit high gear. Captain Bruce Van Velle's mine-countermeasures force moved inshore to clear lanes for the amphibious units through Koronan minefields. This required that CVW-1 finish clearing out the last of the Koronan's coastal antiship and SCUD missile sites before the vulnerable amphibious ships of the *Guam* ARG began to operate close to the Kartunan coastline (actually near Camp Lejeune, North Carolina). Meanwhile, a continuous, twenty-four-hour-a-day CAP had to be flown over the amphibious ships, to protect them and the Marines of the 24th MEU (SOC). In fact, the transition to the amphibious phase of operations meant that there was even more for everyone to do, and even less time to do it in.

Now, because I wanted to find out how the air campaign was actually going "over the beach," I went to the one place where I knew I would hear the truth about such things—the pilot's "Dirty Shirt" mess. Located up forward on the O-2 level, it is a less formal place than Wardroom 3. Also, because of its informality and its more "meat and potatoes" diet, most officers in fact prefer the "Dirty Shirt." Here pilots can wear their flight suits to relax and catch a meal (hence the name). All the meals are served cafeteria-style, and are eaten on tables reserved for each squadron in the air wing. If you're an outsider, you have to ask to join them. They rarely refuse. This day I had an old Navy favorite, ''Sliders and Fries'' (''Navy'' for cheeseburgers and French fries), and a glass of cold milk with some VF-102 ''Diamondbacks,'' flying F-14B Upgrades.

This was an exciting time for the Diamondbacks, who had seen their community rise Phoenix-like from the ashes of their Cold-War interceptor mission to become one of the Navy's preeminent strike and reconnaissance platforms. The addition of the AAQ-14 LANTIRN targeting pod and the new D/TARPS reconnaissance pod has changed the face of the Tomcat community, making them once again the kings of the air wings. You could see the pride in the faces of the young pilots, who are now certain of a mission in the CVWs of the 21st century. They will eventually be the first community to receive the new F/A-18E/F Super Hornet when it arrives in a few years, and the new pods will make the time waiting all the more pleasant. During JTFEX 97-3, they had been heavily tasked, flying some fifteen to twenty missions a day (they had fourteen F-14 airframes aboard)—heavy usage for Tomcats. And since their missions tended to last two to four times longer than those of the Hornets, due to their greater internal fuel load and range, the Tomcat crews were getting more flight hours than the Hornet drivers.

When I asked how things were going, the variety of answers reflected the great range of their activities during the previous two weeks. While they all agreed that the new sensor and targeting capabilities of the LANTIRN pod were terrific, they had to admit that they were still learning how to get everything out of it. In particular, the Navy version of the AAQ-14, which has a GPS/INS unit built in, has opened new targeting possibilities. On the down side, the new pod is currently unable to send images from the LANTIRN back to the carrier in the same near real-time manner as the D/TARPS pod. But this problem is being worked on, and will probably be solved by late 1998. As for the new D/TARPS pod, they had absolutely no reservations (except for their small numbers). The addition of the digital line scanner and near-real-time transmission capability for imagery has given theater commanders their first real ability to find and target mobile high-value targets like SCUD launchers. With only four D/TARPS-capable F-14's per CVW, these are arguably the most valuable aircraft in the air wing.

When I asked about the current exercise, they all agreed that the Diamondbacks and their CVW-1 partners had done very well during JTFEX 97-3. The few air-to-air engagements during the exercise had been decidedly one-sided, with most ending in a hail of AIM-54 Phoenix and AIM-120 AMRAAM shots and the Koronan aircraft going down in flames. The reconnaissance missions had gone equally well, though the classified aspects of their tactics and equipment kept the air crews from discussing the results. The real smiles came when they talked about the bombing results with their LANTIRN pods and Paveway II and III LGBs. Thanks to their superior FLIRs, dedicated RIOs as operators, and excellent weapons, the Tomcats had become the scourge of the mobile targets ashore. Though there was a general feeling that Koronan SCUD and antiship missile launchers were probably dead by now, a few older Tomcat crew members who had flown in Desert Storm doubted this. They would go ''SCUD hunting'' one more time later that night, while others would hunt down Koronan artillery pieces, so they would be out of action before the Marine landing rumored for the following night.

At that point, it was time for me to move on. Deep fatigue was visible in the eyes of these men, and I wanted to intrude no more on their crew rest. All too soon, they would be climbing into their cockpits and heading into the night skies to once again hunt their ''enemies.'' Meanwhile, the USACOM staff had a few more tricks up their sleeves to keep things interesting. And as I stopped by the public affairs office, I learned from Joe Navritril that a contracted civilian Learjet, pretending to be a CNN camera aircraft, had simulated a kamikaze dive into one of the escort ships. The *Kamikaze* attack had been defeated by a short-range SAM shot, though only just barely.

Sunday, August 31st, 1997

> JTFEX 97-3—Day 14: The rumors of an Allied invasion near the Kartunan capital city of Temal continue, with reports of Allied Naval vessels beginning to conduct shore bombardments with gunfire. There is a feeling that the expected Allied invasion by coalition forces may be only a matter of days away.

This morning found everyone on *GW* busy getting things ready for the planned invasion. The actual time of the invasion was a secret to most people on the *GW*, including me. I assumed, like just about everyone else, that the Marines of the 24th MEU (SOC) would hit the Camp Lejeune beaches sometime around midnight of the following evening—a tactical time that had become more or less standard in the last few JTFEXs. Meanwhile, I wanted to head up to Flight Deck Control for a small ceremony that has been a tradition going back centuries. Today, Captain Groothousen, the *GW*'s XO, would leave the ship and officially hand over the job to Commander Chuck Smith, a fast-tracked flier from the S-3 Viking community. In a few months, Groothousen (''Groot'' to his friends) would take over command of the *Shreveport* (LPD-12), over in the *Guam* ARG, which was the next step on the way to command of his own flattop. Around noon, the ship's department heads met in the Flight Deck Control Room, where they said their final goodbyes to ''Groot.'' After he headed across the flight deck to the waiting COD aircraft, the various department heads left the room and went back to their tasks; but I lagged behind to watch the activities.

Flight Deck Control, at the base of the island, monitors and controls the movement of aircraft, personnel, and equipment on the flight and hangar decks. On a pair of scale models of the flight and hangar decks, movable templates show the location of aircraft (with their wings folded) and equipment (such as tractors, firefighting trucks, etc.). Meanwhile, on the walls there are a series of transparent status boards, upon which are noted (in grease pencil) the side numbers of every aircraft aboard the flattop. You can see at a glance on these boards what every aircraft based aboard the ship is doing, how it is loaded, and who is flying it. The models and templates are moved by skilled aircraft handlers, who know just how much room you need to park a line of aircraft in the smallest possible space. Decades of experience have gone into the procedures that run the flight and hangar decks, and it is likely that they will continue for as long as Americans take aircraft to sea.

That night, as the aircraft continued their round-the-clock shuttle over the beach, John, Lieutenant Navritril, and I joined Commander Smith for a short visit in his new quarters. The XO's quarters aboard a *Nimitz*-class carrier are quite pleasant, though the lack of time that he gets to spend there more than makes up for the few pleasures. With Commander Smith sitting at his desk, the rest of us found comfortable spots on the couches, and we talked of how he had come to be here tonight. He talked of the path to command of a carrier, and why he supports nuclear propulsion for future U.S. flattops. He also spent a few minutes talking about the fine people and procedures that Captain Groothousen had left him. As the minutes became an hour, he talked of his experiences on the way to this job, and how many good jobs the folks coming out of the S-3 Viking community were getting. About the only thing missing was a good cigar and a snifter of brandy to go with it. But the U.S. Navy is ''dry'' and smoking is rapidly leaving our ships as an allowable vice. What stimulation Chuck Smith would find aboard the *GW*, he would have to find on his own. As the new mayor of almost six thousand people, he undoubtedly would over the next two years. As we rose, the chaplain came over the 1MC to announce the command to darken ship for the night and say a prayer. Heading up the ladders to my stateroom, I again was reminded why I love the Navy so much. Here were thousands of young men and women, going to sea to preserve the kinds of things I love America for. As I went to sleep, I felt the safety of knowing that good people were around me.

Monday, September 1st, 1997

> JTFEX 97-3—Day 15: There has been a news blackout by the Allied coalition forces, which would seem to indicate that the planned invasion of occupied Kartuna is imminent. Meanwhile, the Koronan government is calling for their forces to prevail in the coming ''Maximum Battle,'' which will determine the fate of this region.

Almost two weeks earlier, Admiral Mullen had mentioned that Colonel Richard Natonski, the CO of the 24th MEU (SOC), was a "sneaky" kind of Marine. He proved it when he decided to invade Camp Lejeune before the sun went down. At 1600 (4 P.M.), the first elements of Battalion Landing Team (BLT) 3-6 began to hit the beaches and landing zones around the town of Temal (actually the communities around the New River inlet), and a battalion of the 82nd Airborne Division began to drop from the skies. I heard later that the Koronan troops (being played by several battalions of the 2nd Marine Division) had been caught getting ready to watch the opening game of Monday Night Football. The truth was that the colonel's bold move had stolen a march on them; and with the Koronan force already heavily depleted by air and missile strikes from the *GW* group, the 24th MEU (SOC) made rapid progress.

Through it all, the round-the-clock flight schedule continued, although you could see the fatigue in the movements of the air crews and flight deck personnel. They had done their jobs well.

Tuesday, September 2nd, 1997

> JTFEX 97-3—Day 16: The Allied coalition forces landed yesterday near Temal, the capital of occupied Kartuna. Elements of the 24th MEU (SOC) and 82nd Airborne Division have seized a bridgehead, and are awaiting the arrival of follow-on forces. The Koronan forces are reportedly in retreat, headed back to their original borders. Other reports indicate that the exiled government of Kartuna will return to Temal sometime late today. . . .

I awoke this last morning of the exercise to the sound of Joe Navritril knocking on my stateroom door. As I opened up, he informed me with a smile that the exercise would be terminated in a few hours. Since the reoccupation of Kartuna was essentially complete, he had already arranged seats for John and me on the midday COD flight back to NAS Norfolk. After packing, I headed to the wardroom for a quick breakfast. Around 1000 hours (10:00 A.M.), Captain Rutheford came up on the 1MC and addressed the ship. "The ENDEX [End of Exercise] time has been declared," he announced, "and we'll be home tomorrow. I am therefore ordering an immediate suspension of flight operations. I hope everyone can take a breather before we take the *GW* back to the carrier dock at Norfolk."

All over the ship, you could feel a collective sigh as the tension of the exercise passed.

During the next hour John and I paid our mess bills (yes, the Navy makes me pay to eat on their ships), bought a few patches and "zap" stickers

The USS *South Carolina* (CGN-37) deploys on October 3rd, 1997. Part of the escort force assigned to the *George Washington* battle group, the nuclear cruiser was on her final deployment. She was decommissioned when she returned.
JOHN D. GRESHAM

from the squadron stores, and made our farewells. As noon approached, we headed up to the ATO, grabbed our float coats and cranial helmets, and waited for the word to move out to the flight deck. Soon after the command came, we marched up the stairs to the flight deck, where we had the opportunity to see one more impressive sight before the C-2A Greyhound taxied to the catapult. Now that the ENDEX time was now a matter of record, the ships of STANAFORLANT had requested to make a parade past the *GW* before they headed home to Europe.

Soon *Witte De With,* Admiral Peter van der Graaf's flag flying on her halyard, came alongside. As the rest of the multinational force passed in review, the crew chiefs buckled us into our seats and raised the cargo ramp. Once again, the familiar sounds of the COD aircraft filled our ears, and we prepared for the thrill of a catapult shot. Two seconds and more than a few heartbeats later, we were airborne, flying northwest toward NAS Norfolk. Our trip aboard the *GW* was over. But for the personnel of the battle group, it was a new beginning. School was out and they were about to graduate to the job they had all sought. A trip to the other side of the world to support American interests overseas.

Deployment: The Acid Test

In October 1997, John and I drove south one more time to say good-bye to the men and women who had been our shipmates the previous month, to walk the flight deck one more time, and see how ready the *GW* was for her six-month cruise. The first thing we noticed when we stepped aboard was the non-skid coating up on the "roof." During JTFEX 97-3, it had been worn to bare metal. Now it was factory fresh and ready to receive Captain Stufflebeem's airplanes. Down below, supplies were being loaded and personnel were bringing aboard the last of their personal items. Most of the crew would stay aboard that night. As the sun set over the James River, we returned to our hotel and asked for an *early* wake-up call, so we could take part in a very moving ritual: the sailing of the *GW* battle group.

Friday, October 3rd, 1997

The gathering began before dawn, as the families and friends came down to the carrier dock at Naval Station Norfolk to see the *GW* off. For most, there was a quick trip to the McDonald's across the street for an Egg McMuffin and some coffee. Most of the crew had stayed aboard the night before, including Lieutenant Joe Navritril, who had said good-bye to his family in Maryland several days earlier. All the officers and men were in their white uniforms, looking distinctly cooler than they had four weeks earlier. The tropical heat of summer had given way to a pleasant fall in the Mid-Atlantic region, and this morning was cool and sweet.

As a rose-colored sunrise began to appear over the eastern sky, the last of the preparations for sailing came to an end. Now it was time for a little public-relations work. Joe Navritril and several of his staff were herding television camera crews and newspaper reporters around the dock, shooting pictures of Captain Rutheford taking a ceremonial sword from a George Washington lookalike. Command Master Chief Kevin Lavin prowled the dock with Commander Smith, herding the last of the late arrivals aboard the ship. Both shook our hands good-bye, and headed aboard to get to their sailing stations. At the same time, the crowd of several hundred families and well-wishers began to hold up their signs of encouragement for their sailors on board the *GW*.

What followed was a duplicate of the sailing a month earlier, with Captain Rutheford again at the helm. With an HH-60G helicopter overhead for security and guidance, Chuck Smith ordered the last lines brought over and the brows raised. At precisely 0800 hours (8 A.M.), a signal was given, the American flag was raised, and over a thousand sailors in their best whites manned the sides. This impressive sight was made even more so by the emotions of the people left behind on the dock—some sobbing, some stoically silent, some talking nervously.

As the tugs pushed the *GW* into the channel, the crowd began to move to various points around the bay to watch the carrier head out. As John and I walked up the dock to our cars, we stopped and chatted with a young woman wearing a cruise jacket that must have belonged to one of the sailors on board. She just sat there watching the ship and her man move into the channel, then walked with us back to the cars. Her Sailor was a member of the CVW-1 staff, and they were planning to meet in Europe for the holidays. As we parted, though, John and I had the terrible feeling that she might not get the chance. The affairs of the world were going their usual chaotic way. Already that morning, the *Nimitz* battle group had been ordered to the Persian Gulf, to show the flag in the face of renewed tension between Iraq and Iran.

And the crisis that would bring the *GW* to the Gulf was only a month away.

Now, though, the *GW* began to move down the channel, followed at ten-minute intervals by the *Normandy*, *Guam*, *South Carolina*, and *Seattle*. Again, at bases up and down the Atlantic coast, other ships of the battle group and ARG were sailing, planning to rendezvous off the Virginia capes the following day. CVW-1 flew aboard that afternoon, and the 24th MEU (SOC) was already loading down at Moorehead City, North Carolina. As I drove out of the Naval station several hours later, I passed by the British aircraft carrier HMS *Invincible* (R 05) and her battle group, which were making a port visit of their own. Ironically, these same ships would also wind up in the Persian Gulf several months later, along with the *Nimitz* and *GW*. Before that, though, there would be some of the planned exercises and port visits that had been scheduled before Saddam's newest troublemaking.

The *GW* battle group and *Guam* ARG took part in Operation Bright Star 97, the annual joint U.S./Egyptian maneuvers in the desert west of Cairo. However, by mid-November, the crisis over the UN weapons inspectors had broken, and the plans to split the battle group were already in the works. The *GW*, *Normandy*, *Carney*, *Annapolis*, and *Seattle* would make a run through the Suez Canal and Red Sea, following a short port visit to Haifa, Israel. The rest of the group would stay in the Mediterranean with the *Guam* ARG to support operations in Bosnia, and generally "show the flag."[86] On the night of November 20th/21st, the *GW* and her escorts went to GQ, ran through the Straits of Hormuz, and joined the *Nimitz* group in flying patrols over Southern Iraq. The men and women of the battle group never did get their Christmas in Europe.

There was a personal cost to the *GW* and CVW-1 during these operations. On February 6th, two VMFA-251 F/A-18's collided while on patrol. While both pilots ejected (albeit with injuries), Lieutenant Colonel Henry Van Winkle, the XO of VMFA-251, was killed. His would be the only life lost in the crisis with Iraq. The *GW* and *Nimitz* continued their vigil, until relieved by the *Stennis* and *Independence* groups. The *Seattle* was left behind for a time because of the need for extra logistics ships in the Persian Gulf. Moving back through the Suez Canal, the *GW* rendezvoused with the *Guam* ARG and her escorts, and headed home.

They arrived home several weeks later, and the eighteen-month cycle began anew. Along the way, more changes took place to the people that we had met. Captain Stufflebeem was relieved in late 1997, and became an aide to Admiral Jay Johnson in the CNO's office. Captain Groothousen took over command of the *Shreveport* about the same time, and continues on the path to command his own carrier someday. Though the various crises continue, the cycle never stops. The battle groups work up, go out, and come back. Let us hope that they continue that way.

---

86 The *GW* group that went to the Gulf had two things in common with the *Nimitz* CVBG. In addition to the carrier and logistics ship, both of the escort ships were Aegis- and Tomahawk-capable, as was the submarine. In short, they took the ships with the most firepower where they would be needed.

Aircraft Carriers in the Real World

As throughout this series, I've reserved a bit of space at the end of this volume to spin a yarn, to try to tell the story of what I think future carrier operations might be like. Though the following story is set some two decades in the future, it is based upon what I believe to be solid plans and ideas. I hope that it also says something about the evolution of our world, and how democratic nations will function in the 21st century.

Birth of a Nation: Sri Lanka, 2016

In the terrible summer of 2015, the great powers of the world—the United States, Russia, and China—all knew that the Indo-Pakistani War was likely to go nuclear at some point. They also knew that there was absolutely nothing that anyone could do to prevent it. Yet when India and Pakistan went to war over a series of escalating border clashes in Kashmir, the suddenness and magnitude of the catastrophe took everyone by surprise.

The roots of the conflict lay in over sixty years of deepening hatred. Border raids and warfare, terrorist actions, fighting on every level had been a part of the landscape since Pakistan's separation from India after the end of British colonial rule. By the time fighting escalated in Kashmir in 2015, the more fanatical elements of the Indian military and political leadership saw no way to resolve the conflict using conventional means. Instead, they chose a do-or-die course. India fired eight nuclear-tipped ballistic missiles at Karachi and Islamabad, the two most important cities in Pakistan. The results were terrible, horrifying beyond the most exaggerated expectations of the almost forgotten Cold War back in the 20th century.

Both Karachi and Islamabad were bracketed by a quartet of five-hundred-kiloton warheads, set to airburst over the cities for maximum damage to buildings and people. In a matter of minutes, both cities were destroyed, with firestorms roaring outward from the explosion epicenters at over sixty miles an hour. Over twenty-two million Pakistanis were killed instantly. Retaliation was automatic and immediate. Though somewhat more limited in their arsenal than the Indians, the Pakistani armed forces also had

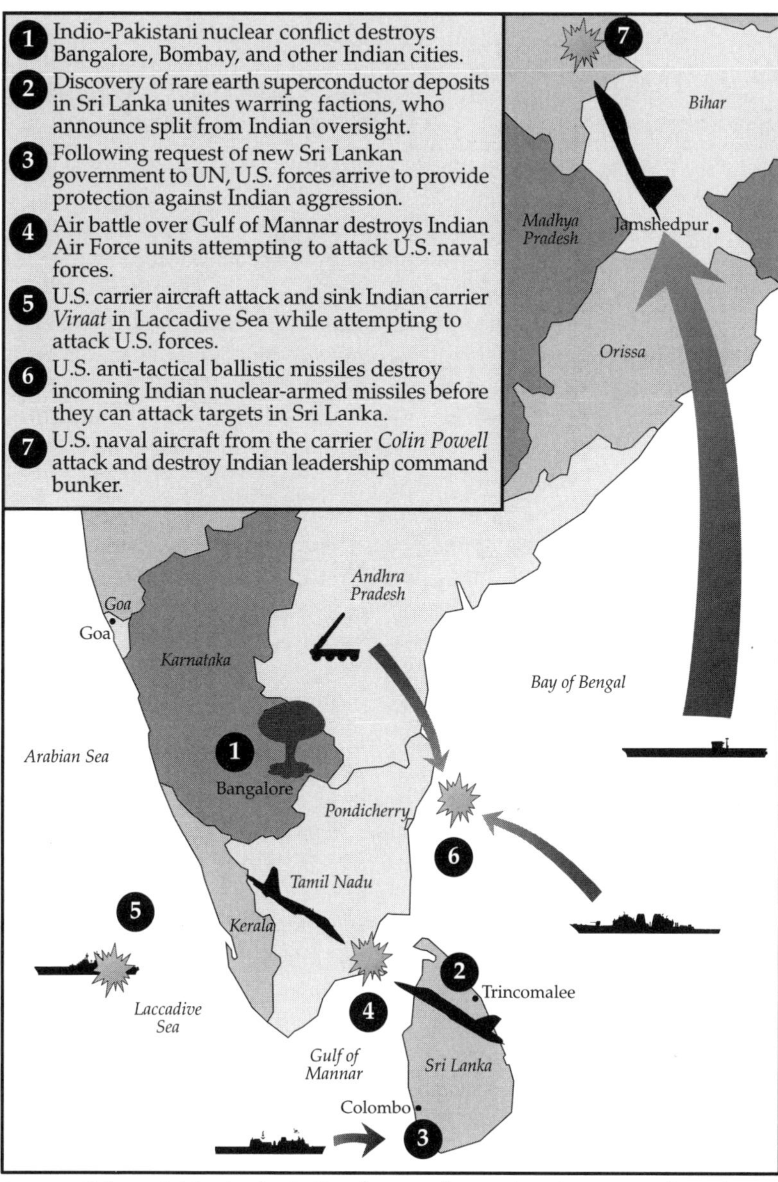

A map of the activities in the Indian theater of operations in 2015 and 2016.

JACK RYAN ENTERPRISES, LTD., BY LAURA DENINNO

missiles with nuclear warheads, and they used them. They fired a dozen missiles at India, each with its own four-hundred-kiloton warhead. The targets they selected were Bombay, New Delhi, and Bangalore—the high-technology center of India's booming military-industrial complex. Over fifty-two million

Indians died in the initial explosions. As prevailing winds carried lethal clouds of fallout over Southeast Asia, an outraged world demanded an immediate cease-fire. The demand was enforced by a unanimous United Nations Security Council resolution. Within days, that demand was backed up by the rapidly growing military presence of its members in the Indian Ocean.

Pakistan's provisional military regime immediately agreed to the cease-fire. They had seen that country's government and fully ten percent of its population snuffed out, and had their hands full dealing with the aftermath of the Indian attack. India's government, evacuated to a command center tunneled deep beneath a Himalayan mountain hours before its capital was vaporized, grudgingly complied. Nevertheless, they continued to denounce ''external interference in our natural and inevitable leadership of South Asia.'' It was clear to everyone in the world that the situation was unstable, likely to explode again at any time. By the time diplomats had ironed out the new cease-fire line in late 2015, the other nations in the region were beginning to consider their options.

Ever since the enforced partition of England's imperial ''Jewel of the Crown'' led to the creation of India and Pakistan in 1947, conflict between the two newly independent nations had never died down. Other nations bordering the Indian Ocean took natural sides, with Muslim states supporting Pakistan, and non-Muslim ones supporting India. Yet after the nuclear holocaust that threatened not only India and Pakistan, but also the entire region, and possibly the world, the states in the region began to distance themselves politically from the two nuclear rogue nations.

Thus the small island nation of Sri Lanka, which had been under virtual Indian control since the partition, took initial steps to remove itself from India's sphere of influence. The reaction of India to Sri Lanka's attempt to declare independence was quick and fierce. India was determined to retain control of the island nation; and might even have managed to do so if the rivalry of the island's Sinhalese and Tamil populations had followed its traditional course. The Indian government had learned the art of ''divide and rule'' all too well during two centuries of English domination. After independence was declared in India, the ruling class put those lessons to good use, playing the divergent interests and goals of many minority groups off against each other in order to keep a firm grip on national affairs. But the current disaster had changed the Indian subcontinent forever. And in the days that followed, India would discover that the old rules had changed.

Aboard the Command Ship USS *Mount McKinley* (LCC-22), Two Hundred Fifty Nautical Miles (NM) Northeast of Diego Garcia, February 4th, 2016

Vice Admiral Matt Connelly was always happiest when he was at sea. His current post as commander of the U.S. Fifth Fleet, and the naval component commander (NAVCENT) for the U.S. Central Command (CENT-

COM), had kept him out at sea for months now, overseeing a mission vital to his county and the world. He was in charge of the Navy's ships and aircraft in a place that was as geographically far as you could go from the miserable climate and politics of Washington, D.C. Even better, he was a *real* fleet commander, in charge of *real* personnel, ships, and aircraft doing a critical mission in an area of great tension. Best of all, his ships and aircraft were the newest and best in the fleet. Given where he was and what he was doing, nothing less was acceptable. India was poised on the brink of another war, possibly even another nuclear war. His success or failure in achieving his mission might determine the fate of this part of the world.

The ship he was aboard, the *Mount McKinley* (LCC-22), was a purpose-built command ship, based upon the design of the *San Antonio* (LPD-17) amphibious landing dockship. Even though it was built as a political concession to keep several shipyards busy following the completion of twelve *San Antonio*-class ships, the *Mount McKinley* was one of the finest fleet flagships ever built. Comfortable and fast, it was a marvelous balance of the complex technologies that make up specialized warships. Other wonderful ships were part of the Fifth Fleet, which Connelly was using to quarantine the Indian subcontinent while the United Nations decided what to do with the Indians and Pakistanis.

Several hundred miles to the east was the new carrier USS *Colin Powell* (CVN-79), another proud ship with a notable namesake. The second of the new class of carriers that was then being constructed, she carried an air wing with ten of the new F-25B joint strike fighters backed up by thirty F-18E and F-18F Super Hornet strike aircraft. These jets were armed with a new family of precision standoff weapons—weapons with amazing new warhead effects.

Also aboard the *Colin Powell* were several new variants of the V-22 Osprey, including the SV-22 ASW/sea-control version, the EV-22 airborne-early-warning/surface-surveillance variant, and the KUV-22 tanker/utility model. Though the *Colin Powell* was only one ship carrying a few dozen aircraft, it was a formidable weapon in the current crisis. The aircraft launched from its deck could maneuver anywhere in the region and hit anything that the National Command Authorities cared to target.

Connelly also had an MEU (SOC) aboard the three ships of his amphibious ready group (ARG), as well as a dozen highly capable escort vessels. Eight of these were Aegis-capable cruisers and destroyers, while the rest were new SC-21-class land-attack and ASW destroyers to protect the underway replenishment train ships. Finally, he had four nuclear submarines prowling about, just in case the Indians decided to get aggressive with their fleet of diesel boats.

A few Allied ships would rotate in and out of what he was calling Task Force 58 (named in honor of Admiral Raymond Spruance's famous World War II force), but by and large this was an American force, protecting American interests and values. Not that Connelly didn't enjoy working with coalition allies.

Over the years he had become known as a master of naval diplomacy. But like any commander, he felt more comfortable with a force whose personnel and capabilities he knew intimately, whose commanders spoke his language without the need to resort to translators, and whose ships and men did what he told them to do without him needing to say "please" first.

His mission was essential, even if it could sometimes grow rather monotonous. He had learned the quarantine game back in 1990 during Desert Shield, and knew how to make it work. Backed up by patrol aircraft out of Diego Garcia and satellite surveillance from the U.S. Space Command (USSPACECOM) warfighting center at Colorado Springs, Colorado, Task Force 58 had the whole region under tight control. His force would keep it that way as long as the equipment, crews, and food held out. He was an American naval officer doing what he had spent a life training to do. Here in the *Mount McKinley*'s Tactical Flag Command Center (TFCC), with the computerized equipment around him constantly monitoring every creature and machine larger than a gnat within the theater of operations, Connelly was exactly where he wanted to be. As he cleared his head for the morning video teleconference with his ship and air unit commanders, he took a deep breath, drank some coffee, and reviewed the computer screen in front of him. So far, it had been a quiet morning. It was his job to be sure that it stayed that way.

University of New Mexico High Energy Physics Laboratory, February 5th, 2016

Jill Jacobs was a lovely blonde. She could have been a college cheerleader in Texas, or possibly a starlet in Beverly Hills. She turned heads wherever she went; she had the kind of looks that made most people assume she got by on body, not brains. Most people would be wrong. She was a well-regarded doctoral candidate in high-energy physical chemistry, exploring rare earth properties for her thesis. It was slow, painstaking work, typically done at night when the lab spaces were open and she could mix and test the bizarre concoctions that were the basis of her ideas about superconductivity. Tonight's work was typical of what she had been doing for almost six months—another apparent failure. It had not generated any of the improvements that her computer models had projected two years earlier. *Oh well*, she thought, *at least this batch didn't explode.*

She stared then at the next batch on her list—samples of a hybrid copper-platinum-scandium mix that represented a sort of cul-de-sac in her projected family of superconducting materials. Always a low-probability set within her computer-modeled group, she had mixed it only because she had the time and materials at hand, and needed to try this particular formula out *sometime*. She took the samples, formed into lengths of wire, to her test bench to measure their resistance and conductivity properties. As she stepped up to the bench, she was tired to her bones. It was discouraging to work so hard without noticeable progress.

She knew the world needed metals that were superconductive at average atmospheric temperatures, but wondered if she would ever find them. If she didn't find them soon, would she ever make a difference with this work? Most likely, she would wind up in a corporate lab somewhere working on improved alloys for jet engines or household appliances. It was the first time she'd even allowed herself to visualize failure, and it surprised her. Maybe the sleep she was losing every night to acquire the lab time for her tests was taking its toll. Or maybe it was the news in the paper every morning. That was enough to depress anyone. But something wasn't right, she decided. She was normally an optimist with a rose-tinted world-view. She needed a break. Perhaps after she finished this test, she would take off for the weekend, and drive to Taos for an overnight visit to a spa, or up into the mountains for a camping trip. If she could get away for a little, maybe she'd feel human again. Maybe.

Turning her attention back to the sample in the test stand, she began to run current through it at a variety of temperatures. At first the readings did not seem out of the ordinary. At -200° Centigrade, the sample had exactly the superconducting properties that one would expect it to have. But as the sample came up past 0° C, it finally hit her what was wrong, or more properly, what was *right*. The sample had stabilized its conductive properties at 98% of their optimum, and held them. She continued to ramp the temperatures up, and the material held up until it finally melted at about 300° C.

She'd done it.

If her eyes and her machinery weren't lying to her, she'd found her material. Stunned, she cleaned up the chamber, recalibrated her equipment, loaded an identical sample into the test rig, and tried it again. Identical result.

"I've really done it," she whispered to herself.

As she fumbled in her purse for her mobile phone to call her faculty adviser, her brain was spinning like a pair of dice in Vegas. Her doctoral thesis was a done deal now. She could finally finish her degree and get on with her life in the *real* world. She'd realized her goals in her current research and could move on to new frontiers. But even as she called her adviser to share the news, she had no idea how crucial her new discovery would be to the rest of the world. She'd just created a practical high-temperature superconductor, and in so doing was about to change the face of civilization. "Power" and "wealth" would never be the same again.

Headquarters of the Liberation Tigers of Tamil Eelam, Near Mankulam, Sri Lanka, February 7th, 2016

Arjuan Ranatunga sat in the place he called his office and contemplated how to change the course of his nation's history. Grand thoughts for a man whose major passion had only recently been playing cricket. But the continued suppression of the Tamil sect by the Indians on the mainland and the Sinhalese on the southern half of Sri Lanka had no end in sight. This repression had drawn him to the Liberation Tigers of Tamil Eelam (LTTE), more commonly referred to as the "Tamil Tigers." The Tigers had always been a part of his country's political landscape—for as long as he could remember, anyway. He was a revolutionary soldier in a battle that had been going on for longer than he had been alive. Now, at age thirty-seven—an age when he should have been coaching a regional cricket team—he had become the leader of the LTTE. When events in his country had spun out of control, he had been unable to turn his back on the needs of the people. The final straw that had made his current occupation inevitable was the death of the previous Tiger leader, his brother Sanath. Sanath had been killed by an Indian helicopter gunship a few weeks earlier.

He was sitting in a tent surrounded by jungle near Route A9. His "desk" was a folding table and his office chair a ration crate. In front of him were a laptop data slate and his encrypted satellite cellular phone. Despite the spartan surroundings, he had the power to control considerable military clout from the humble resources at his fingertips. He could dispatch forces ranging from patrol boats to special assassination teams with just a few taps on his keyboard, or a simple phone call.

And yet, force wasn't doing the job. Decades of active resistance against the Indians and Sinhalese had utterly failed to give the Tamil Tigers the homeland they dreamed of. Already today, he had been advised by his regional commanders to begin a terror campaign in the south to avenge his brother's death. Yet revenge was not his objective today. He knew better than anyone did how futile it was. Nothing would bring back his brother. Instead of planning and setting into motion a campaign of terror, he'd chosen to spend the morning considering his options, and the options of the organization and his people. Though well financed by the Tamil supporters on the mainland, he could see no combination of military action that would ever result in Tamil domination of Sri Lanka. Even if they won the bloody civil war that would be necessary, they would inevitably lose the peace that would follow. The Sinhalese would start their own liberation movement, and the cycle would start again.

What he needed was something different. A new kind of weapon—some new power that would break the rules, that would give his cause an edge that would count for something in a world where large-scale violence was relatively rare, but where the warfare of commerce, corporations, and economics was everywhere. A few days earlier, it came to him that an answer might lie in the rich earth at his feet. Sri Lanka was his home, the mother of his people. Perhaps that mother might provide the milk that would make them powerful enough to win, powerful enough to keep the Indians from crushing them, powerful enough to encourage a superpower like America to support them, as they had Kuwait back in 1990.

Not an easy task.

To catch the attention of the United States and focus it on the sufferings of a handful of people on the far side of the globe would take no less than magic. Luckily, he had recently hired a wizard.

West of the Kokkita Bird Sanctuary, Sri Lanka, March 9th, 2016

The foothills of north Sri Lanka are unique in the South Asian region. While most of the Indian subcontinent is among the newest terrain on the globe, these foothills are some of the oldest. Old things are likely to be valuable, and that was why the geologist was here. The contract to survey this area had been both lucrative and timely. Short of money for his children's school tuition, he had jumped at the chance when the Internet inquiries about his availability had reached his home in Perth, Australia. He had immediately said yes.

Before he'd even started packing, he had commissioned a series of one-meter-resolution multi-spectral satellite photographs from the French SPOT Corporation. Running the images through his desktop workstation in Perth, he had found several promising areas to explore. The commission had been explicit. Find rare and valuable mineral deposits, report them to the commissioning agent, accept the fee, and then deny that he had ever visited Sri Lanka. As an enticement to silence, the agent had promised him a tenth-of-a-percent royalty on anything that he found that was developed during his lifetime. With an offer like that, he had gone to extraordinary efforts for his employers. For almost a month, he had run the tires off of his hired Land Rover, looking for some exceptional mineral deposit to report back to them.

Now he was working the last area on his list of possibilities. So far he had found some promising discoveries, but nothing spectacular. A few days earlier, he spotted what might be a major vein of platinum in the side of a mountain, and he had taken several core samples around it to assay when he got home. Today, his chemical "sniffer" was finding samples of rare earth metals; and there seemed to be particularly large concentrations of scandium. What struck him was the purity of the sample he'd collected here—it exceeded anything he had ever heard reported.

In three days, he would return to Perth and start on his analysis and report. He hoped, for the sake of his future royalties, that the platinum find would pan out. Nobody had ever found a significant use for scandium.

National Press Club, Washington, D.C., April 1st, 2016

April Fool's Day is normally a day for pranks and lies, but this day would go down in the history books as a day when new truths were told. Jill Jacobs and the head of the Sandia Labs stood before a packed house of disbelieving science reporters to announce a breakthrough in superconductor technology, which would allow for the development of electric motors thousands of times more compact, powerful, and efficient than any made previously. A patent for the metal formulation had been applied for, and it would be available for commercial license immediately.

Chuckles broke out among the reporters, and there were cracks about cold fusion—until Jill came to the podium and asked everyone to go down to the street below, where she promised to demonstrate the material. Moments later, the assembled press personnel found what appeared to be a completely normal pair of buses painted with the logos of the University of New Mexico and Sandia Labs. After the reporters were all aboard and seated, Jill stepped onto the first bus, the Sandia chief onto the one behind it. A moment later, the buses accelerated smartly away from the curb, silent as ghosts, the typical diesel roar completely absent. In fact, the street noise outside was deafening by comparison. The stunned reporters sat in silence as they rode to the base of the Washington Monument several miles away.

After everyone got out of the buses and filed onto the sidewalk, Jill and the Sandia Labs chief opened the engine compartments to show the press corps a single car battery running an electric motor the size of a beer keg. All told, they informed the reporters, the two buses had consumed less than an amp-hour of power, less than one percent of what was stored in each battery. Even better, the motors, which had been designed from existing models, had cost less than a thousand dollars to build. Most of the reporters dragged out their cell phones then and there to report in, rather than waiting for the buses to return them to the Press Club building.

Headquarters of the Liberation Tigers of Tamil Eelam, Near Anuradhapura, Sri Lanka, April 3rd, 2016

Today Arjuan Ranatunga's headquarters were located in a gamekeeper's hut, and his table and chair were actually comfortable. Moreover, the news on the data slate before him was like a gift from God himself. For the better part of a week, he'd had the mineral report he'd commissioned, but until now, it had seemed disappointing. It had promised nothing like the riches he had hoped for. But overnight, the news from America had turned the economy of the world upside down. Everyone had gone superconductor-crazy. Oil prices had taken a precipitous drop, and the prices for platinum and scandium had jumped off of the charts. This was hardly surprising. The world's known reserves of scandium could be measured in just a few tons. These would supply a bare handful of the proposed applications for the new superconducting metal formula.

He did not need to be a financial genius to figure out that what had been found in the foothills to the east could make Sri Lanka the superconductor capital of the world. The problem was what India would do when they found out what was sitting in the foothills of Sri Lanka. Once they knew what was there, they would crush both the Tamils and Sinhalese faster than they had nuked the Pakistanis. Even worse, the rest of the world would probably not care, if what he had seen over the Internet on the various news service web pages could be believed. As long as the resource was developed, it didn't matter who was offering it. He had to act quickly if he were to save his people and—ironically—their enemies, the Sinhalese. Taking a deep breath, he tapped out an E-mail message to his Sinhalese counterpart in Colombo.

Indian National Command Bunker, Near the Himalayan Town of Puranpur, April 4th, 2016

Roshan Gandhi was having another in a long string of bad days here in his bunker. The Indian Prime Minister had not seen a ray of sunshine for weeks, and was beginning to wonder if he would ever see sun again. Since the day four months ago when he had authorized the firing of the nuclear-tipped missiles into Pakistan, his fortunes and those of his country had been spiraling out of his control.

Like so many other Indian politicians who shared his name, Gandhi was in no way related to the great man who had led India to independence six decades earlier. It had never seemed to worry the Indian people that the name Gandhi had helped a string of politicians gain power in India over the years. Still, his family did share a political connection with him. Roshan's grandfather had been a follower of the great Gandhi's, and had adopted the name after the assassination in the late 1940s.

The current Gandhi had been a popular provincial governor before he ran for and won his present office. He'd become the political leader of his party, and was then elected to national office because he was an honest man. He'd offered a pleasant contrast to the scandal and graft of the previous administration. Unfortunately (tragically, as it transpired), during all the discussions and analyses of what he was not, nobody had ever thought to ask what kind of leader Roshan would be. It would have been an illuminating question. As Roshan himself was the first to admit, he was a better follower than leader. And, honest man that he was, he'd have admitted that to the press. But no one thought to ask the question. From his first day as Prime Minister of India, with a vast majority in Parliament, Roshan Gandhi had been in over his head.

In the early days of his administration, his Defense Minister had badgered him into ordering a nuclear war with Pakistan. Even after the war was unleashed so catastrophically, the man was still badgering him for more. Roshan wasn't happy about the way matters stood, either for him or for his country. Gandhi was aware of the problems his government's actions had caused. How could he not be, even insulated here in the mountain fortress? There were tens of millions of Indians dead. Even four months after it was over, more were dying every day from the lasting effects of the nuclear exchange with Pakistan. Prevailing winds had swept the fallout to the east, making whole swathes of the land uninhabitable. Uncontaminated water was in critically short supply throughout the country. Plague and famine were rampant. Existing food stores, the crops in the fields, dairy products—all were contaminated by radioactive waste.

Unrest was everywhere in India, in a thousand villages and towns. Over the war, over the lack of food and water, over the destruction of the infrastructure, even over the UN quarantine. Mobs were forming, demanding action. Military units were suppressing the demonstrators and rioters, using deadly force if necessary. Roshan had agreed to that. It was a bad choice, but the only one that might allow India to survive as a nation.

But Roshan's current problem was not centered on India's massive domestic difficulties. Just at the moment, he was worrying about what would happen if any of India's neighbors became too independent. Both Bangladesh and Sri Lanka had been showing signs of slipping away from India's influence. India was not really a "melting pot" like the United States, but a huge patchwork quilt composed of many thousands of distinct language and ethnic groups. Held together now only by the iron force of the Indian military, India might fragment into a hundred little kingdoms and regions—unless Roshan could make the center hold. In Roshan's opinion, a crucial stage in this process would be getting the trade and imports embargo imposed by the UN dropped. Roshan's people were starving, dying of thirst, rotting away from radiation sickness, and succumbing to a long list of ordinary diseases that could be controlled with proper medications.

Roshan wanted the means to repair the damage he'd done. He needed the basics of life—food, water, medical supplies. What the Defense Minister needed—ammunition—he unfortunately had in abundance. Maybe they could use it to buy more time. Right now, India had none. What Roshan *really* needed was a solution to the problems he himself had created by authorizing the launch of nuclear missiles against Pakistan. Such a solution was even less likely to materialize.

A Private Plantation near Colombo, Sri Lanka, April 5th, 2016

The plantation was a hallowed place in Sri Lankan history. It was the former home of a celebrated scientist and science-fiction author who had spent his later years tapping out novels on a computer in the study, and then uploading them to his New York publisher via a personal satellite uplink in the courtyard. A literary shrine for tourists, it was closed today, ostensibly for cleaning and maintenance. Venkatesh Prasad, the Sinhalese Prime Minister, had come here in response to an E-mail he'd received the day before from his counterpart in the LTTE. The unofficial cease-fire between the government and the Tamil Tigers notwithstanding, Prasad was extremely suspicious of this meeting.

But Prasad's suspicion rapidly gave way to astonishment when, a few minutes later, Arjuan Ranatunga arrived, accompanied only by a driver for his Land Rover. Prasad had spent a lifetime fighting the LTTE to preserve Sinhalese control of Sri Lanka. Now he was about to sit down for a private talk with his sworn enemy. About what? He had no idea. Maybe Arjuan would suggest that they settle everything in a nice, civilized way, perhaps with a cricket match. That thought made him smile thinly.

As the two men sat together in the former author's study, the LTTE leader laid out an astonishing offer before Prasad. Arjuan proposed that they just stop fighting. Stop fighting, put down their weapons, and share the most valuable mineral strike in the history of mankind. It was a peace proposal so remarkably simple it was impossible to refuse. Both men could see clearly what would happen if they could just cooperate. Their little island would become the 21st century equivalent of OPEC, with all the wealth, power, advantages, and liabilities that would naturally ensue. They agreed on the need for support from outside, particularly from the Americans. Most of all, they decided that the existence and location of the platinum and scandium would remain secret, until the security of their new nation was assured. Otherwise, that knowledge would bring genocide on both their peoples.

United Nations Security Council Chamber, New York, New York, May 2nd, 2016

The two Sri Lankan leaders had decided to let the British ambassador, rather than the Americans, convey their proposal to the Security Council. The old colonial ties with the British Empire would lend credibility to the proposal, and the American second would almost certainly assure its passage. By nightfall, the following resolution had been passed, with only two abstentions:

RESOLUTION 2209

The Security Council,

Recognizing the desire of the combined peoples of the Island Nation of Sri Lanka for self-determination,
Alarmed by the recent actions by India in the suppression of their own ethnic minorities, as well as the illegal use of weapons of mass destruction against all known international laws and treaties,
Determining that there exists a breach of international peace as a result by the Government of India,
Acting under Articles 39 and 40 of the Charter of the United Nations,

1. *Condemns* the Indian suppression of their region;
2. *Demands* the immediate recognition of the Sri Lankan Republic by the Indian Government;
3. *Calls upon* the Government of the Sri Lankan Re-

public and the Government of India to begin immediately intensive negotiations for the resolution of their differences and supports all efforts in this regard;

4. *Orders* that the Indians shall be the object of a reinforced UN-sanctioned air, ground, and Naval quarantine of all Indian efforts against the Sri Lankan Republic;
5. *Authorizes* that member nations providing forces for the quarantine may use military force consistent with their own security, and the enforcement of the previously mentioned action;
6. *Decides* to meet again as necessary to consider further steps to ensure compliance with the present resolution.

Indian National Command Bunker, near the Himalayan Town of Puranpur, May 4th, 2016

Prime Minister Gandhi was in the middle of another shouting match, this one involving his Defense Minister and the service chiefs. The mysterious union of the two warring factions on Sri Lanka into a single government was puzzling, but irrelevant. More important was the fact that the island was a de facto province of India, and the mainland population would see any attempt by the islanders to go their own way as a sign of weakness on the part of Gandhi's government. But neither Gandhi nor the men around him could decide what to do about it. The Navy and Air Force chiefs were busily trying to explain the suicidal folly of trying to take Sri Lanka in the face of the previous day's UN vote. After the war with Pakistan, the UN was looking for any excuse it could find to turn India into a null-power in South Asia. Within a matter of hours, American Army and Marine pre-position squadrons would sail from Diego Garcia. In just four days, they would unload in the harbors at Colombo and Trincomalee with enough equipment for a 25,000-man joint air/land task force to protect the small island. Already, there were reports of the American ARG beginning to head for Sri Lanka. As if to add to Roshan's troubles, there were reports from the BBC and CNN that units of the 82nd Airborne Division were preparing to deploy to Sri Lanka from Fort Bragg. In less than a week, Sri Lanka would be as free of India's rule as Antarctica.

Gandhi knew this turn of events would mean the end of his government, and he wanted it stopped. Since such an effort would involve amphibious and air operations, it would fall on these two chiefs to make it happen, and they did not want any part of it. Their forces had suffered in the short and bloody fight with Pakistan the previous winter, in humanitarian missions after the war, and in quelling the riots ever since. Both commanders, doing their utmost to hold what was left of their services for better causes, better days, were firmly opposed to Roshan's decision. Predictably, the Defense Minister was in favor of the Sri Lankan expedition. He didn't care about the preservation of the lives of the men in the armed forces; trained men could be replaced or bought.

After a time, further debate was useless. A decision had to be made. Roshan closed his eyes, thought for a moment, and ordered the expedition. It was another bad choice in a seemingly endless line of bad choices, dating from the very moment he'd sought to become Prime Minister.

Aboard the Command Ship USS *Mount McKinley* (LCC-22), Five Hundred Nautical Miles South of Colombo, May 5th, 2016

''Well, Jack, I think we understand what is needed here. I'll get the staff working on it,'' Vice Admiral Connelly said into his video teleconferencing terminal. His satellite-assisted meeting with the Chairman of the Joint Chiefs of Staff had gone as expected, and the official orders from the UN and American National Command Authorities would be on their way via secure FAX in a matter of minutes. Now he could officially begin what he had privately started two days earlier when the chairman had told him of the probable UN resolution. Already, he had begun to concentrate his forces around Sri Lanka, and set up the wall of fire and sensors that would be needed to protect the island nation from what was considered the inevitable Indian response.

Shutting down the terminal, he walked back to his day cabin, pulled out a yellow legal pad and mechanical pencil, and began to sketch an outline of the plan for the defense of Sri Lanka. He knew using pencil and paper was so outdated it was laughable, but he also knew that he did his best thinking while he wrote the old-fashioned way. He smiled as he began, knowing his Fifth Fleet staff would probably take twice as long to argue over what he was about to write as he would to do so. That was after they'd laughed themselves sick over his method of encoding the data. *Well,* he thought, *this is how we did it on the old days before voice-recognition word processors and eye-controlled pointing devices. And it will work under any circumstances, even in a total power outage. I'd like to see them say that about their computers.* Thirty minutes later, he was finished.

The plan was quite simple, actually. The 26th MEU (SOC) would land on the island and establish coastal defenses to keep the Indians from crossing the Gulf of Mannar. Two brigades of the 82nd Airborne Division would begin arriving in thirty-six hours to back up the Marines. He would then create a series of ''missile traps,'' composed of pairs of Aegis ships and land-attack destroyers, to provide fire support and protect against air and ballistic missile attack. Finally, his command ship, the carrier, and the four remaining escorts would establish an operating area southeast of Sri Lanka to provide air cover and support for the ground and naval forces. When the MPS ships arrived in three days, he would land their cargo, and begin flying in the Army and Marine Corps personnel needed to make the island into a fortress. After that, UN peacekeeping personnel with their blue berets would arrive and take over, along with the inevitable multi-national air and Naval force to cover the island from attack. All he had to do was keep the Indians honest for the next few days.

Unfortunately, this particular Indian government was composed of a few irrational people with the ugly habit of launching nuclear weapons when they lost their temper. He was more than a little concerned about whether his Aegis ships and the battle staff at USSPACECOM in Colorado Springs were ready to play for all the marbles. The Indians were using serious firepower. Not modified SCUDs fired like shotgun shells, but IRBMs with nuclear weapons. He found himself wondering if American magic would be better than Indian magic.

Indian Naval Base, Goa, India, May 6th, 2016

After the destruction of Bombay, the major fleet units of the Indian Navy had made Goa their new fleet base. All told, over a dozen warships and a comparable number of submarines lay at anchor, surrounded by the merchant ships being taken up and loaded with men and equipment for the expedition to Sri Lanka. As he looked across the bay at his fleet, Admiral Ajay Jadeja, the Chief of the Indian Navy, contemplated the death ride that his fleet was about to take. He wondered how much he would personally sacrifice in the name of Indian honor, and how many young men on both sides he would have killed as he did so.

He had no doubt of the Americans' ability to destroy his surface force before it rounded Cape Comorin at the southern tip of India. Right now, his most hopeful outcome was for the world to be so appalled by *his* losses that the UN might back away from their resolution to maintain a complete embargo against India. Meanwhile, since much of his submarine force had been destroyed when Bombay had been destroyed, he wanted to be careful with the handful of subs he still had. He was still hopeful that his submarines would get in a few lucky shots against the American ships, though nobody had had much luck on that score since the 1990's.

But in his heart he feared a round of Indian nuclear missile launches against Sri Lanka would cause retaliation in kind against his country. Should that occur, he mused, India, the world's largest democracy, might just have solved its population problems permanently. He was a man of no little integrity; and he had argued against this silly adventure to his superiors. It did no good. They'd simply told him to "be silent and lead your men in their duty." He would follow his orders to the death, he supposed—anything but a glorious death. It would be a slaughter. On the other hand, if he resigned, his replacement would be indifferent to the fears that burned within him. Better to take his fleet to sea, and try to save what he could.

Over the Gulf of Mannar, May 6th, 2016

The first action between the American and Indian forces inevitably took place in the air. In the late afternoon, an Indian force of 24 Su-30 Flanker fighter-bombers armed with antiship missiles launched with a dozen old MiG-29 Fulcrums as escorts. Their targets were the two missile-trap ships on either side of the narrows between Sri Lanka and the Indian mainland. The Indian pilots had no idea they had been detected even before their aircraft had left the ground. Their takeoff was picked up by one of the new EV-22 surveillance aircraft. As they flew toward their destination, they were intercepted by eight F-25B stealth strike fighters from the *Colin Powell,* armed with the newest long-range version of the AIM-120 AMRAAM air-to-air missile. Before the Indian fighters had even formed up, two thirds of their force was vaporized by the first salvo of American missiles. These were followed by a quartet of F-18E Super Hornets that finished off all but three of the survivors. Then came a salvo of standard surface-to-air missiles from one of the Aegis destroyers. When it was all over, only a single MiG-29 pilot made it home to tell about the massacre over the Gulf of Mannar. The Americans would later call it an ''overmatch.'' The Indians called it suicide.

Aboard the Aircraft Carrier USS *Colin Powell* (CVN-79), Fifty Nautical Miles Southeast of Sri Lanka, 2000 Hours, May 6th, 2016

Admiral Connelly had taken a helicopter over to the *Colin Powell* to congratulate the pilots on their intercept of the Indian fighters, and to confer with the captain and air wing commanders on what they would do the next morning when the Indian fleet came into range. They all agreed that what he had in mind was not going to be easy, and could become extremely difficult if the Indian fleet commander tried anything radical with his course or formations.

As things were then proceeding, this appeared unlikely. The Indian commander seemed bent on a death ride. Already, the Fifth Fleet staff analysts had decided that the Indians hoped to shame the Americans with the slaughter—as the Iraqis had done during Desert Storm by drawing media attention to what was falsely called the ''Highway of Death.'' More than one historian had noted that press coverage of that event had caused the war to be stopped at least a day or two earlier than it should have been. The price had been several decades of problems in the Persian Gulf. Connelly did not intend to repeat that mistake.

Over the Lakshadweep Sea, 0700 Hours, May 7th, 2016

The Global Hawk reconnaissance drone was settled safely over the Indian task force, and the live satellite imagery feed was operating perfectly. Launched eighteen hours earlier from Diego Garcia, it would stay in the air for days, feeding data to the American forces. Right now, the main camera was focused upon the Indian aircraft carrier *Viraat*, at one time the British flattop *Hermes*. She carried a dozen modernized Sea Harrier fighter-bombers, which were currently loaded with rather elderly Sea Eagle antiship missiles. Admiral Jadeja figured that he'd been indulging in a bit of wishful thinking when he'd had the Harriers tasked. More than likely, they would never leave the deck of the *Viraat*. His only real question was whether the attack that demolished them all would come from a submarine or from the air. Either way, the death of his fleet might serve to shame the superpower into relaxing its hold on Sri Lanka. In truth, he doubted that.

CNN Center, Atlanta, Georgia, 2000 Hours, May 6th, 2016

The LIVE EVENT graphic went up on the screen followed by an introduction by the news anchors. Viewers worldwide were about to see a live feed from the Indian Ocean where the Sri Lankan quarantine was in effect. The CNN feed was accompanied by a voice-over from the Chairmen of the Joint Chiefs, who began to provide the world's first official play-by-play commentary of an actual battle. What the world saw was the Global Hawk view of the Indian carrier group, with an occasional zoom in on the *Viraat*. What was said next stunned the worldwide audience.

"Since the United States wishes to fulfill its commitment to the Sri Lankan people and its UN partners, but wishes no excess bloodshed in the process, we are about to show the world, especially the Indian government, what will happen to all of their ships if they do not turn back their forces immediately."

He nodded to his assistant, who relayed a signal to Admiral Connelly on the *Mount McKinley*.

Over the Lakshadweep Sea, 0705 Hours, May 7th, 2016

The four F/A-18Es Super Hornets had just downlinked the final targeting templates for their ATA-equipped hypersonic cruise missiles, and fed the image of the *Viraat* into the guidance systems. When Admiral Connelly gave the order, the four pilots salvoed the missiles at fifteen-second intervals, the better for the world to watch the results. Each missile immediately ignited its rocket motor, and climbed at Mach 6 into the upper atmosphere for the two-minute run to the target. When directly over the *Viraat*, each missile began to dive, and scanned the surface below for a shape that matched the image template in its guidance package. The results were stunning even to the people who had planned the strike.

The Global Hawk camera zoomed in on the *Viraat* just before the first missile struck the flight deck on the fantail. The missile penetrated the flight deck before the thousand-pound warhead detonated, blowing chunks of the after flight deck into the air. Seconds later, the next missile arrived, landing about one hundred feet forward of the first missile hit. This time three Sea Harriers were blown apart, the pieces flung into the air. The explosions continued. By the time the last two missiles arrived, the ship was a mass of flames and explosions. Since there was no longer a target to hit, the missiles splashed into the ocean. Almost immediately, the old flattop began to settle. Within ten minutes it was nothing but a pool of burning oil, floating debris, and men fighting for their lives. One of them was Admiral Jadeja.

CNN Center, Atlanta, GA, 2010 Hours, May 6th, 2016

The images of the final moments of *Viraat* shocked even the JCS Chairman, who had to recompose himself before he completed his statement.

''As you can see, the United States has the ability to strike, and destroy at will, any Indian naval unit that it desires. In the interests of humanity, I make the following statement to the Indian National Command Authorities. You may spend the next two hours conducting search and rescue operations. At that time, if your ships have not reversed course, we will begin to sink additional units at our discretion. In the name of decency, please return your fleet to its base at Goa without delay.''

He need not have said anything. As a burned and bruised Admiral Jadeja was pulled from the oily water, he himself ordered the fleet to complete search and rescue operations, and then to return to Goa at best speed. The Indian Sri Lanka expedition was over.

Indian National Command Bunker, near the Himalayan Town of Puranpur, 0900 Hours, May 7th, 2016

Once again, Prime Minister Gandhi was watching a fight between his Defense Minister and his service chiefs. This one had turned uglier than usual. Physical blows had been exchanged even before news of Admiral Jadeja's fleet recall order had been delivered. Far from shaming the Americans with a slaughter, the Indian Navy, the most powerful navy in the region, had been punished and humbled before the world—not just by a show of arms but by a show of mercy.

After they'd watched the broadcast on CNN, the service chiefs had withdrawn, for their own physical security (they feared that the Defense Minister might find a weapon and kill them). In their absence the Defense Minister had turned his wrath on Gandhi. This infamy, the Minister ranted, must be avenged, and the American mission stopped, whatever the cost. It was at this moment that Roshan realized that he was a coward; he lacked both the moral and physical courage needed to defend himself and his country. So when the Defense Minister pressed for a nuclear-missile strike on Sri Lanka, as the madman hung over him threateningly, Gandhi signed the release orders.

As the Defense Minister left to commit another crime against humanity, the Prime Minister lowered his face into his hands to sob, silently praying to his God that someone would stop this man, even if it killed them all. He could only die once. Best for that to happen before the blood of more millions of innocents stained his hands.

North Coast of Sri Lanka near Jaffra, 1200 Hours, May 7th, 2016

Admiral Connelly liked what he saw. The MEU (SOC) was already in its defensive position. The troopers of the 82nd Airborne down at Colombo had volunteered to send them a platoon of engineers with bulldozers and earthmovers to improve the sites. The artillery was already dug in; and the air defense vehicles had excellent engagement arcs. Seeing that their colonel had things well in hand, he walked back to his HH-60R helicopter for the ride back to the *Mount McKinley*.

As they lifted off and headed out to sea, he got a message on his secure satellite phone, which set him immediately on edge. An NSA ferret satellite had picked up indications of commands being issued to an Indian IRBM battalion. Early analysis indicated that the unit had been ordered to erect and fuel their missiles, and prepare them for launch. Estimated time until they would be ready for action was less than three hours. Realizing that his force had very little time to prepare for what might be the world's first duel between nuclear-armed ballistic missiles and theater ballistic-missile defense forces, he ordered his pilot to push the chopper to the limit.

USSPACECOM Theater Battle Management Center, Falcon AFB, Colorado, 0322 Hours, May 7th, 2016

The battle management staff was fully manned, with off-shift personnel crowding in between the workstation terminals and the gallery. An Air Force brigadier general from the 50th Space Wing was in command, and he had his command and control links and satellites fully netted and ready. For years, they had practiced this very scenario on complex computer networks against synthetic missiles. Today, they would be doing it for real, with actual nuclear-tipped missiles as targets, and the lives of several million human beings at stake. The earliest deadline for possible launch of the Indian missiles had passed about twenty minutes earlier. Everyone was getting a little edgy. Just as the general was about to declare an alert break so his people could get some coffee and donuts, the Defense Support Program (DSP) satellite console operator came on the net with a voice that was frighteningly detached.

''We have missile launches in central India. I repeat, we have multiple missile launches in central India. Confidence is high. I repeat, confidence is *high*.''

It took a few seconds for the DSP bird to obtain rough tracking information on what was now looking like six IRBM-type missiles as they climbed away from their launchers near Nagpur. When the information came in, it was fed automatically to the battle management consoles, where software began to send orders to a series of high-resolution targeting satellites in medium Earth orbit. Within thirty seconds of the last Indian missile's launch, each missile was being tracked by a telescope, which was supplying precise fire control information to the battle management network. The general, seeing that there was only a single wave of missiles headed south toward Sri Lanka, quickly made his decision, then spoke over the network.

"This is Silicon Palace to all stations. Werewolf. Werewolf! We have six inbound missile tracks to the Sri Lanka area. Confidence is *high*. I repeat, confidence is *high*. All ships and batteries, I declare *weapons free*. Repeat. I declare *weapons free!* Go get 'um, space rangers!"

He had done his job. Now they all got to see if a few hundred billion dollars had been wasted.

Aboard the Command Ship USS *Mount McKinley* (LCC-22), Five Hundred Nautical Miles (NM) South of Colombo, 1525 Hours, May 7th, 2016

The displays showed the inbound missile tracks, even though the radars of his Aegis ships could not yet see the weapons on their own. Like everyone else, Admiral Connelly had run simulations of missile defense time and time again. But this time, it was terribly real. Right now, the targeting data was being relayed via satellite link from Falcon AFB, and it was good enough to shoot with. The idea was to try to engage the incoming missiles as soon as they came into view of the Aegis ships. He had already given weapons-release authority to the theater ballistic-missile defense officer in the corner console in the TFCC. The young lieutenant commander had an Aegis cruiser and two destroyers to engage with, as well as a pair of Army Patriot batteries from XVIII Airborne Corps on Sri Lanka itself. This gave them two layers of firepower to apply against the incoming missile stream. He hoped it would be enough.

Over on the destroyers *Mahan* (DDG-72) and *Hopper* (DDG-70), as well as the cruiser *Cape St. George* (CG-71), the battle management software from Falcon Field ordered each ship to launch a modified Standard SAM with a miniature homing vehicle as the payload. Because of their limited loadout of ATBM SAMs, the three ships had to fire one at a time at the incoming missiles, so that the chances of a kill would be maximized. The first salvo had been dispatched before the Indian IRBMs had even come over the horizon, but this would increase the number of possible shots against the missile stream.

Admiral Connelly watched transfixed as the six SAM symbols moved across the large-screen display toward the IRBM icons. The flight time was almost two minutes, and the results were gratifying. Three of the Indian missiles were destroyed by direct kinetic energy hits from the SAMs, while the others would require further engagement. Another salvo of three ATBM SAMs erupted from the Aegis ships, this time with a flight time of less than forty-five seconds to their targets. The miniature homing vehicles vaporized two more IRBMs. That left just one targeted on Colombo.

Connelly began to ball his fists when he saw two shots at the final Indian missile miss due to bad engagement geometry, allowing it past the picket line of Aegis ships. This left only their goaltender, the Patriot battery on a hill overlooking Colombo Harbor. The site had originally been the headquarters of Lord Louis Mountbatten during the Second World War, and now had the best firing arc of the Army SAM batteries. The Indian missile was less than two hundred miles out when the battery spat out a pair of PAC-3 ERINT anti-missile SAMs. The Army had deployed this system in great numbers, and a second pair of ERINTs were fired to make sure that this last inbound had no chance.

The problem was that the Indian missile was of a fairly advanced design, with a system for detaching the warhead at apogee. This improved the accuracy of the warhead and made interception more difficult. However, U.S. design teams hadn't been standing still either. Hard-won experience from several decades earlier in the Persian Gulf had taught the software engineers some valuable tricks, and the Patriot radar easily picked out the warhead from the fragments of the missile that were breaking up upon reentry into the atmosphere. As it turned out, the first salvo of ERINTs was enough. The second PAC-3 struck the warhead, vaporizing it into an exploding stream of plutonium and ceramic from the heat shield. On both sides of the world, the winners of the first nuclear-missile/anti-missile battle jumped to their feet and issued a collective victory cry. The American magic had been better.

Indian National Command Bunker, near the Himalayan Town of Puranpur, 1835 Hours, May 7th, 2016

Prime Minister Gandhi sat alone now in the conference room. He'd sent the military chiefs away to their quarters, and put the Defense Minister under arrest. He had finally pulled himself together enough to do the right thing, which was precisely nothing. The failure of the missile strike had given him back his options, and now he was going to limit the retribution on India to this bunker, and probably the missile launch site. He knew that the Americans had probably already targeted both locations, and that they would hit them soon. He ordered all non-essential personnel out of the facility, than sat down and began to pray for his soul. He hoped that it would be over soon.

Flight Deck of the Aircraft Carrier *Colin Powell*, 1925 Hours, May 7th, 2016

They had been forced to wait until the resolution of the Indian missile strike to know which weapons they would upload. Had any of the Indian

IRBMs hit their targets, then the F-25Bs would have been each loaded with a pair of B-61-15 nuclear penetrating gravity bombs targeted on what had been called "strategic" targets. The population density of India meant that the use of any such weapon would kill hundreds of thousands of civilians at a minimum. Thankfully for the ordnance personnel and the pilots, the orders from the National Command Authorities had been explicit. Response in kind. This meant that unless a nuclear detonation had taken place, only convention weapons were authorized for use in the coming strike on the Indian leadership and their nuclear missile depots.

The F-25Bs would each carry a GBU-32 JDAMS with a modified BLU-109 two-thousand-pound penetrating warhead to seal the bunker entrances. Then the F/A-18 Super Hornets would finish the job with 4,700-pound GBU-28 "Deep Throat" bombs armed with BLU-113 warheads to collapse the tunnels. Similar attention would be given to the Indian missile silos near Nagpur.

It took a little over three hours to get the aircraft loaded and the crews briefed. As usual for such things, it would be a precision night strike to help degrade the Indian defenses. As the first pair of F-25Bs taxied up to the catapults at the bow, the deck crews lined the catwalks, cheering the pilots as they launched into a beautiful night sky. It would take a few hours for the planes to reach their targets.

Indian National Command Bunker, near the Himalayan Town of Puranpur, 2242 Hours, May 7th, 2016

Prime Minister Gandhi lay in his bedroom waiting for the end. He had authorized the actions that had resulted in the deaths of tens of millions of human lives. He would be remembered as the first great genocidal despot of the new millennium, and that was a difficult thought to die with. But he knew he was doing the right thing now. Down the corridor he heard the sounds of the first penetrating bombs sealing the exits. At the same time, the air raid sirens went off, an unnecessary distraction. Death was at most a minute or two away.

When the F/A-18s finally arrived overhead thirty seconds after the F-25Bs had done their jobs, it took just a few minutes for the four pilots to set up their laser designators, get the weapons into parameters, and make the drop. Thirty seconds later, eight of the big bombs entered the solid granite protecting the mountain bunker. They split the wet stone for almost a hundred feet before detonating, setting up a shear shock wave in the rock strata. The effect was to collapse the bunkers below, destroying everyone and everything inside instantly. With the destruction of the command bunker, the American aircraft headed home to the *Colin Powell* and an early breakfast.

Aboard the Command Ship USS *Mount McKinley* (LCC-22), Five Hundred Nautical Miles South of Colombo, 0400 Hours, May 8th, 2016

''That's right, Jack,'' Admiral Connelly said over the conference phone to the JCS chairman. ''We got them back safe and with all the targets hit, at least as far as the early BDA can tell. In addition, the two MPS squadrons arrive in the morning, and should be off-loading by midday. What do you hear on your end?''

The JCS chairman was quick and concise, having been up for almost two days holding the President's and National Security Advisor's hands during the short but brutal combat. ''Well, what's left of the Indian government is asking for UN peacekeeping and nation-building teams to reform their government. Pakistan is doing the same thing. My guess is that we'll be able to pull you and your people out within a few weeks, when the permanent UN units arrive. The boss says to tell your people that they did an incredible job out here, and that he'll meet them when they get home next month.''

''Thanks, Jack,'' said Connelly. ''You know, he'll probably want to give me another star or some other damned thing and get me back home again on shore duty.''

''He just might at that. You'll be back to that snoozer work you love so much,'' the JCS chairman replied. Unable to resist that perfect opening, he ended the conversation with, ''Have a nice nap.''

As it happened, Connelly slept for two straight days.

Stockholm, Sweden, February 14th, 2017

The Nobel Prize ceremonies were agreeably short this year, though the significance of the awards made the usually esoteric descriptions of the winners' work absolutely sparkle with excitement. The combined prizes in physics and chemistry went, of course, to Jill Jacobs, who was already a billionaire from her licensing advances on the superconducting-wire formula. She chose to donate the Nobel Prize money to her alma mater at New Mexico. The Peace Prize went jointly to Venkatesh Prasad, the Sinhalese Prime Minister, and his new Interior Minister, Arjuan Ranatunga, for their peaceful forging of a new nation. Both men had decided to donate their prizes, as well as significant funds from their overflowing national coffers, to disaster relief in India and Pakistan, an olive branch to their new customers to the north. Finally, the Nobel Committee had awarded a special peacekeeping award to Admiral Connelly, now the JCS Chairman in Washington, D.C. It was the first time that all of them had met, but their paths had already crossed in the currents of history, and between them they had created a better world.

Conclusion

When I started working on this book in 1997, I had little doubt that I could justify good reasons for America to continue its support for carrier aviation. If I've done nothing else in this book, those reasons should be readily apparent by now. However, at the same time, I went into this book with a real concern about the ability of the U.S. Navy to address the many leadership and material problems that have plagued the service since the end of the Cold War. As it turned out, I need not have been so worried. The U.S. Navy is a resilient institution, which has endured trial, scandal, and other ills many times, and continued to prosper. So too, with the Navy of our times. The simple fact is that as a nation whose trade is primarily maritime based, we need the oceans the way that humans need oxygen. This country was founded on a strong maritime tradition, and will likely be that way for the rest of our existence. Therefore, the question is not whether we need naval forces, but what form and numbers those units will represent. Within our current concept of naval operations, that means that aircraft carriers and their embarked air wings are here to stay.

In fact, after the disastrous years following Desert Storm and the Cold War, naval aviation seems on the verge of a new golden age, with new carriers, aircraft, and weapons on the way, and strong leadership to guide it. Best of all though, the U.S. Navy seems to be moving away from the self-imposed tyranny that has marked the development and use of carrier airpower since the end of World War II. Far from the dreaded expectations, the new ''joint'' method of packaging and deploying U.S. armed forces (as a result of the Goldwater–Nichols reform bill) has actually allowed carrier admirals greater latitude in the use of their flattops. Jay Johnson's innovative use of his carriers during the 1994 Haiti operation would have been unthinkable just five years earlier. Even today, there are many naval aviation leaders who consider his actions heresy. Those voices though, are growing more silent with every new JTFEX and training exercise. Joint warfare is here to stay, and nothing will ensure a strong future for carriers and their aircraft more than regional commanders and joint task force commanders who *want* a carrier battle group as part of their complement of units.

From this is coming new and innovative roles and missions for aircraft carriers and their supporting battle amphibious groups. One of these is the use of "adaptive" air wing organizations, which would allow changing the mix and types of aircraft embarked for a particular mission. Haiti back in 1994 was just a point of departure for what might be possible in the future. Using aircraft, UAVs, and UCAVs from other services as well their own will allow the Navy greater participation in future military operations, and expand the range of possible supporting missions. It also raises the possibility of utilizing the big-deck flattops in disaster relief and humanitarian aide missions, which have become a hallmark of post–Cold War military operations. Ironically, these expanded missions will also help justify future construction of new carriers, since their inherent value and flexibility will become more apparent and valuable to a wider base of users. The idea of Army generals helping to support new warship construction may seem outlandish, but is already happening on Capitol Hill and the Pentagon.

It is with this knowledge that I want to take one last look ahead at what the new century may bring for naval aviators. For starters, there will finally be a new set of carrier designs. The CVX program is committed to transitioning from the existing *Nimitz*-class (CVN-68) ships to a new design that will be oriented toward the power projection missions of the new millennium. Though the program is undergoing a restructuring at the moment, plan on seeing a series of two or three transitional designs while the new design features are ironed out. By that time, around 2020, the future of warship design should be much clearer, given the political/world situation a generation from now. There also is the real possibility of technical breakthroughs that may effect new designs, particularly if low-temperature superconductors or high output fuel cells finally become a reality.

There also will be new aircraft, some so wondrous that I cannot even describe them. JSF and the F/A-18EF Super Hornet I have already shown to you. However, the new generation of Uninhabited Combat Aerial Vehicles (UCAVs) is likely to appear sooner than later, given the rapidly escalating costs of manned aircraft. The performance of those aircraft are likely to be unimaginable by today's standards, with maneuvering capabilities more like that of air-to-air missiles than 20th-century manned aircraft. The fighter pilots of tomorrow may not even need to be flight qualified. Flown from consoles aboard ships or transport aircraft, they will be able to fly missions that today's manned aircraft would not even be considered for. Best of all, a lost aircraft will just be money, and not human lives. Before you call this science fiction, it is useful to remember that the Navy ran maneuvering trials between an F-4 Phantom and an unmanned Firebee drone in the early 1970s, and the drone consistently won!

These, though, are matters for another generation of Americans, some of whom have not even been born yet. Today the issue is finding the money to make the transition to these wondrous new ships, aircraft, and weapons, and this is the real challenge. Since the end of the Cold War, both we and our allies have downsized the armed forces to the point where their credibility is now coming into question. For the Navy, this means that the dozen carrier and amphibious groups that are being retained are the absolute minimum if we are to maintain the current rotation policies. It also has meant that the personnel are now at the breaking point, as Admiral Johnson indicated in his interview. U.S. military personnel have been exiting the services in growing numbers for the booming civilian job market. Long deployments and eroding salaries are a formula for disaster, and must be dealt with if our forces are to remain strong and credible. The answer of course is more money, and that is going to require leadership. Leadership from an elected administration and Congress, which currently is more interested in political squabbling than national security issues. It also will take military leaders willing to put their own careers on the line to tell the truth to those civilian leaders, even if they do not want to hear. Fine men like Jay Johnson and Chuck Krulak are leading this fight, but cannot do it themselves. All of us must accept the fact that the current economic boom, which has been powering the 1990s, has been accomplished in a time of virtually no military threats to America or its Allies. To assume that this happy set of circumstances will continue is folly, given the eruption of nationalism since the end of the Cold War. The threats are out there, and I have no doubt that they will find us without difficulty. Let us hope that our sea services continue to have the necessary support to protect us all from them. We're going to need it.

Glossary

A-12 General Dynamics/McDonnell Douglas A-12 Avenger, a 1990's Navy program for a stealthy carrier strike aircraft, canceled due to cost overruns and program mismanagement.

AAA Antiaircraft artillery also called "triple-A" or "flak" (from the German "fliegerabwehrkanone" or air defense gun).

AAQ-13/14 LANTIRN Low Altitude Navigation Targeting Infrared for Night. A pair of sensor pods mounted on the F-15E and certain F-16C/D aircraft. The AAQ-13 Navigation Pod combines a Forward Looking Infrared sensor and a terrain-following radar. The AAQ-14 Targeting Pod combines a Forward Looking Infrared and Laser Target Designator. Entire system is built by Martin Marietta (now Lockheed Martin) and tightly integrated with the aircraft's flight control and weapons delivery software. A version of the AAQ-14 targeting pod with an internal GPS/INS system is used aboard the F-14 Tomcat.

ACC Air Combat Command. Major command of the USAF formed in 1992 by the merger of Strategic Air Command (bombers and tankers) and Tactical Air Command (fighters).

ACES II Standard U.S. ejection seat built by Boeing, based on an original design by the Weber Corporation. ACES is a "zero-zero" seat, which means that it can save the crew person's life (at the risk of some injury) down to zero airspeed and zero altitude, as long as the aircraft is not inverted. Humorously known as the "hostage delivery system."

ACM Air Combat Maneuvering, the art of getting into position to shoot the other guy, preferably from behind, before he can shoot you. A vital but expensive part of advanced flight training for fighter pilots, ACM is most effective on an instrumented radar range with "playback" facilities for debriefing.

Aegis Advanced automated tracking and missile fire-control system on modern U.S. Navy cruisers and destroyers. Key components are the SPY-1 phased-array radar and the SM-2 missile. Named for the shield of Zeus in Greek mythology.

AEW Airborne Early Warning. Specifically used to describe aircraft like the

Northrop Grumman E-2C Hawkeye and Boeing E-3 Sentry, but also used generically to describe similar types used by other Air Forces.

AFB Air Force Base.

Afterburner Device that injects fuel into the exhaust nozzle of a jet engine, boosting thrust at the cost of greater fuel consumption. Called "Reheat" by the British.

AGM-62 Walleye AGM-62 television-guided glide bomb with 2,000-lb warhead. Maximum range of about 20 miles, depending on speed and altitude of launch aircraft. Used in Vietnam War; obsolescent but still in stock.

AGM-65 Maverick Family of air-to-surface missiles, produced since 1971 by Hughes and Raytheon with a variety of guidance and warhead configurations. Range about 14 nm. Navy versions carried by S-3, P-3, F/A-18, and other aircraft use imaging infrared guidance.

AGM-84 Harpoon/SLAM AGM-84, turbojet powered antiship missile, up to 120 miles range with 488-lb/220-kg explosive warhead. AGM-84E version ("SLAM") uses Maverick IIR seeker and GPS-aided guidance.

AGM-154 JSOW Joint Standoff Weapon. Low-cost 1,000-pound glide bomb with 25-mile range, using INS/GPS guidance. Carries 145 BLU-97 bomblets. A version carrying a 1,000-pound unitary warhead is under development.

AH-1W "Cobra" attack helicopter found in Marine light-attack squadrons. Nicknamed "Whiskey Cobra" or "Snake."

AIM-9 Sidewinder Heat-seeking missile family, used by the Air Force, Navy, Marines, Army, and many export customers. A letter, such as AIM-9M or-9X, designates variants.

AIM-120 AMRAAM AIM-120 Advanced Medium Range Air to Air Missile (AMRAAM). First modern air-to-air missile to use programmable microprocessors with active radar homing (missile has its own radar transmitter, allowing "fire and forget" tactics). Currently carried by Navy and Marine F/A-18's.

ARG Amphibious Ready Group.

ATO Air Tasking Order. A planning document that lists every aircraft sortie and target for a given day's operations. Preparation of the ATO requires careful "deconfliction" to ensure the safety of friendly aircraft. During Desert Storm the ATO ran to thousands of pages each day.

Avionics General term for all the electronic systems on an aircraft, including radar, communications, flight control, navigation, identification, and fire-control computers. A "data bus" or high-speed digital network increasingly interconnects components of an avionics system.

BDA Bomb Damage Assessment. The controversial art of determining from fuzzy imagery and contradictory intelligence whether or not a particular target has been destroyed or rendered inoperative.

BVR Beyond visual range; usually used in reference to radar guided air-to-air missiles. "Visual range" depends on the weather, how recently the windscreen was cleaned and polished, and the pilot's visual acuity, but against a fighter-sized target rarely exceeds 10 miles (16 km).

C-130 Hercules Lockheed medium transport aircraft. Four Allison T56 turboprops. Over 2,000 of these classic aircraft have been built since 1955, and it is still in production.

C³I Command, Control, Communications, and Intelligence; the components and targets of information warfare. Pronounced "C-three-I."

Call Sign (1). An identifying name and number assigned to an aircraft for a particular mission. Aircraft in the same flight will usually have consecutive numbers. (2). A nickname given to an aviator by his/her squadron mates and retained throughout his/her flying career, often humorous.

Canopy Transparent bubble that covers the cockpit of an aircraft. Usually made of Plexiglas, or polycarbonate, sometimes with a microscopically thin layer of radar-absorbing material or gold. Easily scratched or abraded by sand or hail. Ejection seats have a means of explosively jettisoning or fracturing the canopy to reduce the chance of injury during ejection.

CAP Combat Air Patrol, a basic fighter tactic that involves cruising economically at high or medium altitude over a designated area searching for enemy aircraft.

CBU Cluster Bomb Unit. An aircraft munition that is fused to explode at low altitude, scattering large numbers of "submunitions" over a target area. Submunitions can be explosive grenades, delayed-action mines, antitank warheads, or other specialized devices.

CENTAF Air Force component of U.S. Central Command, including units deployed to bases in Kuwait, Saudi Arabia, and other states in the Gulf region. The commander of CENTAF is an Air Force lt. general, who typically also commands Ninth Air Force based at Shaw AFB, SC.

CENTCOM United States Central Command, a unified (joint service) command with an area of responsibility in the Middle East and Southwest Asia. Headquartered at McDill AFB, FL, and generally commanded by an Army four-star general. CENTCOM normally commands no major combat units, but in a crisis situation it would rapidly be reinforced by units of the Army's XVIII Airborne Corps, the U.S. Marine Corps, and allied forces.

Chaff Bundles of thin strips of aluminum foil or metallized plastic film that are ejected from an aircraft to confuse hostile radar. A chaff cloud creates a temporary "smoke screen" that makes it difficult for radar to pick out real targets. The effectiveness of chaff depends on matching the length of the chaff strips to the wavelength of the radar.

Chine A sharp-edged projection running along the fuselage of an aircraft, often as an extension of the leading-edge wing root. Particularly prominent on the F/A-18 Homet.

CinC Commander in Chief. Used to designated the senior officer, typically a four-star general or admiral in charge of a major command, such as CINCPAC (Commander in Chief of the U.S. Pacific Command).

CIWS Mk. 15 "Close-in Weapons System." Pronounced "Sea-Whiz." The Phalanx automatic gun and radar system, as installed on Navy ships of many classes.

CONOPS Concept of Operations. The commander's guidance to subordinate units on the conduct of a campaign.

CSAR Combat Search and Rescue. Recovery of downed air crew evading captures in an enemy-held area. Typically a helicopter mission supported by fixed-wing aircraft.

CTAPS Contingency Tactical Air Control System Automated Planning System. A transportable network of computer workstations, linking various databases required for the generation of an Air Tasking Order.

CVW Carrier Air Wing; a force of Navy aircraft organized for operation from an aircraft carrier, Typically includes one fighter squadron, two attack squadrons, and small units of helicopters, antisubmarine, electronic warfare, and early warning radar planes.

DoD Department of Defense. U.S. Government branch created in 1947, responsible for the four armed services and numerous agencies, program offices, and joint projects.

Drag The force that resists the motion of a vehicle through a gaseous or liquid medium. The opposite force is lift.

E-2C Hawkeye U.S. Navy carrier-based twin-turboprop airborne early warning aircraft built by Northrop Grumman. Entered service in 1964. Also operated by France, Israel, Egypt, Taiwan, Singapore, and Japan.

E-8 JSTARS Joint Surveillance and Targeting Attack Radar System. An Army/Air Force program to deploy about 20 Boeing E-8C aircraft equipped with powerful side-looking synthetic-aperture radars to detect moving ground targets at long range.

E/O Electro-optical. A general term for sensors that use video, infrared, or laser technology for assisting navigation or locating, tracking, or designating targets.

ECM Electronic Countermeasures. Any use of the electromagnetic spectrum to confuse, degrade, or defeat hostile radars, sensors, or radio communications. The term ECCM (electronic counter-countermeasures) is used to describe active or passive defensive measures against enemy ECM, such as frequency-hopping or spread-spectrum waveforms.

ELINT Electronic Intelligence. Interception and analysis of radar, radio, and other electromagnetic emissions in order to determine enemy location, numbers, and capabilities.

ESM Electronic Security Measures. Usually refers to systems that monitor the electromagnetic spectrum to detect, localize, and warn of potential threats.

Exocet French-built antiship missile, widely exported in air-launched (AM-39), ship-launched (MM-38/40), and submarine-launched (SM-39) versions. Two AM-39 Exocets fired by an Iraqi aircraft damaged the U.S. Navy frigate *Stark* (FFG-31) in the Persian Gulf on May 17, 1987.

F/A-18 Hornet Boeing "Hornet" carrier-capable fighter-bomber, operated by Navy and Marine squadrons, and the Air Forces of Canada, Kuwait, Malaysia, Spain, Switzerland, Australia, and Finland. Improved F/A-18E/F model under development.

FADEC Full Authority Digital Engine Control, a computer that monitors jet-engine performance and pilot-throttle inputs and regulates fuel supply for maximum efficiency.

Flameout Unintended loss of combustion inside a jet engine, due to a disruption of airflow. This can be extremely serious if the flight crew is unable to restart the affected engine.

Flap A hinged control surface, usually on the trailing edge of a wing, commonly used to increase lift during takeoff and drag during landing.

Flare (1). A pyrotechnic device ejected by an aircraft as a countermeasure to heat-seeking missiles. (2). A pitch-up maneuver to bleed off energy performed during landing, just before touching down.

FLIR Forward Looking Infrared: an electro-optical device similar to a television camera that "sees" in the infrared spectrum rather than visible light. FLIRs display an image based on minute temperature variations, so that hot engine exhaust ducts, for example, appear as bright spots.

G-Force One G is the force exerted by Earth's gravity on stationary objects at sea level. High-energy maneuvers can subject the aircraft and pilot to as much as 9 Gs. Some advanced missiles can pull as much as 60 Gs in a turn.

GBU Guided Bomb Unit. General U.S. term for precision-guided munitions.

GBU-29/30/31/32 JDAM Joint Direct Attack Munition. A general-purpose bomb or penetration warhead, with inertial/GPS guidance package in the tail cone. Initial operational capability planned for 1999. Navy requirement is 12,000 bombs; Air Force requirement is for 62,000.

Geosynchronous Also called "geostationary." A satellite in equatorial orbit at an altitude of 35,786 km (about 22,230 miles) will take 24 hours to circle the Earth. In 24 hours the Earth rotates once on its axis, so the satellite will appear to be "fixed" over the same point on the earth.

"Glass" Cockpit Design that replaces individual flight gauges and instruments with multi-function electronic display screens. A few mechanical gauges are usually retained for emergency backup.

Goldwater-Nichols Common name for the Military Reform Act of 1986, which created a series of Unified Commands cutting across traditional service boundaries and strengthened the power of the Chairman of the Joint Chiefs of Staff.

GPS Global Positioning System. A constellation of 24 satellites in inclined earth orbits, which continuously broadcast navigational signals synchronized by ultra-precise atomic clocks. At least four satellites are usually in transit across the sky visible from any point on Earth outside the Polar Regions. A specialized computer built into a portable receiver can derive highly accurate position and velocity information by correlating data from three or more satellites. An encoded part of the signal is reserved for military use. A similar, incomplete, Russian system is called GLONASS.

HARM AGM-88 High Speed Anti-Radiation Missile, produced by Texas Instruments. Mach 2+. 146-lb blast-fragmentation warhead. Typically fired 35 to 55 miles from target, but maximum range is greater. First used in combat in April 1986 raid on Libya; 40 missiles fired.

Have Blue Original Lockheed "Skunk Works" prototype for the F-117 Stealth fighter. Considerably smaller than the production aircraft, and still highly classified.

HEI High Explosive Incendiary, a type of ammunition commonly used with air-to-air guns.

HOTAS Hands on Throttle and Stick. A cockpit flight control unit that allows the pilot to regulate engine power settings and steering commands with one hand.

HS Helicopter Antisubmarine Squadron.

HSL Helicopter Antisubmarine Squadron, Light.

HUD Heads-Up Display: a transparent screen above the cockpit instruments on which critical flight, target, and weapons information is projected, so that the pilot need not look down to read gauges and displays during an engagement. Current HUD technology provides wide-angle display of radar and sensor data.

IIR Imaging Infrared. An electro-optical device similar to a video camera that "sees" small differences in temperature and displays them as levels of contrast or false colors on a operator's display screen.

ILS Instrument Landing System. A radio-frequency device installed at some airfields that assists the pilot of a suitably equipped aircraft in landing during conditions of poor visibility.

INS Inertial Navigation System. A device that determines location and velocity by sensing the acceleration and direction of every movement since the system was initialized or updated at a known point. Conventional INS systems using mechanical gyroscopes are subject to "drift" after hours of continuous operation. Ring-laser gyros sense motion by measuring the frequency shift of laser pulses in two counter-rotating rings, and are much more accurate. The advantage of an INS is that it requires no external transmission to determine location.

Interdiction Use of airpower to disrupt or prevent the movement of enemy military units and supplies by attacking transportation routes, vehicles, and bridges deep in the enemy rear.

IOC Initial Operational Capability. The point in the life cycle of a weapon system when it officially enters service and is considered ready for combat, with all training, spare parts, technical manuals, and software complete. The more complex the system, the greater the chance that the originally scheduled IOC will slip.

IRBM Intermediate Range Ballistic Missile. A rocket (typically two-stage) designed to deliver a warhead over regional rather than intercontinental distances. This class of weapons was eliminated by treaty and obsolescence from U.S. and Russian strategic forces, but is rapidly proliferating in various world trouble spots, despite international efforts to limit the export of ballistic missile technologies.

JCS Joint Chiefs of Staff. The senior U.S. military command level, responsible for advising the President on matters of national defense. The JCS consists of a Chairman, who may be drawn from any service, the Chief of Naval Operations, the Chief of Staff of the Army, the Commandant of the Marine Corps, and the Chief of Staff of the Air Force.

JFACC Joint Forces Air Component Commander. The officer who has operational control over all air units and air assets assigned to a theater of operations. The JFACC is typically drawn from the service that has the greatest amount of airpower in the area of operations, and reports directly to the theater Commander in Chief.

JP-5 Standard U.S. jet fuel. A petroleum distillate similar to kerosene.

JTF Joint Task Force: a military unit composed of elements of two or more services, commanded by a relatively senior officer. JTFs may be organized for a specific mission, or maintained as semi-permanent organizations, such as the anti-drug JTF-4 based in Florida.

KC-10 Extender Heavy tanker/transport based on Boeing DC-10 wide-body commercial airliner. 59 aircraft in service, some modified with drogue refueling hose reel as well as tail boom. Three CF6 turbofan engines. Maximum takeoff weight is 590,000 lbs.

KC-130 Lockheed ''Hercules'' four-engine turboprop, used as a transport and aerial tanker by Marine air units.

Knot Nautical miles (6,076 feet) per hour. Often used by U.S. Air Force and Navy to measure aircraft speeds, particular in the subsonic range. One knot equals one nautical mile per hour.

LGB Laser-guided bomb, such as the Paveway-series LGBs produced by Raytheon.

LHA Large amphibious assault ship designed to operate helicopters and STOVL aircraft, with a well deck for landing craft.

LHD Amphibious assault ship with flight deck and well deck.

LPD Amphibious ship with well deck.

LPH Amphibious assault ship designed to operate helicopters.

LRIP Low Rate Initial Production. A phase in the development of a new weapon system in which the ''bugs'' are worked out of manufacturing techniques, tooling, and documentation before shifting to full-rate production.

LSD Amphibious landing dockship.

M-61 Vulcan Six-barreled rotary (''Gatling'') 20mm cannon used as standard weapon on U.S. aircraft. Very high rate of fire. Also mounted on Army vehicles and Navy ships for short-range antiaircraft defense.

Mach The speed of sound at sea level (760 feet per second). An aircraft's Mach number is dependent on altitude, since sound travels faster in a denser medium. Named for Ernst Mach (1838–1916), Austrian physicist.

MAG Marine Aircraft Group.

MAGTF Marine Air-Ground Task Force.

MAW Marine Aircraft Wing.

MCAS Marine Corps Air Station.

MEB Marine Expeditionary Brigade.

MEF Marine Expeditionary Force.

MEU Marine Expeditionary Unit.

MEU (SOC) Marine Expeditionary Unit (Special Operations Capable).

MFD Multi-function Display. A small video monitor or flat panel display on an aircraft control panel that allows the operator to display and manipulate different kinds of sensor information, status indications, warnings, and system diagnostic data.

MiG Russian acronym for the Mikoyan-Gurevich Design Bureau, developers of some of the greatest fighter aircraft in history. Survived the breakup of the Soviet Union, and is actively competing in the global arms market.

MOS Military Occupational Specialty.

MPF Maritime Prepositioning Force.

MPS Maritime Prepositioning Ship.

MPSRON Maritime Prepositioning Ship Squadron.

MRC Major Regional Contingency Pentagon euphemism for small war or large crisis requiring a significant intervention of U.S. military forces as directed by the President.

MRE Meals, Ready to Eat. Military field ration in individual serving packs. Eaten by Marines on deployment until regular dining facilities can be constructed. Humorously known as "Meals Rejected by Everyone."

NAF Naval Air Facility (typically a smaller base than a Naval Air Station).

NAS Naval Air Station.

Nautical Mile (nm) 6,076 feet. Not to be confused with Statute Mile, of 5,280 feet.

NAVAIR Naval Air Systems Command. Organization that procures and manages aircraft and related systems and equipment for the Navy and Marine Corps. Formerly called Bureau of Aeronautics (BuAir)

NAVSEA Naval Sea Systems Command. Organization that procures and manages ships and related systems for the Navy. Formerly called Bureau of Ships (BuShips).

NBC Nuclear-Biological-Chemical. General term for weapons of mass destruction, including nuclear bombs or weapons designed to disperse radioactive material, toxic gases, liquids, or powders, infectious microorganisms or biological toxins. Forbidden by many international treaties that have been widely ignored.

NCO Non-Commissioned Officer. An enlisted soldier, sailor, or airman with supervisory responsibility or technical qualifications. There are nine standardized enlisted pay grades, but each service has its own complex nomenclature for NCO ranks. In the navy these are petty officer, chief petty officer, senior chief petty officer, and master chief petty officer. The senior NCO on a vessel, regardless of rank, is informally known as "Command Master Chief."

NEO Non-combatant Evacuation Operations. Use of military force to rescue American and foreign citizens, diplomatic personnel, and relief workers endangered by civil unrest or factional fighting.

NORAD North American Air Defense Command. Joint U.S.-Canadian headquarters located inside Cheyenne Mountain, CO, responsible for air defense of North America. CINCNORAD is also the Commander of U.S. Space Command.

NRO National Reconnaissance Office. Formerly super-secret intelligence agency established in early 1960's within the Department of Defense. Not officially acknowledged to exist until 1990's. Responsible for procurement, operation, and management of various types of reconnaissance satellites. A separate organization, the Central Imagery Office, is responsible for processing, interpretation, and dissemination of satellite imagery.

NS Naval Station, typically a larger base that includes shore facilities, airfields, and logistic installations.

"Nugget" Pilot jargon for an inexperienced aviator, or new person in the squadron.

O&M Operations and Maintenance. A major budget item for all military units.

OpTempo Operational tempo—subjective measure of the intensity of military operations. In combat high OpTempos can overwhelm the enemy's ability to respond, at the risk of burning out your own forces. In peacetime a high OpTempo can adversely affect morale and exhaust budgeted funds.

Ordnance Weapons, ammunition, or other consumable armament. Frequently misspelled.

OTH Over the Horizon. Used in references to sensors and targeting. Distance to the visual horizon may be 20 miles from the masthead of a ship, or more than 200 miles from an aircraft at high altitude.

PAA Primary Aircraft Authorized—the number of planes allocated to a unit for the performance of its operational mission. PAA is the basis for budgeting manpower, support equipment, and flying hours. In some cases, a unit may have fewer aircraft because of delivery schedule slippage or accidents. Units may also have more aircraft than their PAA, such as trainers, spare "maintenance floats," or inoperable "hangar queens."

PAO Public Affairs Officer. Military staff officer responsible for media relations, coordination with civil authorities, VIP escort duties, and similar chores. The PAO of an aircraft carrier is typically a Navy lieutenant, supervising a small team of enlisted writers and media specialists.

Paveway Generic term for laser-guided bombs. Made by Raytheon/Texas Instruments since 1968. Latest Paveway III version provides standoff range of 3-5 nm.

PGM Precision Guided Munition. Commonly called a "smart bomb," any weapon that uses electronic, electro-optical, inertial, or other advanced forms of terminal guidance to achieve a very high probability of hitting its target.

Pitch Change of an aircraft's attitude relative to its lateral axis (a line drawn from left to right through the center of gravity). Pitch up and the nose rises; pitch down and the nose drops.

"Pucker Factor" Flight crew anxiety level. Typically related to highly stressful combat situations such as major aircraft system malfunctions while under fire from enemy missiles.

Pylon A structure attached to the wing or fuselage of an aircraft that supports an engine, fuel tank, weapon, or external pod. The pylon itself may be removable, in which case it is attached to a "hard point" that provides a mechanical and electrical interface.

RAM Radar Absorbing Material. Metal or metal-oxide particles or fibers embedded in synthetic resin applied as a coating or surface treatment on radar-reflective areas of a vehicle in order to reduce its radar cross section. A particular RAM formulation may be specific to a narrow band of the radar frequency spectrum.

RC-135V Rivet Joint Program name for electronic reconnaissance aircraft, operated by 55th Wing based at Offut AFB, NE. Used in Saudi Arabia during Desert Shield/Desert Storm.

RH-53E Sikorsky "Sea Dragon" found in mine-countermeasure units.

RIM-116A RAM RIM-116A Rolling Airframe Missile. Development began in 1975 as a joint U.S./German/Danish program. Entered service in June 1993 on USS *Peleliu* (LHA-5). Combines seeker head of Stinger SAM with motor, warhead, and fuse from AIM-9 Sidewinder. Angled tail fins cause the missile to spin in flight for stability. Range of around 5 nm, and carried in a 21-round box launcher.

RO-RO Roll-on/Roll-off. A cargo ship with vehicle parking decks, flexible ramps, and special ventilation, allowing loaded vehicles to drive on or off under their own power.

ROE Rules of Engagement. Guidance often determined at the highest levels of national government, regarding how and when warriors may employ their weapons. In air-to-air combat, ROE usually specify specific criteria for declaring a non-friendly aircraft as hostile. In air-to-ground combat, ROE usually forbids attacking targets likely to involve significant collateral damage to civilian populations or religious sites. Regardless of the ROE, the right of self-defense against direct armed attack is *never* denied.

Roll Change of attitude relative to the longitudinal axis (a line drawn from nose to tail through the center of gravity). Roll to port and an aircraft tilts to the left; roll to starboard and it tilts to the right. Roll also describes a class of aerobatic maneuvers, such as the barrel roll.

RWR Radar Warning Receiver. An electronic detector tuned to one or more hostile radar frequencies and linked to an alarm that alerts the pilot to the approximate direction, and possibly the type of threat. Similar in concept to automotive police radar detectors. Also known as a RHAW (Radar Homing and Warning Receiver).

SAM Surface to Air Missile. A guided missile with the primary mission of engaging and destroying enemy aircraft. Most SAMs use rocket propulsion and some type of radar or infrared guidance.

SAR Synthetic Aperture Radar. An aircraft radar (or operating mode of a multi-function radar) that can produce highly accurate ground maps.

SCUD Western reporting name for the Soviet R-11 (SCUD-A) and R-17 (SCUD-B) short-range ballistic missiles. Based largely on captured WW II German technology. Range of 110-180 miles with 900-kg/1,980-lb warhead, with inaccurate inertial guidance. Can be transported and erected for launch by a large truck. Widely exported to Iraq, North Korea, and other Soviet client states. Iraq modified basic SCUD-B design to produce longer-ranged Al Abbas and Al Hussein missiles with much smaller warheads.

SEAD Suppression of Enemy Air Defenses. This requires enticing the enemy to ''light up'' search and tracking radars, launch SAMs, or fire antiaircraft guns, which can then be targeted for destruction or neutralization by jamming and other countermeasures.

SIGINT Signal Intelligence. Interception, decoding, and analysis of enemy communications traffic.

Skunk Works® Lockheed's Burbank, California, Advanced Development group, created during WW II by engineer Clarence ''Kelly'' Johnson. Developed the U-2, SR-71, F-117, and other secret aircraft. Lockheed Martin copyrights the name and skunk cartoon logo.

Slat A long, narrow, moveable control surface, usually along the leading edge of the wing, to provide additional life during takeoff.

Sortie The basic unit of airpower: one complete combat mission by one aircraft. ''Sortie generation'' is the ability of an air unit to re-arm, refuel, and service aircraft for repeated missions in a given period.

Sparrow AIM-7 family of long-range radar-guided air-to-air missiles produced by Raytheon. Variants include the ship-launched Sea Sparrow.

Stall Sudden lost of lift when the airflow separates from the wing surface; may be caused by a variety of maneuvers, such as climbing too steeply with insufficient thrust. ''Compressor stall'' is a different phenomenon that occurs inside a turbine engine.

Stealth A combination of design features, technologies, and materials, some highly classified, designed to reduce the radar, visual, infrared, and acoustic signature of an aircraft, ship, or other vehicle. This can be taken to the point where effective enemy detection and countermeasures are extremely unlikely before the vehicle has completed its mission and escaped. The F-117A is the best-known modern example.

STOVL Short Takeoff-Vertical Landing. Capability of certain vectored-thrust aircraft, notably the Harrier and variants of the future Joint Strike Fighter. Short takeoff is assisted by a fixed "ski-jump" ramp.

T-38 Talon Twin-turbojet advanced trainer, over 1,100 built by Northrop. Entered service in 1961. First supersonic aircraft specifically designed as a trainer.

T-3A Firefly Lightweight two-seat propeller-driven trainer based on British Slingsby T67. Used by U.S. Air Force for screening of prospective pilots. Top speed 178 mph, ceiling 19,000 ft.

TARPS Tactical Air Reconnaissance Pod System. A 1,700-pound/770 kg pod built by Naval Avionics Center and fitted to four F-14 fighters in each carrier air wing. Pod carries a 9-inch panoramic camera, 12-inch frame camera, and infrared line scanner.

TDY Temporary Duty. A military assignment to a location away from one's normal duty station. TDY generally involves separation from family and entitles personnel to supplementary pay and allowances.

TERCOM Terrain Contour Matching, a cruise-missile-guidance concept that relies on a radar altimeter and a stored digital map of elevations along the line of flight. Flight plans require detailed and lengthy preparation, and cannot be generated for relatively flat, featureless terrain.

Top Gun U.S. Navy Fighter Weapon School at NAS Fallon, NV. Responsible for training fleet pilots in air-combat maneuvering.

TRAP Tactical Recovery of Aircraft and Personnel.

U-2 High-altitude (over 90,000 ft/27,430 m) reconnaissance aircraft originally developed in 1950 for the U.S. Central Intelligence Agency by Lockheed. Single J57, later J75 turbojet. Many variants with diverse sensors operated by the USAF and NASA (civilian research).

UAV Unmanned Aerial Vehicle. Also known as a drone or RPV (remotely piloted vehicle). A recoverable pilotless aircraft, either remotely controlled over a radio-data link, or preprogrammed with an advanced autopilot. The U.S. Air Force has tended to resist any use of UAVs, except as targets, because they take jobs away from pilots. There are also real safety concerns about operating UAVs and manned aircraft in the same airspace, since they are usually small and hard to see.

UCAV Unmanned Combat Aerial Vehicle. Also known as a drone or RPV (remotely piloted vehicle). A recoverable pilotless aircraft, either remotely controlled over a radio-data link, or preprogrammed with an advanced autopilot.

UH-1N ''Huey'' light utility helicopter found in Marine light-attack squadrons and support units.

UH-46 Aging Boeing Vertol ''Sea Knight'' twin-rotor helicopter found in Navy utility and logistic squadrons. Nicknamed ''Bullfrog.''

UPT Undergraduate Pilot Training

V-22 Osprey Bell-Boeing-built twin-engine tilt-rotor aircraft, combining the agility of a helicopter with the speed and range of a fixed-wing turboprop. Joint Marine/Navy/Air Force program, delivery began in 1997.

VA Navy Attack Squadron, previously equipped with A-7 or A-6 aircraft. With the retirement of these types, this squadron designation is no longer used.

VAQ Navy Tactical Electronic Warfare Squadron, equipped with EA-6 Prowler aircraft.

Variable geometry Ability of an aircraft to change the sweep of its wings in flight, to optimize performance for a given speed and altitude.

VAW Navy Carrier Airborne Early Warning Squadron, equipped with E-2C Hawkeye aircraft.

VF Navy Fighter Squadron, typically equipped with F-14 Tomcats.

VFA Navy Strike Fighter Squadron, typically equipped with F/A-18 Hornets.

Viewgraph An overhead projector transparency or slide used in briefings or presentations. Also spelled Vu-graph.

VMA Marine Attack Squadron (e.g., VMA-211). Typically equipped with AV-8B Harrier.

VMAT Marine Attack Training Squadron (e.g., VMAT-203). Typically equipped with AV-8B Harrier.

VMFA Marine Fighter Attack Squadron (e.g., VMFA-115). Typically equipped with F/A-18 Hornet.

VMFAT Marine Fighter Attack Training Squadron (e.g., VMFAT-101). Typically equipped with F/A-18 Hornet.

VMGR Marine Aerial Refueler Transport Squadron (e.g., VMGR-252). Typically equipped with KC-130.

VP Navy Patrol Squadron, typically equipped with land-based P-3 Orion aircraft.

VRC Fleet Logistics Support Squadron (Carrier onboard delivery) equipped with C-2 Greyhound aircraft.

VS Navy Sea Control Squadron equipped with S-3 Viking aircraft.

V/STOL Vertical/Short Takeoff and Landing.

VT Navy Training Squadron. Equipped with a wide variety of aircraft—carrier training squadrons typically fly the T-2 Buckeye or T-45 Goshawk.

Wild Weasel An aircraft configured with Radar Homing and Warning (RHAW) gear and Anti-Radiation Missiles (ARMs) operated to suppress enemy surface-to-air-missile sites. Originally performed by F-100F, F-105F, and F-4G Phantom II aircraft, this mission will increasingly be borne by specially trained and equipped F-16Cs, F/A-18Cs, and EA-6Bs.

XO Executive Officer, second in command of a squadron, vessel, or equivalent unit.

Bibliography

Books:

Adams, James. *Bull's Eye: The Assassination and Life of Supergun Inventor Gerald Bull.* Times Books, 1992.

Albrecht, Gerhard, ed. *Weyers Flotten Taschenbuch (Warships of the World) 1994/96.* Bernard & Graefe Verlag, Bonn, Germany, 1994.

Alexander, Joseph H., and Merrill L. Bartlett. *Sea Soldiers in the Cold War.* Naval Institute Press, 1995.

Allen, Thomas, and Norman Polmar. *Codename Downfall: The Secret Plan to Invade Japan and Why Truman Dropped the Bomb.* Simon & Schuster, 1995.

———. *Merchants of Treason.* Dell Publishing, 1988.

Allen, Thomas B. *War Games: The Secret World of Creators, Players, and Policy Makers Rehearsing World War III Today.* McGraw-Hill, 1987.

Amett, Peter. *Live from the Battlefield: From Vietnam to Baghdad.* Simon & Schuster, 1994.

Atkinson, Rick. *Crusade: The Untold Story of the Persian Gulf War.* Houghton Miffen, 1993.

Baker, Arthur D., III. *Combat Fleets of the World, 1998–1999.* Naval Institute Press, 1998.

———. *Japanese Naval Vessels of World War Two.* Naval Institute Press, 1987.

Ballard, Jack S. *The United States Air Force in Southeast Asia.* U.S. Government Printing Office, 1974.

Barker, A. J. *The Yom Kippur War.* Ballatine Books, 1974.

Barron, John. *MiG Pilot: The Final Escape of Lt. Belenko.* Avon Books, 1980.

Basel, G. I. *Pak Six.* Associated Creative Winters, 1982.

Bathurst, Robert B. *Understanding the Soviet Navy: A Handbook.* U.S. Government Printing Office, 1979.

Baxter, William P. *Soviet Air Land-Battle Tactics.* Presidio Press, 1986.

Beach, Edward L., Captain, USN (Ret.). *The United States Navy: 200 Years.* Holt, 1986.

Beaumont, Roger A. *Joint Military Operations: A Short History.* Greenwood, 1993.

Bennett, Christopher. *Supercarrier: USS George Washington.* Motorbooks, 1996.

Berry, F. Clifton, Jr. *Gadget Warfare: The Vietnam War.* Batnam Books, 1988.

———. *Strike Aircraft: The Vietnam War.* Bantam Books, 1987.

Beschloss, Michael R. *At the Highest Levels.* Little, Brown, 1993.

———. *May Day.* Harper & Row, 1986.

Bin Sultan, Khaled. *Desert Warrior: A Personal View of the Gulf War by the Joint Forces Commander.* HarperCollins, 1995.

Bishop, Chris, and David Donald. *The Encyclopedia of World Military Power.* The Military Press, 1986.

Blackwell, James. *Thunder in the Desert: The Strategy and Tactics of the Persian Gulf War.* Bantam Books, 1991.

Blair, Arthur, H., Colonel, U.S. Army (Ret.). *At War in the Gulf.* A&M University Press, 1992.

Blair, Clay. *The Forgotten War: America in Korea, 1950–1953.* Times Books, 1987.

Bodansky, Yossef. *Crisis in Korea: The Emergence of a New & Dangerous Nuclear Power.* SPI Books, 1994.
Bonnanni, Pete. *Art of the Kill: A Comprehensive Guide to Modern Air Combat.* Spectrum Holobyte, 1993.
Bosnia: Country Handbook. U.S. Department of Defense, 1995.
Bowers, Peter M. *Boeing Aircraft Since 1916.* Naval Institute Press, 1989.
Boyne, Walter J. *Clash of Titans: World War II at Sea.* Simon & Schuster, 1995.
———. *Clash of Wings: World War II in the Air.* Simon & Schuster, 1994.
Bradin, James W. *From Hot Air to Hellfire: The History of Army Attack Aviation.* Presidio Press, 1994.
Bradnock, Robert. *India Handbook,* 5th ed. Passport Books, 1996.
———. *Sri Lanka Handbook.* Passport Books, 1996.
Braybrook, Roy. *British Aerospace Harrier and Sea Harrier.* Osprey/Motorbooks International, 1984.
———. *Soviet Combat Aircraft.* Osprey, 1991.
Brown, David F. *Birds of Prey: Aircraft, Nose Art, and Mission Markings of Desert Storm/ Shield.* U.S. Government Printing Office, 1993.
Brown, Eric M., Captain, RN. *Duels in the Sky: World War II Naval Aircraft in Combat.* Naval Institute Press, 1988.
Brugioni, Dino A. *Eyeball to Eyeball: The Cuban Missile Crisis.* Random House, 1991.
Buderi, Robert. *The Invention That Changed the World.* HarperCollins, 1996.
Buell, Thomas B. *The Quiet Warrior: A Biography of Admiral Raymond E. Spruance.* Naval Institute Press, 1987.
Burrows, William E. *Deep Black.* Random House, 1988.
———. *Exploring Space: Voyages in the Solar System and Beyond.* Random House, 1990.
Burrows, William E., and Robert Windham. *Critical Mass.* Simon & Schuster, 1989.
Butowski, Piotr, with Jay Miller. *OKB MiG: A History of the Design Bureau and Its Aircraft.* Speciality Press, 1991.
Bywater, Hector, C. *The Great Pacific War,* reprint of 1925 ed. Naval Institute Press, 1991.
Campen, Alan D., ed. *The First Information War.* AFCEA, 1992.
Chadwick, Frank. *Gulf War Fact Book.* Game Designers Workshop, 1992.
Chant, Christopher. *Encyclopedia of Modern Aircraft Armament.* IMP Publishing Services Ltd., 1988.
Chetty, P. R. K. *Satellite Technology and Its Applications,* 2nd ed. McGraw Hill, 1991.
Chinnery, Philip D. *Life on the Line.* St. Martin's Press, 1988.
Clancy, Tom. *Airborne: A Guided Tour of an Airborne Task Force.* Berkley Books, 1997.
———. *Armored Cav: A Guided Tour of an Armored Cavalry Regiment.* Berkley Books, 1994.
———. *Clear and Present Danger.* G. P. Putnam's Sons, 1999.
———. *Debt of Honor.* G. P. Putnam's Sons, 1994.
———. *Executive Orders.* G. P. Putnam's Sons, 1996.
———. *Fighter Wing: A Guided Tour of an Air Force Combat Wing.* Berkley Books, 1995.
———. *Marine: A Guided Tour of a Marine Expeditionary Unit.* Berkley Books, 1996.
———. *Patriot Games.* G. P. Putnam's Sons, 1986.
———. *Red Storm Rising.* G. P. Putnam's Sons, 1985.
———. *Submarine: A Guided Tour Inside a Nuclear Warship.* Berkley Books, 1993.
———. *The Cardinal of the Kremlin.* G. P. Putnam's Sons, 1988.
———. *The Hunt for Red October.* Berkley Books, 1985.
———. *The Sum of All Fears.* G. P. Putnam's Sons, 1991.
———. *Without Remorse.* G. P. Putnam's Sons, 1993.
Cline, Ray S. *Secrets, Spies and Scholars.* Acropolis, 1976.
Clodfelter, Mark. *The Limits of Air Power: The American Bombing of North Vietnam.* Free Press, 1989.
Cochran, Thomas B., William M. Arkin, Robert S. Norris, and Jeffrey I. Sands. *Soviet Nuclear Weapons.* Harper & Row, 1989.
Cohen, Dr. Elliot A. *Gulf War Air Power Survey Summary Report.* U.S. Government Printing Office, 1993.

———. *Gulf War Air Power Survey*, Vol. I. U.S. Government Printing Office, 1993.
———. *Gulf War Air Power Survey*, Vol. II. U.S. Government Printing Office, 1993.
———. *Gulf War Air Power Survey*, Vol. III. U.S. Government Printing Office, 1993.
———. *Gulf War Air Power Survey*, Vol. IV. U.S. Government Printing Office, 1993.
———. *Gulf War Air Power Survey*, Vol. V. U.S. Government Printing Office, 1993.
Cohen, Dr. Elliott A. and John Gooch, *Military Misfortunes: The Anatomy of Failure in War.* Free Press, 1990.
Cohen, Eliezer "Cheetah," Colonel, 10F. *Israel's Best Defense: The First Full Story of Israeli Air Force.* Orion Books, 1993.
Conduct of the Persian Gulf War. U.S. Government Printing Office, 1992.
Cooling, Benjamin F., ed. *Case Studies in the Achievement of Air Superiority.* Center for Air Force History, 1991.
Cooling, Benjamin F., ed. *Case Studies in the Development of Close Air Support.* Office of Air Force History, 1990.
Coonts, Stephen. *Flight of the Intruder.* Naval Institute Press, 1986.
Cormier, Zeke, Wally Schirram, Ohil Wood, with Barrett Tillman. *Wildcats to Tomcats: The Tailhook Navy.* Phalanx, 1995.
Corum, James S. *The Luftwaffe: Creating the Operational Air War, 1918–1940.* University Press of Kansas, 1997.
Coyne, James P. *Airpower in the Gulf.* Air Force Association, 1992.
Crickmore, Paul F. *Lockheed SR-71: The Secret Missions Exposed.* Osprey Aerospace, 1993.
Crowe, Admiral William J., Jr. *The Line of Fire: From Washington to the Gulf, the Politics and Battles of the New Military.* Simon & Schuster, 1993.
Cunningham, Randy, with Dr. Jeffery Ethell. *Fox Two: The Story of America's First Ace in Vietnam.* Chaplin Fighter Museum, 1984.
Darwish, Adel, and Gregory Alexander. *Unholy Babylon: The Secret History of Saddam's War.* St. Martin's Press, 1991.
David, Peter. *Triumph in the Desert.* Random House, 1991.
Davis, Larry. *MiG Alley: Air to Air Combat over Korea.* Squardron/Signal Publications, 1978.
Dawood, N.J., ed. *The Koran.* Penguin Books, 1956.
De Jomini, Baron Antoine Henri. *The Art of War.* Green Hill Books, 1992.
Devereaux, Tony. *Messenger Gods of Battle: Radio, Radar, Sonar: The Story of Electronics in War.* Brassey's, 1991.
Divine, Robert A. *The Sputnik Challenge.* Oxford, 1993.
Doleman, Edgar C., Jr. *The Vietnam Experience: Tools of War.* Boston Publishing Company, 1985.
Donald, David, and Jon Lake, eds. *U.S. Navy and Marine Corps Air Power Directory.* Aerospace Publishing, Ltd., 1992.
Doolittle, Jimmy A., with Carol V. Glines. *I Could Never Be So Lucky Again: An Autobiography by General James H. "Jimmy" Doolittle.* Bantam Books, 1991.
Dorr, Robert F. *Air War Hanoi.* Blanford Press, 1988.
———. *Desert Shield: The Build Up: The Complete Story.* Motorbooks, 1991.
———. *Desert Storm: Air War.* Motorbooks, 1991.
———. *F-86 Sabre: History of the Sabre and FJ Fury.* Motorbooks, 1993.
———. *Vietnam MiG Killers.* Motorbooks International, 1988.
Dorr, Robert F. and Chris Bishop. *Vietnam Air War Debrief.* Aerospace Publishing Ltd., 1996.
Dorr, Robert F. and Norman E. Taylor. *US Air Force Nose Art: Into the 90's.* Speciality Press, 1993.
Dull, Paul S. *The Imperial Japanese Navy.* Naval Institute Press, 1978.
Dunnigan, James F. and Austin Bay. *A Quick and Dirty Guide to War.* 3rd ed. Morrow, 1996.
———. *From Shield to Storm.* Morrow Books, 1992.
Dunningan, James F., and William Martel. *How to Stop a War: Lessons on Two Hundred Years of War and Peace.* Doubleday Books, 1987.
Dupuy, Trevor N., Colonel, USA (Ret.). *Attrition: Forecasting Battle Casualties and Equipment Losses in Modern War.* Hero Books, 1990.
———. *Future Wars: The World's Most Dangerous Flashpoints.* Warner Books, 1993.

———. *Numbers, Predictions & War: The Use of History to Evaluate and Predict the Outcome of Armed Conflict.* Hero Books, 1985.
———. *Options of Command.* Hippocrene Books, Inc., 1984.
———. *Saddam Hussein: Scenarios and Strategies for the Gulf War.* Warner Books, 1991.
———. *The Evolution of Weapons and Warfare.* Bobbs-Merrill, 1980.
———. *Understanding Defeat: How To Recover from Loss in Battle To Gain Victory in War.* Paragon House, 1990.
———. *Understanding War: History and Theory of Combat.* Paragon House, 1987.
Dzhus, Alexander M. *Soviet Wings: Modern Soviet Military Aircraft.* Greenhill Books, 1991.
Edwards, John E., Major, USA (Ret.). *Combat Service Support Guide*, 2nd ed. Stackpole Books, 1993.
Eliot, Joshua, ed. *Indonesia, Malaysia & Singapore Handbook.* Passport Books, 1994.
Eshel, David. *The U.S. Rapid Deployment Forces.* Arco Publishing, Inc., 1985.
Ethell, Dr. Jeffrey L., and Clarence Simonsen. *The History of Aircraft.* Ethall and Simonsen, 1991.
Ethell, Dr. Jeffrey, and Dr. Alfred Price. *Air War South Atlantic.* Macmillan, 1983.
———. *One Day in a Long War: May 10th, 1972, Air War, North Vietnam.* Random House, 1989.
———. *World War II Fighting Jets.* Naval Institute Press, 1994.
Ethell, Dr. Jeffrey, and Robert T. Sand. *Fighter Command: American Fighters in Original WWII Color.* Motorbooks International, 1991.
Faulkner, Keith. *Jane's Warship Recognition Guide.* HarperCollins, 1996.
Fisher, David E. *A Race on the Edge of Time.* McGraw Hill, 1988.
Flagherty, Thomas J. *Air Combat.* Time-Life Books, 1990.
———. *Carrier Warfare.* Time-Life Books, 1991.
Flintham, Victor. *Air Wars and Aircraft: A Detailed Record of Air Combat, 1945 to Present.* Facts on File, 1990.
Ford, Brian. *Allied Secret Weapons, the War of Science.* Ballatine Books, 1971.
———. *German Secret Weapons, Blueprint for Mars.* Ballatine Books, 1969.
Foster, Simon. *Hit The Beach!* Arms and Armour Press, 1995.
Francillon, Rene, J. *Lockheed Aircraft since 1913.* G. P. Putnam's Sons, 1982.
———. *Tonkin Gulf Yacht Club: US Carrier Operations off Vietnam.* Naval Institute Press, 1988.
———. *World Military Aviation, 1997–1998.* Naval Institute Press, 1997.
Frank, Richard B. *Guadalcanal: The Definitive Account of the Landmark Battle.* Random House, 1990.
Fricker, John. *Battle for Pakistan.* Ian Allen, 1979.
Friedman, George and Meredith Friedman. *The Future of War.* Crown, 1996.
———. *The Intelligence Edge*, Crown, 1997.
Friedman, Norman. *Desert Victory: The War for Kuwait.* U.S. Naval Institute Press, 1991.
———. *Naval Institute Guide to World Naval Weapons Systems 1997–98.* Naval Institute Press, 1997.
———. *US Naval Weapons.* Naval Institute Press, 1985.
Furtrell, Robert F. *The United States Air Force in Korea.* U.S. Government Printing Office, 1983.
———. *The United States Air Force in Southeast Asia: The Advisory Years to 1965.* U.S. Government Printing Office, 1981.
Gandt, Robert. *Bogeys and Bandits: The Making of a Fighter Pilot.* Viking, 1997.
Gann, Ernest K. *The Black Watch: The Men Who Fly America's Secret Spy Planes.* Random House, 1989.
Garrett, Dan. *Wings of Freedom.* Lockheed-Ft. Worth, 1988.
Geoffrey Perrett. *Winged Victory: The Army Air Forces In World War II.* Random House, 1993.
Giangreco, D. M. *Stealth Fighter Pilot.* Motorbooks, 1993.
Gibson, James William. *The Perfect War: Technowar in Vietnam.* Atlantic Monthly Press, 1986.
Godden, John, ed. *Shield & Storm: Personal Recollections of the Air War in the Gulf.* Brassey's, 1994.

Goldstein, Donald L., et. al. *D-Day, Normandy: The Story and the Photographs*. Brassey's, 1994.

Gordon, Michael R., and Bernard E. Trainor. *The General's War: The Inside Story of the Conflict in the Gulf*. Little, Brown, 1995.

Gordon, Yefim, and Vladimir Rigmant. *MiG-15*. Motorbooks, 1993.

Gorshkov, S. G., Admiral of the Fleet of the Soviet Union. *The Sea Power of the State*. Pergamon Press, 1979.

GPS—A Guide to the Next Utility. Trimble Navigation, 1989.

Graham, Christian Garnett. *Against All Odds: The Battle of Britain*. The Rococo Group, 1990.

Graham, T. Allison. *Essence of Decision*. Little, Brown, 1971.

Gray, Colin S. *The Leverage of Seapower: The Strategic Advantage of Navies in War*. Free Press.

Gribkov, General Anatoli I., and General William Y. Smith. *Operation Anadyr: US and Soviet Generals Recount the Cuban Missile Crisis*. Edition q, Inc., 1994.

Grove, Erio. *Battle for the Fiørds: NATO's Forward Maritime Strategy in Action*. Naval Institute Press, 1991.

———. *Sea Battles in Close Up: World War II*, Vol. 2. Naval Institute Press, 1993.

Gumble, Bruce L. *The International Countermeasures Handbook*. EW Communications Inc., 1987.

Gunston, Bill. *Grumman: Sixty Years of Excellence*. Orion, 1988.

Gunston, Bill. *Mikoyan MiG-21*. Osprey Publishing Limited, 1986.

Gunston, Bill, with Peter Gilchrist. *Jet Bombers: From the Messerschmitt Me 262 to the Stealth B-2*. Osprey Aerospace, 1993.

Halberstadt, Hans. *Desert Storm: Ground War*. Motorbooks International, 1991.

———. *F-15E Strike Eagle*. Windrow & Greene, 1992.

Hallion, Dr. Richard P. *Air Power Confronts an Unstable World*. Brassey's, 1997.

———. *On the Frontier: Flight Research at Dryden, 1946–1981*. Richard P. Hallion, National and Aeronautics and Space Administration, 1984.

———. *Rise of the Fighter Aircraft 1914–1918*. The Nautical & Aviation Publishing Co., 1988.

———. *Storm over Iraq: Air Power and the Gulf War*. Smithsonian Books, 1992.

———. *Strike from the Sky: The History of Battlefield Air Attack 1911–1945*. Smithsonian Books, 1989.

———. *The Literature of Aeronautics, Astronautics and Air Power*. U.S. Government Printing Office, 1984.

Hammel, Eric. *Guadalcanal: Decision at Sea*. Crown, 1988.

———. *Guadalcanal: Starvation Island*. Crown, 1987.

———. *Guadalcanal: The Carrier Battles*. Crown, 1986.

Hanak, Walter. *Aces & Aerial Victories*. U.S. Government Printing Office, 1979.

Hansen, Chuck. *US Nuclear Weapons: The Secret History*. Orion Books, 1988.

Hartcup, Guy. *The Silent Revolution: Development of Conventional Weapons 1945–85*. Brassey's, 1993.

Hartmann, Frederick H. *Naval Renaissance: The US Navy in the 1980s*. Naval Institute Press, 1990.

Hartmann, Gregory K. and Scott C. Truver. *Weapons That Wait: Mine Warfare in the US Navy*. Naval Institute Press, 1991.

Heatley, C. J., III. *Forged in Steel: US Marine Corps Aviation*. Howell Press, Charlottesville, Virginia, 1987.

Heinlein, Robert A. *Starship Troopers*. Ace Books, 1959.

Heinmann, Edward H., Rosario Rausa, and K. E. Van Every. *Aircraft Design*. Nautical & Aviation Publishing Co., 1995.

Hill, J. R., Rear Admiral, RN (Ret.). *Anti-Submarine Warfare*. Naval Institute Press, 1985.

Hilsman, Roger, *George Bush vs Saddam Hussein: Military Success! Political Failure?* Lyford Books, 1992.

Holley, I. B., Jr. *The US Special Studies: Ideas and Weapons*. U.S. Government Printing Office, 1983.

Honan, William H. *Visions of Infamy*. St. Martin's Press, 1991.

Hudson, Heather E. *Communication Satellites: Their Development and Impact.* Free Press, 1990.
Hurley, Alfred, Colonel, USAF, and Major Robert C. Ehrhart, USAF. *Air Power and Warfare.* U.S. Government Printing Office, 1998.
India and Bangladesh: Travel Atlas. Lonely Planet, 1995.
Inoguchi, Rikihei, Captain, IJN, and Tadashi Nakajima, Commander, IJN. *The Divine Wind.* Bantam, 1978.
Isby, David. *Fighter Combat in the Jet Age.* HarperCollins, 1997.
———. *Weapons and Tactics of the Soviet Army.* Jane's, 1981.
Isenberg, Michael T. *Shield of the Republic: The United States Navy in an Era of Cold War and Violent Peace, 1945–1962.* St. Martin's Press, 1993.
Jablonski, Edward. *America in the Air War.* Time-Life Books, 1982.
Japan at War. Time-Life Books, 1980.
Jessup, John E., Jr. and Robert W. Coakley. *A Guide to the Study and Use of Military History.* U.S. Government Printing Office, 1991.
Johnson, Brian. *Fly Navy:The History of Naval Aviation.* William Morrow, 1981.
Joss, John. *Strike: US Naval Strike Warfare Center.* Presidio Press, 1989.
Kahn, David. *Seizing the Enigma: The Race to Break the German U-Boat Codes, 1939–1943.* Houghton Mifflin Company, 1991.
———. *The Codebreakers: The Comprehensive History of Secret Communication from Ancient Times to the Internet.* Scribner, 1996.
Keany, Thomas A., and Eliot A. Cohen. *Revolution in Warfare? Air Power in the Persian Gulf.* Naval Institute Press, 1995.
Keegan, John. *A History of Warfare.* Alfred A. Knopf, 1993.
———. *The Illustrated Face of Battle.* Viking, 1988.
———. *The Second World War.* Viking, 1989.
Kelly, Mary Pat. *''Good to Go'': The Rescue of Scott O'Grady from Bosnia.* Naval Institute Press, 1996.
Kelly, Orr. *Hornet: The Inside Story of the F/A-18.* Presidio Press, 1990.
Kenney, Geore C. *General Kenney Reports.* Office of Air Force History, 1987.
Kershaw, Robert J. *D-Day: Piercing the Atlantic Wall.* Naval Institute Press, 1994.
Kindersley, Dorling. *The Ultimate Visual Dictionary.* Dorling Kindersley, 1994.
Kinzey, Bert. *The Fury of Desert Storm: The Air Campaign.* McGraw-Hill, 1991.
———. *US Aircraft & Armament of Operation Desert Storm.* Kalmbach Books, 1993.
Kissinger, Henry. *Henry Kissinger: Diplomacy.* Simon & Schuster, 1994.
Knott, Richard, C., Capt., USN. *The Naval Aviation Guide*, 4th ed. Naval Institute Press, 1985.
Kohn, Richard H. and Joseph P. Haraan. *Air Interdiction in World War II, Korea, and Vietnam.* U.S. Government Printing Office, 1990.
———. *Air Superiority in World War II and Korea.* U.S. Government Printing Office, 1983.
———. *USAF Warrior Studies.* Coward McCann, Inc., 1942.
Krulak, Victor H., Lieutenant General, USMC. *First to Fight: An Inside View of the US Marine Corps.* Naval Institute Press, 1984.
Kyle, James H., Colonel, USAF (Ret.). *The Guts to Try.* Orion Books, 1990.
Lake, Donald, David, and Jon, eds. *US Navy and Marine Corps Air Power Directory.* Aerospace Publishing, Ltd., 1992.
Lake, Jon. *McDonnell Douglas F-4 Phantom: Spirit in the Skies.* Aerospace Publishing Ltd., 1992.
———. *MiG-29: Soviet Superfighter.* Osprey Publishing, 1989.
Lambert, Mark. *Jane's All the World's Aircraft, 1991–92.* Jane's Publishing Group, 1991.
———. *Jane's All the World's Aircraft, 1992–93.* Jane's Information Group, 1992.
Lavalle, A.J.C., Major. USAF. *Airpower and the 1972 Spring Invasion.* Airpower Research Institute, 1979.
———. *The Tale of Two Bridges and The Battle for the Skies over North Vietnam.* Office of Air Force History, 1978.
———. *The Vietnamese Air Force, 1951–1975, and Analysis of its Role in Combat and Fourteen Hours at Koh Tang.* Office of Air Force History, 1978.
Lehman, John H. *Command of the Seas.* Scribners, 1988.
———. *Making War.* Scribners, 1992.

Levinson, Jeffrey L. *Alpha Strike Vietnam: The Navy's Air War 1964–1973*. Presidio Press, 1989.
Liddell Hart, B. H. *Strategy*. Frederick A. Praeger, Inc., Publishers, 1967.
Lord, Walter. *Day of Infamy*. Holt Rinehart, 1957.
———. *Incredible Victory: The Battle of Midway*, reprint of 1967 ed. Harper Collins, 1993.
Lundstrom, John B. *The First Team and the Guadalcanal Campaign*. Naval Institute Press, 1994.
———. *The First Team: Pacific Naval Air Combat from Pearl Harbor to Midway*. Naval Institute Press, 1984.
Luttwak, Edward, and Stuart L. Koehl. *The Dictionary of Modern War: A Guide to the Ideas, Institutions and Weapons of Modern Military Power*. HarperCollins, 1991.
Macedonia, Raymond M. *Getting it Right*. Morrow Publishing, 1993.
Macksey, Kenneth. *Invasion: The German Invasion of England, July 1940*. Macmillan, 1980.
Makower, Joel. *The Air and Space Catalog: The Complete Sourcebook to Everything in the Universe*. Vintage Tilden Press, 1989.
Manning, Robert. *The Vietnam Experience: The North*. Boston Publishing Company, 1986.
Mark, Edward. *Aerial Interdiction in Three Wars*. Center for Air Force History, 1994.
Marolds, Edward J. *Carrier Operations: The Vietnam War*. Bantam Books, 1987.
Maroon, Fred J., and Edward L. Beach. *Keepers of the Sea*. Naval Institute Press, 1983.
Mason, Francis K. *Battle over Britain*. Aston Publications, 1990.
Mason, John T., Jr. *The Pacific War Remembered: An Oral History Collection*. Naval Institute Press, 1986.
Mason, Tony, Air Vice Marshal, RAF. *Air Power: A Centennial Appraisal*. Brassey's, 1994.
———. *To Inherit the Skies: From Spitfire to Tornado*. Brassey's, 1990.
May, Ernest R., and Philip Zelikow, ed. *The Kennedy Tapes: Inside the White House During the Cuban Missile Crisis*. Belknap Harvard, 1997.
McConnell, Malcolm. *Just Cause: The Real Story of America's High-Tech Invasion of Panama*. St. Martin's Press, 1991.
McFarand, Stephen L., and Wesley Phillips Newton. *The Command of the Sky*. Smithsonian, 1991.
McKinnon, Dan. *Bullseye—Iraq*. Berkley, 1987.
Meisner, Arnold. *Desert Storm: Sea War*. Motorbooks International, 1991.
Melhorn, Charles M. *Two Block Fox: The Rise of the Aircraft Carrier, 1911–1929*. Naval Institute Press, 1974.
Melson, Charles D., and Paul Hannon. *Marine Recon, 1940–90*. Osprey, 1994.
Mersky, Peter B., and Norman Polmar. *The Naval War in Vietnam*. Kennsington Books, 1981.
Michel, Marshall L. *Clashes: Air Combat over North Vietnam 1965–1972*. Naval Institute Press, 1997.
Micheletti, Eric. *Operation Daguet: French Air Force in the Gulf War*. Concord Publications Company, 1991.
Middlebrook, Martin. *Task Force: The Falklands War, 1982*. Penguin Books, 1987.
Miller, Edward S. *War Plan Orange: The US Strategy to Defeat Japan 1897–1945*. Naval Institute Press, 1991.
Moore, John, Captain RN. *Jane's American Fighting Ships of the 20th Century*. Modern Publishing, 1995.
Morrocco, Jon. *The Vietnam Experience: Rain of Fire*. Boston Publishing Company, 1985.
———. *The Vietnam Experience: Thunder from Above*. Boston Publishing Company, 1984.
Morse, Stan, ed. *Gulf Air War Debrief*. Aerospace Publishing Ltd., 1991.
Moskin, J. Robert. *The U.S. Marine Corps Story*. McGraw-Hill, 1987.
Musciano, Walter A. *Messerschmitt Aces*. TAB/Aero Books, 1990.
Navy League Program—1998. Hornet Team, 1993.
Neafeld, Jacob. *Ballistic Missiles in the United States Air Force*. U.S. Government Printing Office, 1993
Neufeld, Michael J. *The Rocket and the Reich: Peenemunde and the Coming of the Ballistic Missile Era*. Free Press, 1995.

Neustadt, Richard E., and Ernest R. May. *Thinking in Time: The Uses of History for Decision Makers*. Free Press, 1986.
Newhouse, John. *War and Peace in the Nuclear Age*. Alfred Knopf Publications, 1989.
Nichols, John B., Commander, USN (Ret.), and Barrett Tillman. *On Yankee Station: The Naval Air War over Vietnam*. Naval Institute Press, 1987.
Nissen, Jack. *Winning the Radar War*. St. Martins Press, 1987.
Nordeen, Lon O., Jr. *Air Warfare in the Missile Age*. Smithsonian Books, 1985.
———. *Fighters over Israel: The Story of the Israeli Air Force from the War of Indenpendence to the Bekaa Valley*. Orion Books, 1990
Nordeen, Lon O., Jr., and David Nicolle. *Phoenix over the Nile: A History of Egyptian Air Power, 1932–1994*. Smithsonian 1996.
O'Ballance, Edgar. *No Victor, No Vanquished*. Presidio Press, 1978.
Ogley, Bob. *Doodlebugs and Rockets*. Froglets Publications, 1992.
O'Grady, Scott, Captain, USAF. *Return with Honor*. Doubleday, 1995.
O'Neill, Richard *Suicide Squads of World War II*. Salamander Books, 1981.
Pagonis, William G., Lieutenant General, USA, with Jeffrey L. Crulkshank. *Moving Mountains: Lessons in Leadership and Logistics from the Gulf War*. Harvard Business School Press, 1992.
Paloczi-Horvath, George. *From Monitor to Missile Boat: Coast Defence Ships and Coastal Defence since 1860*. Naval Institute Press, 1996.
Pape, Robert A. *Bombing to Win: Air Power and Coercion in War*. Cornell, 1996.
Parrish, Thomas. *The American Codebreakers: The US Role in Ultra*. Scarborough Publishers, 1986.
———. *The Cold War Encyclopedia*. Henry Holt, 1996.
Parsons, Dave, and Derek Nelson. *Bandits!* Motorbooks, 1993.
———. *Hell-Bent for Leather*. Motorbooks International Publishers & Wholesalers, 1990.
Peebles, Curtis. *Dark Eagles: A History of Top Secret US Aircraft Programs*. Presidio Press, 1995.
———. *Guardians: Strategic Reconnaissance Satellites*. Presidio Press, 1987.
Penkovskiy, Oleg. *The Penkovskiy Papers*. Avon Books, 1965.
Perla, Peter P. *The Art of Wargaming*. Naval Institute Press, 1990.
Peterson, Philip A. *Soviet Air Power and the Pursuit of New Military Options*. United States Air Force, 1979.
Pocock, Chris. *Dragon Lady: The History of the U-2 Spyplane*. Motorbooks International, 1989.
Polmar, Norman. *Aircraft Carriers*. Doubleday, 1969.
———. *Chronology of The Cold War at Sea, 1945–1991*. Naval Institute Press, 1998.
———. *Naval Institute Guide to the Ships and Aircraft of the US Fleet*, fifteenth ed. Naval Institute Press, 1993.
Polmar, Norman, and Floyd D. Kennedy. *Military Helicopters of the World*. Naval Institute Press, 1981.
Polmar, Norman, and Peter Mersky. *The Naval Air War in Vietnam*. Nautical & Aviation Publishing, 1981.
Polmar, Norman, and Thomas Allen. *Spy Book: The Encyclopedia of Espionage*. Random House, 1997.
Potter, Michael C., Captain, USNR. *Electronic Greyhounds: The Spruance-Class Destroyers*. Naval Institute Press, 1995.
Price, Dr. Alfred. *Air Battle Central Europe*. Warner Books, 1986.
———. *Battle of Britain: 18 August 1940: The Hardest Day*. Granada Books, 1980.
———. *Harrier at War*. Ian Allen, 1984.
———. *Instrument of Darkness: The History of Electronic Warfare*. Peninsula Publishing, 1987.
———. *The History of US Electronic Warfare*, Vol. I & II. Association of Old Crows, 1989.
Rapport, Anatol, ed. *Carl Von Clausewitz on War*. Penguin Books, 1968.
Rendall, David. *Jane's Aircraft Recognition Guide*. HarperCollins, 1995.
Rentoul, Ian, and Tom Wakeford. *Gulf War: British Air Arms*. Concord, 1991.
Reynolds, Clark G. *The Carrier War*. Time-Life Books, 1982.
———. *The Fast Carriers: The Forging of an Air Navy*. Naval Institute Press, 1968.
Rich, Ben, and Leo Janos. *Skunk Works*. Little, Brown, 1994.

Richelson, Jeffrey T. *American Espionage and the Soviet Target.* William Morrow and Company, 1987.
———. *America's Secret Eyes in Space.* Harper & Row Publishers, 1990.
———. *Sword and Shield: Soviet Intelligence and Security Apparatus.* Ballinger Publishing Company, 1986.
———. *The US Intelligence Community.* Ballinger Publishing Company, 1985.
Rogers, Will and Sharon, with Gene Gregston. *Storm Center: The USS Vincennes and Iran Air Flight 655.* Naval Institute Press, 1992.
Santoli, Al. *Leading the Way: How Vietnam Veterans Rebuilt the U.S. Military.* Ballantine Books, 1993.
Schmitt, Gary. *Silent Warfare: Understanding the World of Intelligence.* Brassey's (U.S.), 1993.
Schwarzkopf, General H. Norman, with Peter Petre. *General H. Norman Schwarzkopf: The Autobiography: It Doesn't Take a Hero.* Bantam Books, 1992.
Sharp, Admiral U.S.G. *Strategy for Defeat.* Presidio Press, 1978.
Shaw, Robert L. *Fighter Combat: Tactics and Maneuvering.* Naval Institute Press, 1985.
Shawcross, William. *Sideshow-Kissinge: Nixon and the Destruction of Cambodia.* Simon & Schuster, 1977.
Sheehan, John W., Jr., *Gunsmoke: USAF Worldwide Gunnery Meet,* Motorbooks, 1990
Sheehan, Neil. *The Pentagon Papers.* Batnam Books, 1971.
Sherrod, Robert. *History of Marine Corps Aviation in World War II.* Nautical and Aviation Publishing, 1987.
Shlomo Nakdimon. *First Strike: The Exclusive Story of How Israel Foiled Iraq's Attempt To Get the Bomb.* Summit Books, 1987.
Simonsen, Erik. *This is Stealth: The F-117 and B-2 in Color.* Greenhill Books, 1992.
Sims, Edward H. *Fighter Tactics and Strategy 1914–1970.* Harper & Rowe Publishers, 1972.
Smallwood, William L. *Strike Eagle: Flying the F15E in the Gulf War.* Brassey's, 1994.
———. *Warthog: Flying the A-10 in the Gulf War.* Brassey's (U.S.), 1993.
Smith, Gordon. *Battles of the Falklands War.* Ian Allen, 1989.
Spector, Ronald, H. *Eagle Against the Sun: The American War with Japan.* Free Press, 1985.
Spick, Mike. *All Weather Warriors: The Search for the Ultimate Fighter Aircraft.* Arms and Armour, 1994.
———. *The Ace Factor.* Avon War, 1988.
Spick, Mike, and Barry Wheeler. *Modern Aircraft Markings.* Salamander Books Ltd., 1992.
Stafford, Edward P. *The Big E: The Story of the USS Enterprise.* Naval Institute Press, 1962.
Stephen, Martin. *Sea Battles in Close Up: World War II.* Naval Institute Press, 1991.
Stevens, Paul D., ed. *The Navy Cross: Vietnam.* Sharp & Dunnigan, 1987.
Stevenson, James P. *Grumman F-14 "Tomcat."* Aero Publishers, 1975.
———. *The Pentagon Paradox.* USNI Press, 1993.
Stockdale, Jim and Sybil. *In Love and War.* Naval Institute Press, 1990.
Straubel, James H. *Crusade for Airpower.* Aerospace Education Foundation, 1982.
Summers, Harry G., Jr., Colonel USA (Ret.). *A Critical Analysis of the Gulf War.* Dell Publishing, 1992.
———. *The New World Strategy.* Simon & Schuster, 1995.
Swanborough, Gordon, and Peter Bowers. *United States Military Aircraft since 1909.* Smithsonian, 1989.
———. *United States Navy Aircraft since 1911.* Naval Institute Press, 1990.
Talbott, Strobe. *Deadly Gambits.* Alfred A. Knopf, Inc., 1984.
Terraine, John. *A Time for Courage: The Royal Air Force in the European War, 1939–1945.* MacMillan Publishing Company, 1985.
The Intelligence Revolution—U.S. Air Force Academy. U.S. Government Printing Office, 1988.
The World's Missile Systems. General Dynamics, 1988.
Thomborough, Anthony. *Sky Spies: The Decades of Airborne Reconnaissance.* Arms and Armour, 1993.
Tilford, Earl H., Jr. *Search and Rescue.* Center for Air Force History, 1992.
Tillman, Barrett. *Corsair.* USNI Press, 1984.
———. *Dauntless.* USNI Press, 1976.

———. *Dauntless: A Novel of Midway and Guadalcanal.* Bantam, 1992.
———. *Hellcat.* USNI Press, 1979.
———. *Hellcats: A Novel of War in the Pacific.* Brassey's, 1996.
———. *MiG Master.* USNI Press, 1985.
———. *On Yankee Station.* USNI Press, 1988.
Toffler, Alvin and Heidi. *War and Anti-War: Survival at the Dawn of the 21st Century.* Little, Brown, 1993.
Toliver, Colonel Taymond F., and Trevor J. Constable. *Fighter General: The Life of Adolf Galland.* AmPress, 1990.
Townsend, Peter. *Duel of Eagles.* Simon & Schuster, 1969.
TRW Space Data, 4th ed. TRW, 1992.
Ulanoff, Stanley M., Brigadier General, USAR, and David Eshel, Lieutenant Colonel, IDF (Ret.). *The Fighting Israeli Air Force.* Arco Publishing, 1985.
U.S. News and World Report Staff. *Triumph without Victory: The Unreported History of the Persian Gulf War.* Random House, 1992.
Valenzi, Kathleen D. *Forged in Steel: US Marine Corps Aviation.* Howell Press, 1987.
Van der Vat, Dan. *The Pacific Campaign, World War II.* Simon & Schuster, 1991.
Venkus, Colonel Robert E. *Raid on Qaddafi.* St. Martin's Press, 1992.
Volkman, Ernest, and Blaine Baggett. *Secret Intelligence: The Inside Story of America's Espionage Empire.* Doubleday, 1989.
Von Hassell, Agostino. *Strike Force: US Marine Special Operations.* Howell Press, Charlottesville, Virginia, 1991.
Wagner, William. *Fireflies and other UAV's.* Midland Publishing Limited, 1992.
———. *Lightning Bugs and Other Reconnaissance Drones.* Aero Publishers, 1982.
Walker, Bryce. *Fighting Jets.* Time-Life Books, 1983.
Waller, Douglas C. *The Commandos: The Inside Story of America's Secret Soldiers.* Simon & Schuster, 1994.
Ward, Commander Nigel ''Sharkey,'' DSC, AFC, RN. *Sea Harrier over the Falklands: A Maverick at War.* Naval Institute Press, 1992.
Warden, John A., III, Colonel, USAF. *The Air Campaign: Planning for Combat.* Brassey's Publishing, 1989.
Ware, Lewis B. *Low Intensity Conflict in the Third World.* U.S. Government Printing Office, 1988.
Warnock, A. Timothy. *The Battle against the U-Boat in the American Theater.* The U.S. Army Air Forces in World War II, 1992.
Watson, Bruce W., Bruce George, MP, Peter Tsouras, and B. L. Cyr. *Military Lessons of the Gulf War.* Greenhill Books, 1991.
Wedertz, Bill. *Dictionary of Naval Abbreviations.* Naval Institute Press, 1977.
Weinberg, Gerhard. *A World at Arms: A Global History of World War II.* Cambridge, 1994.
Weinberger, Caspar. *Fighting for Peace: Seven Critical Years in the Pentagon.* Warner Books, 1990.
Weinberger, Caspar, and Peter Schweizer. *The Next War.* Regnery, 1996.
Weisgall, Jonathan M. *Operation Crossroads: The Atomic Tests at Bikini Atoll.* Naval Institute Press, 1994.
Weissman, Steve, and Herbert Krosney. *The Islamic Bomb.* Times Books, 1981.
Werrell, Kenneth P. *The Evolution of the Cruise Missile.* Air University Press, 1985.
Westerfield, H. Bradford, ed. *Inside CIA's Private World: Declassified Articles from the Agency's Internal Journal.* Yale University Press, 1995.
Whipple, A. B. *To the Shores of Tripoli: The Birth of the US Navy and Marines.* Morrow, 1991.
Wilcox, Robert. *Scream of Eagles.* John F. Wiley & Sons, 1990.
———. *Wings of Fury.* Pocket Books, 1996.
Winnefeld, James A., and Dana J. Johnson. *A League of Airmen: US Air Power in the Gulf War.* Rand Project Air Force, 1994.
———. *Joint Air Operations: Pursuit of Unity in Command and Control 1942–1991.* Naval Institute Press, 1993.
Winter, Frank H. *The First Golden Age of Rocketry.* Smithsonian Institution, 1990.

Winton, John. *ULTRA in the Pacific*. Naval Institute Press, 1993.
Wood, Derek. *Jane's World Aircraft Recognition Handbook*, 5th ed. Jane's Information Group, 1992.
———. *Project Cancelled: The Disaster of Britain's Abandoned Aircraft Projects*. Jane's Publishing Inc., 1986.
Woodward, Robert. *The Commanders*. Simon & Schuster, 1991.
Woodward, Sandy, Admiral, RN. *One Hundred Days: The Memoirs of the Falklands Battle Group Commander*. Naval Institute Press, 1992.
Yergin, Daniel. *The Prize: The Epic Quest for Oil, Money and Power*. Simon & Schuster, 1991.
Yonay, Ehud. *No Margin for Error*. Pantheon, 1993.
Yoshimura, Akira. *Build the Musashi! The Birth and Death of the World's Greatest Warship*. Kodansha, Tokyo, 1996.
———. *Zero Fighter*. Praeger, 1996.
Zaloga, Steven J. *Target America: The Soviet Union and the Strategic Arms Race, 1945–1964*. Presidio Press, 1993.
Zumwalt, Elmo, Admiral, USN (Ret.). *On Watch*. Admiral Zumwalt Associates, Arlington, Virginia, 1976.
Zuyev, Alexander, with Malcolm McConnell. *Fulcrum: A Top Gun Pilot's Escape from the Soviet Empire*. Warner Books, 1992.

Pamphlets:

GPS: A Guide to the Next Utility. Trimble Navigation, 1989.
Measuring Effects of Payload and Radius Differences of Fighter Aircraft. Rand, 1993.
Reaching Globally Reaching Powerfully: The United States Air Force in the Gulf War. Department of the Air Force, 1991.
Space Log, 1993. TRW, 1994.
TRW Space Data, 4th ed. TRW, 1992.

Magazines:

Air and Space Smithsonian. Smithsonian Institute.
Air Force. United States Air Force Association.
Air Forces Monthly. Key Publishing, Ltd.
Air International. Expediters of the Printed Word, Ltd.
Airman. Air Force News Agency.
Airpower Journal. United States Air Force.
Aviation Week and Space Technology. McGraw Hill Publications.
Code One. Lockheed Martin.
Command: Military History, Strategy and Analysis. XTR Corporation.
Naval History. United States Naval Institute.
The Economist. The Economist.
The Hook. The Tailhook Association.
USAF Weapons Review. Commandant, USAF Weapons School.
U.S. News and World Report. U.S. News and World Report.
USNI Proceedings. United States Naval Institute.
Wings of Fame. Aerospace Publishing Ltd., Airtime Publishing, Inc.
World Airpower Journal. Aerospace Publishing Ltd., Airtime Publishing, Inc.

Videotapes:

AGM-137 (ISSAM). U.S. Air Force, 9/6/94.
A New Legacy. Northrop Television Communications, 1994.
BLU-109B: Penetrate and Destroy. Lockheed Missiles and Space Company, 1992
CIA: The Secret Files, parts 1–4. A&E Home Video, 1992.
F/A-18 Hornet '94. McDonnell Douglas, Northrop Grumman, General Electric, Hughes, 1994.
Fighter Air Combat Trainer. Spectrum HoloByte, 1993.
Fire and Steel. McDonnell Douglas, 1992

Heroes of the Storm. Media Center, 1991
It's about Performance. Sight & Sound Media, 1994
JSOW Update 1994. Texas Instruments, 1994.
Loral Aeronutronic-Pave Tack Exec. Version. Loral, 1991
MAG-13 Music Video, long version. McDonnell Douglas, 1992
Navy League—1992. McDonnell Douglas & Northrop, 1992.
Navy League—1993. McDonnell Douglas, Northrop, General Electric, Hughes, 1993.
New Developments in the Harpoon and Slam. Media Center, 1996
Night Strike Fighter F/A-18. McDonnell Douglas, Northrop, General Electric, Hughes, 1992
Night Hawk F/A-18 Targeting FLIR Video. Loral Aeronutronic, 1995
Nobody Does It Better. McDonnell Douglas, 1996
OM94008 Lantim Turning Night into Day///OM94154 Lantim/Pathfinder Cockpit Display. Martin Marietta, 9/29/94.
On the Road Again. McDonnell Douglas, Northrop, General Electric, Hughes, 1995
Operation Desert Storm Night Hawk and Pave Tack FLIR Video for IRIS. Loral Aeronutronic, 1991
Paveway Stock Footage. Defense Systems & Electronics Group, 1991
Slam/Slam ER Product Video. Media Center, 1994
Slam Video Composite. Media Center, 1992
Stealth and Survivability, revision 5. Television Communications, 1994
Storm from the Sea. Naval Institute, 1991.
The Canadian Forces in the Persian Gulf. DGPA-Director General Public Affairs, 1991
War in the Gulf Video, Series 1–4. Video Oradance Inc., 1991.
Wings of the Red Star, volume 1, 2, and 3. The Discovery Channel, 1993.
Wings over the Gulf, volume 1, 2, and 3. Discovery Communications Inc., 1991.

Games:

Ace of Aces of Jet Eagles. NOVA Game Designs, Inc., 0-917037-07-3.
Ace of Aces Wingleader. NOVA Game Designs, Inc., 0-917037-06-5.
Ace of Aces WWI Air Combat Game. NOVA Game Designs, Inc., 0-917037-00-6.
Air Strike: Modern Air-to-Ground Combat. Game Designers Workshop, 1987, 0-943580-30-7.
Air Superiority: Modern Jet Air Combat. Game Designers Workshop, 1987, 0-943580-19-6.
Captain's Edition Harpoon. GDW Games, 1990, 1-55878-054-8.
Dawn Patrol: Role-Playing Game of WWI Air Combat. TSR Hobbies, 1980.
Flight Leader: The Game of Air-to-Air Jet Combat Tactics, 1950–Present. The Avalon Hill Game Company, 1985, 0-911605-22-3.
Harpoon. Game Designers Workshop, 1987, 0-943580-12-9.
Over the Reich: WWII Air Combat over Europe. Clash of Arms Games, 1995.
The Speed of Heat: Air Combat over Korea and Vietnam. Clash of Arms Games, 1993.